T0229529

Insect-Plant Interactions

Volume IV

Elizabeth Bernays, Ph.D.
Professor and Head
Department of Entomology
University of Arizona
Tucson, Arizona

CRC Press
Taylor & Francis Group
Boca Raton London New York

CRC Press is an imprint of the
Taylor & Francis Group, an **informa** business

First published 1992 by CRC Press
Taylor & Francis Group
6000 Broken Sound Parkway NW, Suite 300
Boca Raton, FL 33487-2742

Reissued 2018 by CRC Press

A Library of Congress record exists under LC control number: 88035918

ISBN 13: 978-1-138-57617-9 (hbk)
ISBN 13: 978-1-138-55313-2 (pbk)
ISBN 13: 978-1-351-27100-4 (ebk)

Visit the Taylor & Francis Web site at http://www.taylorandfrancis.com and the CRC Press Web site at http://www.crcpress.com

PREFACE

This is the fourth volume of a unique book series concentrating on in-depth reviews of topics that are particularly important at the present time or are in research areas that are currently breaking new ground. The research areas generally lack recent reviews of any kind elsewhere. For those wishing to be up-to-date on some of the new and controversial elements of insect-plant interactions this series is a must. For those needing access to the literature, surveys, and informed value judgements by leaders in the field, or evaluation of research directions, Insect Plant Interactions will be extremely useful. Volume IV contains six chapters covering six diverse areas that are each of interest and importance to a wide readership.

J. L. Bronstein reviews the status of mutualisms. Using the fig-insect interaction, she examines the meaning of mutualism in insect-plant interactions and the ease with which mutualisms may shift to the cost or benefit of either partner. She puts the study in a modern evolutionary context, with unusual clarity and impeccable logic.

E. A. Bernays presents the case for phytosterols being important components of adaptive syndromes in herbivorous insects. She reviews the patchy literature on plant sterols and insects and indicates especially the work that is now needed to understand the importance of plant sterol profiles in host plant affiliations of insects.

L. M. Schoonhoven, W. M. Blaney, and M. S. J. Simmonds provide a scholarly update on how plant-feeding insects detect compounds that deter feeding: the various codes, how and why they vary. Since deterrents play the larger part in determining host plant selection by insect herbivores, this is an important chapter for all researchers interested in crop protection, as well as for those interested in the physiological basis of behavior.

S. Koptur reviews the nature and significance of extrafloral nectaries in plants. She brings together a disparte literature in a field with many anecdotal studies as well as high quality research studies, to give us exactly what a review should be — a thorough survey, with solid evaluation of the results presented by others.

M. Wink explains the varied roles of a group of alkaloids found in plants — the quinolizidines. Unlike the pyrrolizidines which have been thoroughly reviewed recently, the quinolizidines have been relatively neglected. Yet they are shown to be important in a variety of interactions. They are plant defenses against a wide variety of organisms, and yet as has occurred in other cases, there are insects for whom these chemicals have become indispensible.

G. L. Waring and N. S. Cobb review the difficult and controversial arena of plant stress and insect performance. This is a contribution to the general theories of insect-plant interactions because of the importance placed by previous authors on plant stress as a cause of herbivorous insect outbreaks. These authors provide evidence from a summary of the many published papers that plant stress per se is not generally beneficial for insects. The way in which drought relates to increases in insect numbers must be reexamined.

CONTRIBUTORS

Elizabeth A. Bernays, Ph.D.
Regent's Professor and Head
Department of Entomology
University of Arizona
Tucson, Arizona

W. M. Blaney, Ph.D.
Professor and Head
Department of Biology
Birkbeck College
London, England

Judith L. Bronstein, Ph.D.
Assistant Professor
Department of Ecology and
Evolutionary Biology
University of Arizona
Tucson, Arizona

Neil Cobb
Graduate Research Assistant
Department of Biological Sciences
Northern Arizona University
Flagstaff, Arizona

Suzanne Koptur, Ph.D.
Associate Professor
Department of Biological Sciences
Florida International University
Miami, Florida

L. M. Schoonhoven, Ph.D.
Professor
Department of Entomology
Agricultural University
Wageningen, The Netherlands

M. S. J. Simmonds, Ph.D.
Principal Scientist
Jodrell Laboratory
Royal Botanic Gardens
Surrey, England

Gwendolyn L. Waring, Ph.D.
Research Associate
Department of Biology
Museum of Northern Arizona
Flagstaff, Arizona

M. Wink, Ph.D.
Professor
Institute for Pharmaceutical Biology
University of Heidelberg
Heidelberg, Germany

ADVISORY BOARD

THE EDITOR

Elizabeth A. Bernays, Ph.D., is Regents' Professor of Entomology and Head of the Department of Entomology, and Adjunct Professor of Ecology and Evolutionary Biology, at the University of Arizona, Tucson.

Dr. Bernays graduated in 1962 from the University of Queensland, Australia, with a B.Sc. with honors in zoology and entomology. After a period of high school teaching, she obtained an M.Sc. in 1967 and then a Ph.D. in 1970 from the University of London. The same university awarded her a D.Sc. in 1991 for her contributions to biology. After receiving the Ph.D. degree she became a scientist in the British government service and in 1983 was appointed Professor of Entomology and Adjunct Professor of Zoology at the University of California, Berkeley. She took up her present position at the University of Arizona in 1989.

Dr. Bernays is a member of several societies, including the Royal Entomological Society, the American Zoological Society, and the Entomological Society of America. Among several awards, she won the 1987 gold medal of the Ponifical Academy of Science and has published 130 scientific papers. Her research is funded by the National Science Foundation. She regularly presents research and review papers at national and international meetings, as well as at universities in various countries. She is assistant editor of three research journals, two books, and the review series, "Insect Plant Interactions", of which this is volume IV.

Dr. Bernays is best known for her work on the physiological mechanisms underlying insect/plant relationships. This area of research began with her studies on the regulation of food intake in locusts. Many of the novel laboratory experiments on plant/insect interactions are based on field situations, so that they combine physiological understanding with appropriate behavior or ecology. Through this work she has brought a new understanding to the process of host-plant selection by insect herbivores, ranging from the analysis of chemical stimuli, through acceptance for feeding, to the nutritional implications of the resultant diet. In particular, she has demonstrated that many plant secondary substances, while critical as behavioral cues, are often surprisingly inactive metabolically, and many have subtle benefits — shedding new light on such general questions as diet breadth in insects, and the costs and benefits of different host-plant ranges. Most recently her elegant experiments on predators have highlighted the significance of higher trophic levels in the maintenance of specialized feeding habits.

TABLE OF CONTENTS

1 Seed Predators as Mutualists: Ecology and Evolution of the Fig/Pollinator Interaction

Judith L. Bronstein
Department of Ecology and Evolutionary Biology
University of Arizona
Tucson, Arizona

TABLE OF CONTENTS

I. INTRODUCTION

The concept of coevolution was first discussed explicitly in the 1960s,[54] and studies of apparently coevolved systems proliferated in the 1970s. Since that time, however, researchers have increasingly realized that an impressively close match between traits of interacting species is insufficient evidence to conclude that a long-term process of reciprocal change has occurred. Various constraints have been identified that probably prevent coevolution within most two-species interactions.[13,61,85,91,151,183] For example, both the relative and absolute effects of different potential partners must often vary over space and time, reducing the persistence of selective effects necessary for long-term change. Furthermore, rates of evolution can differ substantially between interacting lineages, reducing the likelihood of reciprocal change, even if strong selection does in fact persist. Heritable variation for traits relevant to the interaction may be rare in any case. While most researchers would not at present deny that coevolution can occur, they increasingly discount its broad importance in shaping species interactions. This has left a few really good examples of coevolution that are repeatedly cited, but only very superficially understood. However, these are precisely the cases that need intensive study if we intend to understand the evolutionary processes by which coevolution *can* occur and the ecological patterns it can produce.

Perhaps the classic case of plant/insect coevolution is the obligate mutualism between figs (*Ficus* spp., Moraceae) and their pollinator wasps (Hymenoptera, Agaonidae). Almost every fig species is pollinated by a different agaonid wasp species; the wasps can only reproduce in the inflorescences of that fig, wherein their larvae feed on some developing seeds. Its extreme specificity, in particular, has led this interaction to be considered as a "limiting case" of what coevolution can produce, and has provided a cornerstone to which other interactions have been compared. The fig/fig wasp relationship is particularly intriguing because it is simultaneously a coevolved plant/pollinator and plant/seed-predator interaction. Although this combination is unusual, it is certainly not unique to this system; the best-understood parallels are the interactions between yuccas and yucca moths,[1] and between the globeflower (*Trollius* spp., Ranunculaceae) and anthomyid flies.[136] Furthermore, coevolved interactions combining seed predation with mutualistic seed dispersal are well known (reviewed by vander Wall[174]). The existence of such interactions raises many questions about how mutualistic benefits can arise and persist within essentially antagonistic interactions. This topic will be one of the major themes of this review.

It is only recently that we have had sufficient knowledge of the fig/pollinator interaction to pose informed questions about the coevolutionary processes that have shaped it. The relationship was poorly understood as recently as the middle of this century. Extensive taxonomic work, first by G. Grandi and later by J. T. Wiebes, had shown that fig wasps were usually fig-species specific, and they had long been implicated in seed formation (reviewed by Wiebes[188]). However, it was unclear what the larvae ate or whether adults did in fact carry pollen. During the 1960s and early 1970s, three lines of research were pursued that provided the foundation for our current understanding of the mutualism. First, detailed field studies of fig-pollinator interactions were conducted, primarily by Galil and his students (e.g., Galil and Eisikowitch[65,66]). Second, studies by Ramirez,[141] Galil and Eisikowitch,[68] and Chopra and Kaur[35] identified the means by which fig wasps collect, transport, and transfer pollen. Finally, continuing taxonomic studies provided essential information on fig/pollinator specificity and the phylogenetic distribution of traits central to the interaction.[90,141,142,144,186] This work was reviewed in detail in an influential review paper published in 1979 by Janzen,[95] which also posed many of the central questions about the evolution of this system that researchers continue to address today.

After reviewing the natural history of the fig/pollinator interaction, I will focus in turn on a particularly critical aspect of fig/pollinator coevolution, one of the population-level consequences of it, and on its community-level consequences. I first consider how mutualistic benefits might be able to persist within this seed predation interaction. It has increasingly been recognized that the evolutionary interests of the partners within a coevolved mutualism can differ or even conflict.[163] However, the consequences of those conflicts have received very little attention, although it is their outcome that will determine the coevolutionary dynamics of an interaction. In the mutualism between the monoecious fig species (half of the species of *Ficus*) and their pollinators, one conflict arises because of the following imbalance: while it is in the interests of the fig to produce both seeds and seed-feeding wasp offspring (the only possible pollen vectors), wasps aim only to leave offspring. If higher wasp fecundity could evolve, it would be advantageous to wasps but increasingly costly to the fig. The outcome of this conflict will, over evolutionary time, determine whether fig wasps will remain mutualists or become antagonists of their hosts.

These evolutionary conflicts have direct consequences for fig and pollinator population dynamics. Fig wasps only benefit figs by transferring pollen between trees, and fig flowering phenology has evolved to increase the likelihood that this will happen. One consequence is that within a fig population, trees flower out of synchrony with each other across the year. I will show how a gap in flower production within a fig population can cause local extinction of the fragile pollinators. This risk carries distinct consequences for individual reproductive success and the short-term persistence of the interaction.

Finally, I discuss the community-level consequences of fig/pollinator coevolution. I review the ecology of several classes of organisms that interact with and may influence the outcome of this mutualism, and examine whether they, in turn, may have coevolved with the mutualists. I then consider the growing opinion among conservationists that the fig/pollinator interaction can play a central role in structuring tropical vertebrate communities.

Throughout this review my primary focus is on monoecious figs, those species in which each individual reproduces both as a female (seed producer) and male (pollen donor). Monoecious figs make up about half of 700 species within the genus *Ficus*, including all the New World species.[11] At this point their natural history is much better known than the gynodioecious figs; of the latter, only the cultivated edible fig, *Ficus carica*, has been studied in depth.[108,171,172] However, gynodioecious figs are fascinating in their own right, particularly with regard to evolutionary conflicts with their pollinators,[108,109] and are deserving of much more attention than they have received to date.

II. NATURAL HISTORY

The fig pollination interaction involves all members of one plant genus (*Ficus*, Moraceae) and all members of one insect family (Agaonidae, superfamily Chalcidoidea, Hymenoptera). Evolutionary relationships within each are still incompletely understood. The taxonomy of *Ficus* is currently being reworked to take into account morphological traits related to reproduction and pollination.[11] Meanwhile, the agaonids associated with many figs remain undescribed.[46,191] The sketchiness of these phylogenies unfortunately makes it premature to evaluate the likelihood that figs and agaonids have speciated in parallel, although evolutionary biologists have often speculated upon this possibility.[164,185] Furthermore, although the interaction seems in general to be remarkably species specific (one fig species associated with one agaonid pollinator), exceptions are beginning to accumulate in the literature.[38,121,140,191] In this section, I review the natural history of the pollination interaction, first for the better known monoecious fig species, then for the more derived gynodioecious members of the genus.

A. THE MONOECIOUS FIG/POLLINATOR INTERACTION

To summarize the pollination mutualism in monoecious figs, I will use the example of *Ficus sycomorus*, a widely distributed African species that has been studied in exceptional detail by Jacob Galil and his colleagues.[65,66,68,69,70-72] Following this description, I discuss the extent to which other fig/pollinator pairs deviate from the *F. sycomorus* pattern. For convenience, I will divide the reproductive cycle into the five phases defined by Galil and Eisikowitch.[65] These stages are defined and described in Table 1.

1. Pollinator Arrival

At unpredictable intervals, a *F. sycomorus* tree initiates several thousand syconia on short panicles on the trunk and main branches; this is the beginning of the *A phase*. The syconium is an infolded receptacle bearing many small, unisexual flowers on its internal surface, accessible only through a tight bract-covered pore, the ostiole. (The syconium is colloquially referred to as the "fig", but in this review I will usually refer to it by the botanical term to avoid any confusion with the fig *tree*.) The functional anatomy and development of the syconium has recently been discussed in detail by Verkerke,[179] and I will summarize only the salient points here. Figure 1 shows a cross-section through the syconium of a monoecious species such as *F. sycomorus*. The internal wall is lined with several hundred female flowers borne on pedicels of varying heights and possessing styles of different lengths. (In *F. sycomorus*, style lengths range from 0.8 to 1.5 mm.) In combination, this pedicel and style-length variation results in the formation of a level platform of stigmas within the cavity of the syconium. Depending on the subgenus of *Ficus*, the few anthers are either clustered at the ostiole (as in *F. sycomorus*) or dispersed among the female flowers.[11]

A few weeks after syconia have been initiated, the pollinators, female *Ceratosolen arabicus* in the case of *F. sycomorus*, arrive at the tree in large numbers. There is general agreement in the literature, and extensive circumstantial evidence, that the wasps have followed species-specific odors released by the trees at the point when female flowers are receptive.[7,23,27,175] This is the *B phase*. Each wasp has earlier been inseminated and, in addition, is carrying thousands of pollen grains within several

Table 1
DEVELOPMENTAL PHASES OF MONOECIOUS FIGS AS DEFINED BY GALIL AND EISIKOWITCH[65]

Phase A (prefemale):	Young syconium prior to the opening of the ostiole
Phase B (female):	Ostiolar scales loosen, female flowers ripen, pollinators penetrate into the syconium and oviposit into the ovaries
Phase C (interfloral):	Wasp larvae and fig embryos develop within their respective ovaries
Phase D (male):	Male flowers mature, wasps reach the imago stage, fertilized female wasps leave the syconia via tunnels bored by the males
Phase E (postfloral):	Both the syconia and the seeds inside them ripen

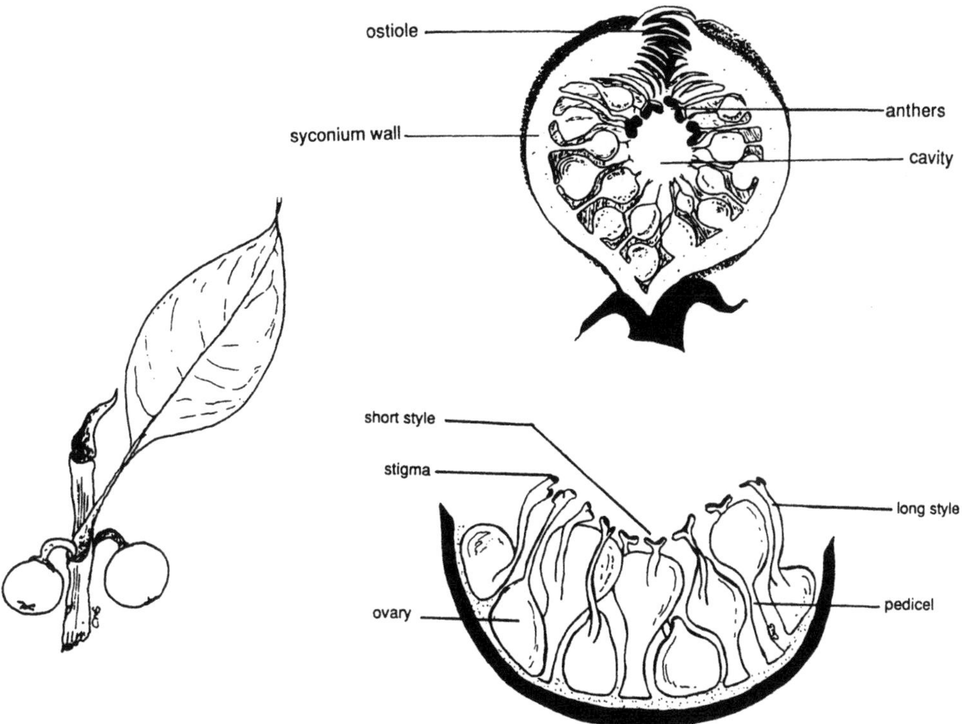

FIGURE 1. Morphology of the syconium of a typical monoecious fig species, such as *Ficus sycomorus*.

pockets on the ventral portion of the mesothorax. Between 1 and 10 wasps enter each syconium by laboriously working their way through the bracts of the ostiole. A few reach the cavity, generally losing their wings and distal segments of the antennae in the process, while the rest are crushed between the bracts of the ostiole. Even the successful individuals become permanently trapped within the syconium. The B phase lasts only a few days. Any unentered syconia are abscised from the tree within about a week of wasp arrival.

2. Oviposition and Pollination

Once inside and on the stigmatic surface, each *C. arabicus* attempts to oviposit into the female flowers. (The anthers are still immature at this phase.) A female first inserts her ovipositor through the stigma and down the style of a flower. However, the ovipositor is shorter than the longest-styled of the flowers. If the style she probes is too long to allow the tip of her ovipositor to contact the ovary, she withdraws it after several seconds. If the style is relatively short, however, she lays one egg between the inner integument and nucellus of the ovary. She then withdraws her ovipositor and the process is repeated in adjacent flowers. Many of her oviposition attempts are followed by a quick movement in which she removes pollen from the thoracic pockets with her forelegs and deposits it upon the stigma of that flower. Consequently, *as a rule*, seeds develop within longer styled flowers, while the seed-eating wasp offspring tend to mature within shorter styled flowers. Detailed discussion of this phenomenon will be deferred until the following section.

3. Seed and Wasp Maturation

The seed and wasp developmental period (*C Phase*) lasts 3 to 4 weeks in *F. sycomorus*. During this period, *C. arabicus* larvae presumably feed upon proliferating endosperm, as has been documented for other pollinator species.[179]

Some fig ovaries contain neither developing seeds nor *C. arabicus* offspring. During development, many *F. sycomorus* ovaries can be seen to be vacant: some of these appear not to have been probed by the pollinators, whereas others ("bladders") have been probed and are swollen but empty. Egg or larval mortality may be responsible for this latter class of vacancies. Additionally, as in all fig species, several chalcidoid wasps that do not transfer pollen can be found to be developing within some ovaries. The ecology of these "interloper" wasps and their interactions with the fig and pollinator are discussed in detail later.

4. Mating, Pollen Loading, and Departure

The fig then enters the comparatively brief *D phase,* the stage at which pollen is dispersed.

C. arabicus males mature about a day before the much more numerous females, leave the fig ovaries in which they matured, and enter the syconium cavity. Using their strong jaws, the wingless, eyeless male wasps puncture the walls of the ovaries that contain females, and inseminate them. They then pull loose the newly mature anthers, which fall into the syconium cavity. A group of males then tunnels through the syconium wall, providing an exit for the females.

Meanwhile, inseminated females leave their fig ovaries and move to the cut anthers. They widen the dehiscence slits with their mandibles and antennal scapes, then shovel pollen into their mesothoracic pockets. Females then depart the syconium through the tunnel excavated by the males in search of B-phase syconia in which to oviposit. The males die, generally still within the syconium.

Finally, during *E phase*, the syconia soften, swell with water, and turn reddish. They are dispersed by a variety of vertebrates, including birds, bats, and monkeys. Note that although technically the syconium is a fleshy peduncle containing many individual achenes or drupes,[48] ecologically it acts as a fruit.

5. The Free-Living Stage

As yet, almost nothing is known about what happens between the time when the pollen-laden females leave their natal tree and when they appear at a receptive syconium. While fig wasps have the *potential* to travel many kilometers between trees,[135,142] it is not known how far they *typically* travel. Adult agaonids do not feed, and their lifespan has usually been estimated to be between a few hours and a few days.[102,110] Mortality in flight is probably extremely high.[27,87]

There is one consequence of the fragility of adult fig wasps that is critical for understanding the mutualism as a whole. For any monoecious fig individual to reproduce, i.e., to either attract wasps carrying non-self pollen or to donate pollen successfully, *that tree must flower out of synchrony with its conspecific neighbors.* If at any point in the year all female agaonid larvae in the population mature, emerge, and *cannot* locate a receptive fig near enough in space and time to reach, pollinators will become locally extinct. In that case, no individual in the fig population can reproduce again until pollinators recolonize the area.

One can see evidence for this phenology and its reproductive consequences by looking across the range of *F. sycomorus.* In Kenya and Namibia, at least some *F. sycomorus* trees are in flower at all times of year,[65,181] although in the more seasonal Namibian forests, relatively few trees fruit in the winter.[181] In contrast, in Israel, *F. sycomorus* is nearly dormant during the winter, resulting in a gap of several months in the flowering sequence of trees.[66] Not surprisingly, the pollinator is absent in Israel.[66,74] I elaborate on the causes and consequences of fig flowering phenology in a later section.

6. Interspecific Diversity

Bronstein and McKey[31] have discussed the fact that there is no single typical "fig/pollinator interaction": although the mutualism is relatively stereotyped, important differences do exist among the 700 or more partners. Because so many wasp and fig traits have consequences for their interaction, interspecific variation in one mutualist taxon is usually matched by traits of the partner, providing clear evidence of coevolution and rich material for comparative studies. Research on this interaction has recently shifted towards documenting this variability and speculating on the evolutionary processes that have produced it.[12,27,86,87,195] Here I will point out just a few of the ways in which coadaptations of other monoecious fig/pollinator pairs deviate from the *F. sycomorus/C. arabicus* pattern; since this was the first well-studied example, it remains, literally as well as figuratively, the textbook fig/pollinator interaction.

Pollination behaviors vary strikingly among agaonids, although at present our knowledge is based on studies of very few species.[35,59,68,70,73,133,141] *C. arabicus*, the agaonid associated with *F. sycomorus*, is an "active" (or ethodynamic[63]) pollinator: it exhibits specific behaviors for collecting and unloading pollen from its pollen pockets. The other agaonid species are passive, or topocentric, pollinators. These species have no pollen pockets; they become covered with pollen as they depart the natal fig, and subsequently deposit it passively as they attempt to oviposit. Certain fig traits are correlated with the different modes of pollinator behavior[144] (but see Herre[87]). For instance, actively pollinated figs tend to contain fewer anthers and less pollen, probably because their mutualists are relatively more efficient at transporting pollen. Individual sections of the genus *Ficus* are associated with pollinators of one of the two types.[11] Passive pollination is believed to be the ancestral condition for fig wasps; however, whether or not the mode of pollen transport and behaviors associated with it should be major characters in classification of the Agaonidae has been a matter of contention.[145,146,189]

Interspecific variation in syconium traits also has direct consequences for interactions with pollinators. Syconium sizes vary greatly among fig species, from less than 0.5 to over 6 cm in diameter (the *F. sycomorus* syconium is about 3 cm at the time of pollinator entry). Data from Hill,[90] Herre,[87] Berg,[12] and others indicate that flower number increases with syconium size; hence, for example, the 0.4-cm syconia of *F. mathewsii* have about 50 flowers, while the 3-cm syconia of *F. cyathistipuloides* contain about 2500. In an important series of papers, Herre[86,87] has documented some of the ecological correlates of flower number within an assemblage of Panamanian fig species. Across species, average numbers of pollinators entering an individual syconium increases with average syconium size.[87] Herre[85] and others[58,107] have shown that sex ratios of the agaonid offspring become increasingly male biased as the number of pollinators per syconium increases, consistent with predictions from theories of local mate competition;[81] thus larger syconia (i.e., those with more flowers) mature more wasps, but proportionately fewer females (the only sex that carries pollen). Species with more flowers per syconium, however, produce proportionately more seeds per syconium than do those with fewer flowers.[87] Apparently, then, the distribution of flower numbers between and within syconia has complex but direct consequences for the reproductive success of the fig species via the female (seed-producing) and male (pollen-donor) functions. It should be pointed out that flower numbers also can vary considerably *within* fig species.[12, 87] The consequences of this variation may be great, but remain largely unexplored.

B. THE GYNODIOECIOUS FIG/POLLINATOR INTERACTION

The most profound interspecific difference among fig/pollinator interactions is found between fig species with different breeding systems. In the monoecious species, which I have discussed up to this point, every syconium on each tree contains both female and male flowers. In contrast, the other species are made up of some individuals with both female and male flowers in each syconium (hermaphrodites) and others with female flowers only (females). Hermaphrodites function exclusively as males; consequently, these species have commonly been referred to in the literature as dioecious, although morphologically they are gynodioecious. As currently delineated, *Ficus* is made up of two groups of subgenera containing exclusively monoecious species, two containing exclusively gynodioecious species, and one containing both.[12] Berg, the current authority on fig taxonomy, believes that gynodioecy has evolved from monoecy at least twice[10] (see also Ramirez[147]).

In the gynodioecious figs, separation of the sexual functions is effected by two traits: the absence of functional anthers in syconia on female trees, and a difference in style length between female flowers within the hermaphroditic and the female syconia. The importance of these traits can be seen by contrasting what happens when a pollinator enters a syconium of a gynodioecious, as opposed to monoecious, species. Figure 2A shows the style length difference between hermaphroditic and female *F. carica* trees in France (Bronstein and Kjellberg, unpublished data); contrast this with Figure 2B for the monoecious *F. pertusa* in Costa Rica.[24] Recall that once within the monoecious syconium, the pollinator encounters female flowers with a wide range of style lengths. Only a portion of these are accessible for oviposition, resulting in the maturation of both pollen-carrying offspring and uneaten seeds within every syconium. In contrast, in a gynodioecious fig the pollinator encounters either only accessible or only inaccessible ovaries, depending on the sex. If she has entered a syconium on a female tree, she confronts ovaries that are entirely inaccessible, because all styles are much longer than her ovipositor. However, she attempts to oviposit anyway, depositing pollen on the stigmas in the process. The result is a syconium that matures only seeds. No eggs are laid, so no offspring develop; even if they did, there are no anthers, so these wasps could transport no pollen. If the wasp enters a syconium on a hermaphroditic tree, however, she encounters only very short-styled female flowers and has access to every one. Consequently, few or no seeds are produced. At maturity, the offspring behave as

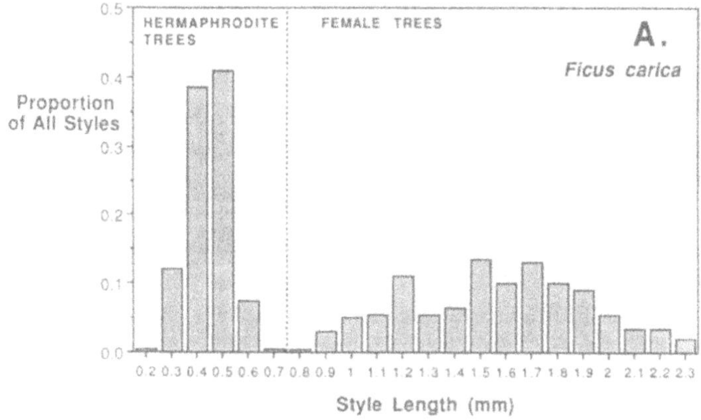

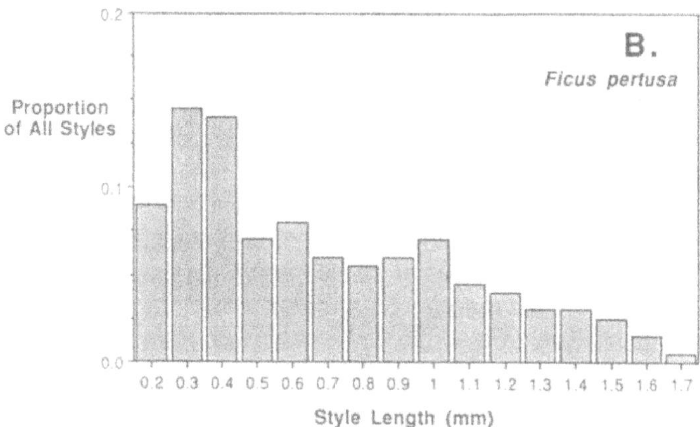

FIGURE 2. Average style-length distributions within individual syconia in a monoecious and a gynodioecious fig species. (A). Average style-length distribution of gynodioecious *Ficus carica*, calculated from six hermaphrodite and six female trees of identical parentage in an experimental plantation in Montpellier, France (Bronstein and Kjellberg, unpublished data). (B). Average style length distribution of monoecious *Ficus pertusa*, calculated from 26 syconia from 3 Costa Rican trees.[25]

described above for the monoecious species: mating, collecting pollen, and departing in search of another syconium.

A number of other fig and wasp traits vary in association with the difference in the breeding system.[11,12,176,180] Many of these remain obscure because so few gynodioecious fig species have yet been studied in the field.[44,64,108,128,129,171,176] Perhaps the most striking difference between monoecious and gynodioecious figs is in flowering phenology. Clearly, it is in the interests of the pollinator of a gynodioecious fig to avoid female trees, in which they cannot reproduce. There is some evidence that sexual differences in flowering time have evolved in a way that prevents pollinators from discriminating between the sexes.[108] I will return to this issue in the subsequent section.

III. CONFLICTS BETWEEN FIGS AND POLLINATORS

The fig/fig wasp association is one of the best-studied examples we have of an obligate mutualism, i.e., a mutually beneficial interaction in which neither participant can persist without the other. At one time, mutualisms were idealized as altruistic, cooperative interactions (reviewed by Boucher[17]). More recent research has, however, started from the view that traits benefiting the mutualist partner at a cost to the bearer generally cannot evolve.[3,8,18,62,78,132,156] The result has been a shift towards regarding mutualisms as "reciprocal parasitisms"[98] in which conflicting individual interests are balanced, at least for the moment.

If this is the case, then some mutualisms should be capable of shifting to increasingly unilateral exploitation when the balance of reciprocal benefits is disrupted.[163] This implies that there may be no sharp discontinuity between mutualism, on the one hand, and antagonism (i.e., predation, parasitism), on the other, as ecologists often assume. Rather, a continuum may exist among outcomes of a given interaction, either over time or in space. This important idea has, however, received little empirical study to date. It has proven difficult to identify ecological systems with simple, measurable traits whose variation could shift an interaction along the putative continuum (but see Clay,[36] Ewald,[55] and Turner et al.[167]).

The fig/pollinator interaction is perhaps an ideal system for studying the mutualism/antagonism continuum. The origins of the association are still obscure, but agaonids probably evolved from chalcidoid wasps that visited the pre-*Ficus* inflorescences to feed on seeds. The transition from antagonism to mutualism may have begun with incidental and occasional passive pollen transfer by the wasp. This would have benefited both the fig and the wasp (since it would guarantee retention of the inflorescence and provide a food source for its offspring). Subsequently, traits associated with increasing interdependence may have arisen. But, a central challenge to the biologist is to explain why figs today produce any *uneaten* seeds at all. We generally assume that, to a fig wasp, a seed only represents a failed oviposition attempt. A mutant genotype that could lay more eggs should then be at an evolutionary advantage. Furthermore, given the huge disparity in generation times between figs and fig wasps, figs may not be able to evolve fast enough to counter the advantage held by increasingly exploitative wasps. In fact, much smaller disparities in generation time have been used to argue that coevolution is unlikely to occur.[88] How can this interaction coevolve at all, and how can it persist as a mutualism?

In the monoecious figs, the evolutionary conflict can be stated as follows: while the fig wasps presumably aim to leave as many offspring as possible, an individual fig tree aims to produce not only the pollen-carrying wasp offspring, but some undestroyed seeds as well. Style-length variation has been generally assumed to be the mechanism by which figs guarantee that they will produce at least some uneaten seeds. "Long-styled" flowers supposedly mature seeds, whereas "short-styled" ones contain seed-eating wasps.

There is at least some evidence that the longest styled flowers do generally mature seeds[65,130] (but see Compton and Nefdt[40]). It does not necessarily follow, however, that style-length variation is sufficient to account for the persistence of fig seed production through evolutionary time, counter to the interests of the wasps. Consider this style-length hypothesis as pictured schematically in Figure 3. Clearly, fig ovaries in a given syconium are only long styled or short styled in relation to an individual wasp's ovipositor length. Now imagine that a wasp with a slightly longer ovipositor entered and laid eggs within this syconium. More wasp offspring, but fewer intact seeds, would mature. As Murray[126] was first to point out, strong directional selection on wasp ovipositor length should result, and since wasps undergo more than a hundred generations per fig generation, longer ovipositors should rapidly evolve (assuming the trait is heritable). This would bias fig reproduction progressively away from seed production and towards the male function of pollen donation, ultimately to the point where seed production would cease. At that point, fig wasps would be antagonists rather than mutualists of the fig. Thus, style-length variation per se cannot explain the persistence of the monoecious fig/pollinator interaction as a mutualism.

What then could? There are two general possibilities. First, some mechanism may have evolved by which figs counter the inevitable evolution of ever-longer pollinator ovipositors, while maintaining a set of ovaries inaccessible due to long style lengths. The challenge in this case is to identify a mechanism that would work and to find evidence for it in nature. Alternatively, style lengths may not, in fact, be the critical constraint on wasp fecundity, as the style-length hypothesis (Figure 3) would predict. Wasps with longer ovipositors would not then experience any fitness advantage; this would allow seed production to continue in longer styled flowers. If so, it is necessary to identify the factors that keep wasp fecundity so low as to assure the continued production of seeds. Below I consider these two possibilities in turn. There is no necessity for one explanation to be universal to all fig/pollinator pairs, although it is tempting to think that this would be the simplest explanation for the mutualism's persistence.

A. FIG CONSTRAINTS ON WASP FECUNDITY

Several mechanisms can be envisioned that could keep fig-wasp ovipositors short enough to guarantee that seed production would continue in the longest-styled flowers of the syconium.

1. "Unbeatable" Style-Length Distribution

The relative advantage of a marginally longer ovipositor is a function of the style-length distribution within the syconium that the wasp enters (Figure 4). If style lengths are continuously distributed, a wasp with a longer ovipositor has access to more oviposition sites (Figure 4A); this is the assumption behind the fig/pollinator conflict pictured in Figure 3. But if style lengths were *bimodally* distributed, and wasp ovipositors were of intermediate length, then a longer ovipositor would confer no advantage to the bearer (Figure 4B). Could a bimodal style-length distribution account for the persistence of seed production?

In fact, every monoecious fig examined so far has had a unimodal rather than bimodal distribution of style lengths.[6,24,40,65,100,130] For example, I found[24] that style lengths in Costa Rican *F. pertusa* were always unimodally distributed within syconia and were skewed towards very short lengths (Figure 2B). Figure 5 superimposes the distribution of pollinators' ovipositor lengths (from 50 *Pegoscapus silvestrii* trapped while ovipositing) onto the average *F. pertusa* style length distribution (calculated from 26 syconia of 3 trees). It can easily be seen that for *P. silvestrii*, a relatively longer ovipositor would permit access to more fig ovaries. Figure 6 shows this advantage quantitatively: an individual with an ovipositor 10% (0.1 mm) longer than average should have access to 8% (or 15) more oviposition sites.

The unimodal style-length distribution seems to be a proximate consequence of the dense packing of flower primordia and competitive style growth during early development within the syconium; styles of different length have similar cell numbers.[177-179] Evolutionarily, it may be an indirect effect of selection for increased numbers of flowers per syconium. Logically, it should be possible to pack more ovaries into the monoecious syconium if the lengths of the pedicels connecting the ovaries to the syconium wall are continuously rather than bimodally distributed. Since stigmas must form a uniform surface within the cavity for pollination to occur, a unimodal style-length distribution would be the inevitable, correlated result. In any case, the tentative generalization

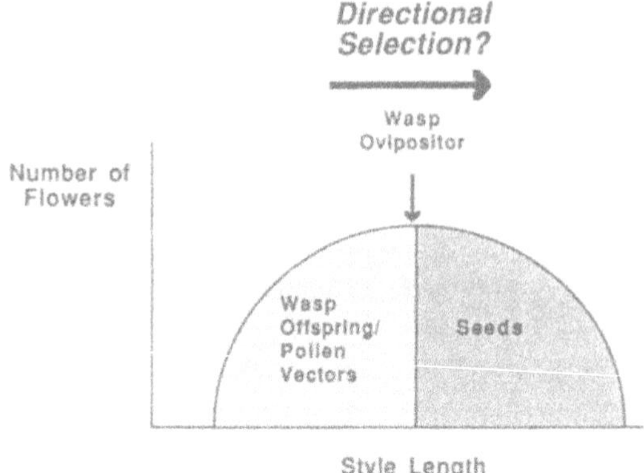

FIGURE 3. Hypothetical scenario of the evolutionary conflict between monoecious figs and their pollinator wasps. If wasp oviposition is regulated by the number of fig ovaries with styles shorter than or equal to ovipositor length, strong directional selection can be expected on wasp ovipositor length. This would lead to the production of progressively more wasp offspring and fewer seeds, ultimately to the point at which seed production stopped.

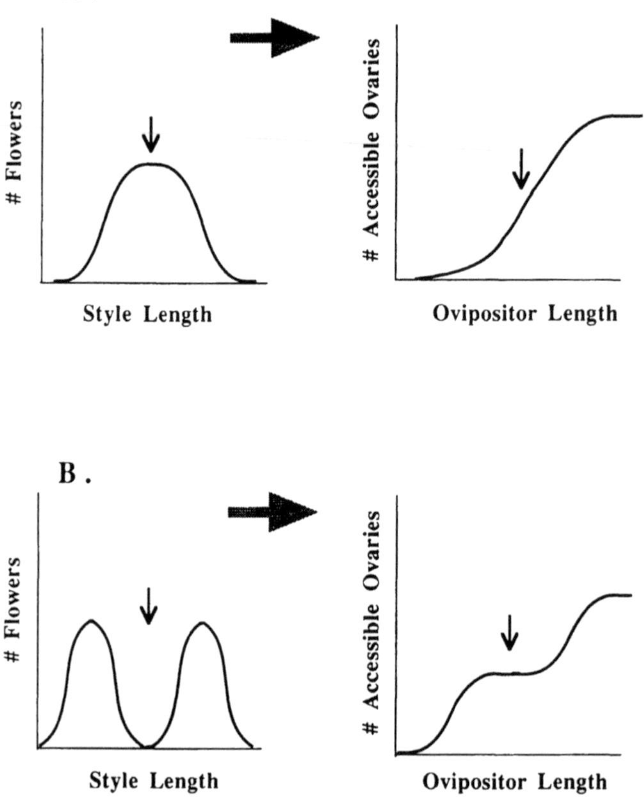

FIGURE 4. Hypothetical consequences of style-length distribution for directional selection on pollinator ovipositors. (A). If the style length distribution is unimodal, a marginally longer ovipositor can reach more fig ovaries (= oviposition sites) and should be favored by selection. (B). If the style length is bimodal, and ovipositors are of intermediate length, a marginally longer ovipositor has access to no more oviposition sites.

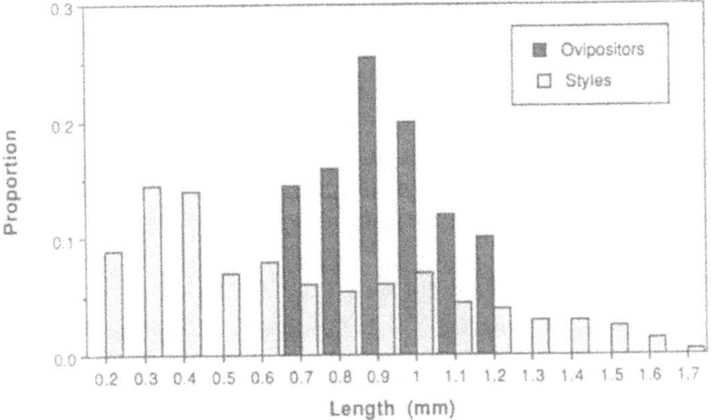

FIGURE 5. Average distribution of style length per *Ficus pertusa* syconium and the distribution of pollinator (*Pegoscapus silvestrii*) ovipositor lengths in Monteverde, Costa Rica. See text for sample sizes.

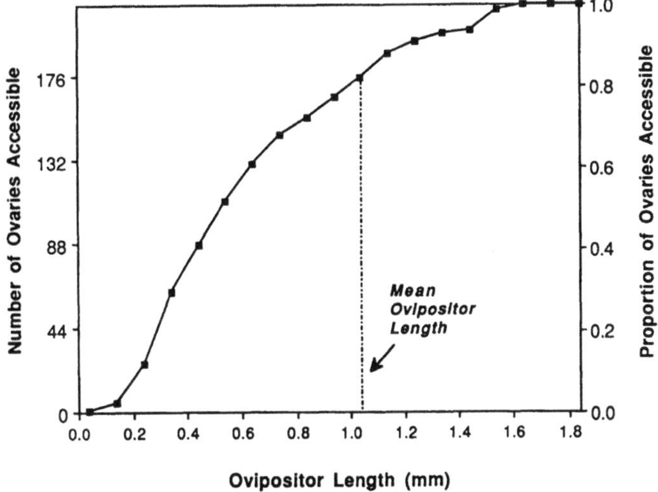

FIGURE 6. Consequences of style-length distribution in the *Ficus pertusa–Pegoscapus silvestrii* interaction. The average pollinator can reach 82% of all fig ovaries; an ovipositor 10% (0.1 mm) longer than average should have access to 8% (15) more ovaries (= oviposition sites).

emerges that in monoecious figs pollinators with longer ovipositors should in fact have access to more oviposition sites, exacerbating rather than alleviating conflicts between the mutualists.

2. Escalating Race Between Style and Ovipositor Lengths

Alternatively, selection for longer wasp ovipositors may select for ever longer style lengths in figs, resulting in an escalating coevolutionary "arms race" (cf. Dawkins and Krebs[53]). Murray[126] has argued cogently against this hypothesis. Its most obvious weakness is that a coevolutionary race is difficult to envision between organisms that differ so radically in generation time. It could be valid if styles lengthened more quickly under selection than ovipositors did, but there is no reason to believe that this is the case, particularly given natural variation for ovipositor length within agaonid populations (e.g., Figure 5) and the apparent value of a longer ovipositor (Figure 6).

3. Disadvantages of Longer Ovipositors

If there were strong selection against longer ovipositors (or a linked trait) during another stage of fig-wasp life history, they may not evolve, despite the advantage they confer at the moment of oviposition. Such selection might or might not be imposed by the fig itself in a coevolutionary response to overexploitation by the wasps. For example, selection may have favored narrower passages through the ostiole, restricting entry to smaller individuals with shorter ovipositors;[126] alternatively, it may simply be more difficult for a fig wasp to fly with a longer ovipositor.

No published data are available that bear on this intriguing hypothesis. However, in a small unpublished study of *F. aurea* and its pollinator *Pegoscapus assuetus* in Florida, I found no evidence for a disadvantage of longer ovipositors. In this population, ovipositor length and body size (using femur length as an index) were uncorrelated ($r^2 = 0.01$, $n = 45$, $p > 0.25$). There was very little variation in either ovipositor length or body size, comparing among wasps departing their natal trees, arriving at receptive trees, and successfully entering receptive syconia. However, this hypothesis deserves further study, using larger sample sizes and more precise measures of wasp size.

4. Abortion of Overexploited Syconia

The hypothesis that Murray[126] favored to explain the persistence of the mutualism was the ability of the fig to selectively abort "overexploited" syconia. Syconia entered by fig wasps with longer ovipositors should contain a high ratio of developing larvae to developing seeds. Murray argued that in periods when the wasps are very common, natural selection should favor figs that increase their investment in seed production. This could occur if young syconia with the highest wasp-to-seed ratio were dropped and the resources were reinvested in more seed-rich syconia, in the same or the subsequent crop.[4,95,126] Fig wasps might subsequently be selected to purposefully pollinate inaccessible ovaries; this would lower the wasp-to-seed ratio, and thus the abortion rate, at no cost to the individual wasp.[126]

Many plant species do abort a large proportion of developing fruits, often showing some ability to discriminate among them according to seed number, pollen source, and/or extent of insect damage.[113,159,192] However, empirical evidence to date provides no support for this phenomenon in figs. Published reports indicate that figs typically mature every pollinated syconium.[25] The exception seems to be during times of unusually strong resource limitation, regardless of local pollinator abundance. For instance, *F. pertusa* aborts developing syconia only when they are pollinated at an exceptionally small size.[25] In Florida, *F. aurea* does the same; it also aborts most of the few syconia pollinated during the winter, when both available pollinators and resources for fruit maturation are scarce (J. Bronstein, unpublished data). We have compared aborted and retained syconia on both *F. aurea* and *F. citrifolia* at one study site and have detected no differences in pollinator number per syconium, syconium size, or ovary number (J. Bronstein and A. Patel, unpublished data).

It is a much more difficult problem to determine whether figs might abort some developing wasps within heavily exploited but retained syconia. Certainly, many fig ovaries are vacant at maturity, suggestive of high wasp mortality (see below). But is this mortality the product of some evolved strategy of the fig to prevent overexploitation by fig wasps? This question leads into the general issue of whether constraints on this exploitation might in fact be *extrinsic* to the conflict between the mutualists, rather than intrinsic to it, as all of the hypotheses discussed above have assumed.

B. ALTERNATIVE CONSTRAINTS ON WASP FECUNDITY

Thus, although some of the above-mentioned mechanisms by which figs could prevent the evolution of longer wasp ovipositors do make intuitive sense, there is as yet no evidence that they occur in nature. I believe that two assumptions underlie all of these hypothetical restraints to pollinator fecundity. First, these hypotheses assume that access to fig ovaries is the primary, or even the only, limit to fig-wasp fecundity. They then assume that a tradeoff is thereupon established between the number of wasps and the number of seeds that mature within an individual syconium. If these assumptions are incorrect, then a pollinator with a longer ovipositor has no clear fitness advantage and inflicts no cost to the fig as simple as that shown in Figure 3. In this case, alternative factors must be sought to explain why wasp fecundity is so low as to assure continued seed production in figs.

1. Does Style Length Actually Regulate Ovary Fate?

Pollinators with longer ovipositors do have access to more fig ovaries in which to oviposit (FIgure 6), but this does not necessarily mean that they leave more offspring. To determine this, one would ideally allow wasps of known ovipositor length individually into syconia, and then count the number of offspring that mature successfully. No one has yet attempted these tedious experiments. However, two other methods have been used to study this issue, and neither indicates that style length is the central constraint on fig wasp fecundity.

I used the data shown in Figure 5 to calculate the frequency of seeds and wasps that would mature in *F. pertusa* if relative style-to-ovipositor lengths were the only constraint on pollinator fecundity.[24] The prediction was that in the average syconium entered by the average wasp, 82% of all ovaries would mature wasp offspring and the remaining 18% would mature seeds (Figure 7A). In reality, these figures vastly overestimated wasp occupancy in the 600 mature syconia sampled from 12 trees (Figure 7B). On average, only 11% of all ovaries, rather than 82%, contained pollinator offspring. Furthermore, seed production was somewhat higher than the predicted value (24% instead of 18% of all ovaries). Another 12% of ovaries contained offspring of three nonpollinator wasps in the family Torymidae; these species feed on sterile tissue in otherwise unoccupied ovaries, rather than on seeds or pollinator larvae.[28] Most importantly, over half the ovaries in the average syconium were vacant at maturity. Note that Figure 7B is based on maturation within the *average* syconium. However, percent occupancy by pollinator larvae ranged from 0 to 60%, so that even in the most thoroughly exploited syconium, wasp occupancy was below the prediction based on relative style and ovipositor lengths.

Compton and Nefdt[40] examined more directly whether utilization of *F. burtt-davyii* by its pollinator *Elisabethiella baijnathi* was a function of access to ovaries. The pattern of wasp utilization was determined, in this case, by dissecting flowers at the stage when wasp pupae were present and the styles still intact. The ovipositor length distribution within the *E. baijnathi* population indicated that the average individual can reach 83% of all *F. burtt-davyii* ovaries, with some individuals able to reach every one. However, pollinator offspring matured in only 7% of all ovaries; seeds matured in 10%, nonpollinator offspring in 11%, and the remaining 72% of all ovaries were vacant at maturity. Of the vacant ovaries, 50% appeared to be infertile, while the rest showed evidence of some development, possibly an indication of early larval mortality.

These results do not seem to be unusual for monoecious figs,[67,100,130] although

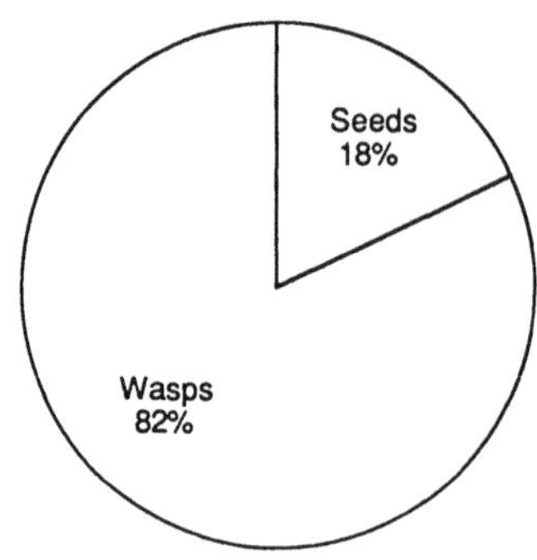

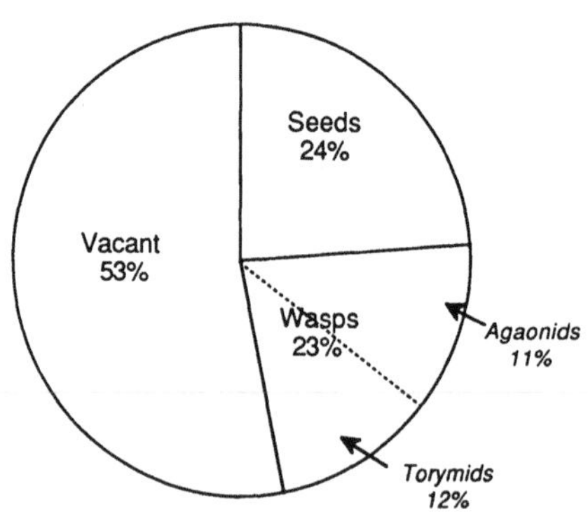

FIGURE 7. Occupancy of ovaries within *Ficus pertusa* syconia in Costa Rica.[25] (A). Prediction from data on relative style to ovipositor lengths (see Figure 5). (B). Average contents of about 600 syconia sampled from 1982 to 1984.

interspecific comparisons are difficult, since there are almost no published data on wasp ovipositor lengths or on fig style lengths and ovary numbers. The insufficiency of style lengths as a predictor of ovary utilization might have been anticipated, however, from some of Galil and Eisikowitch's[65,69] early observations of *F. sycomorus*, in which (1) ovary vacancies at maturity are fairly common, (2) nonpollinator wasps that feed within otherwise unoccupied ovaries are abundant, especially in relatively short-styled flowers; and (3) about 10% of all seeds and pollinator wasps mature in a flower of "inappropriate" style length. In a study of 15 species of Australian figs, Addicott and Bronstein (unpublished data) found that 44% (range 13 to 71%) of all ovaries were vacant at maturity and 21% (range 0 to 58%) contained nonpollinator wasps; as in *F. pertusa* and *F. burtt-davyii*, over half the ovaries contributed neither to fig nor to pollinator reproduction.

Clearly, then, although pollinators with longer ovipositors potentially can exploit more fig ovaries, ovary access is a very inadequate predictor of seed-to-wasp production in this system. From the wasp's perspective, some other constraint or constraints are acting to prevent offspring from maturing within every available oviposition site. From the fig's perspective, these unknown constraints act to skew reproductive output away from the strong male bias predicted by relative style to ovipositor lengths, and towards the female function. It is tempting to conclude that those unaccounted-for constraints are evolved adaptations by which figs prevent overexploitation by wasps; for instance, figs may abort a certain proportion of developing wasps in heavily occupied syconia. However, we must ask at this point whether a reduction in the number of wasp offspring to reach maturity results in increased seed production. If it does not, then any "adaptation" to reduce wasp fecundity could only reduce the fig's *total* reproductive output, putting the individual plant at an immediate selective disadvantage.

2. Is There a Tradeoff Between Seed and Wasp Maturation?

Seeds and wasp larvae are alternative occupants of fig ovaries, and thus the assumption that seed and wasp production are negatively correlated seems eminently reasonable. In fact, there is as yet *no* evidence for a negative correlation between them! In *F. pertusa*, I found seeds and agaonid offspring to be strongly *positively* correlated, comparing among fruit crops (Figure 8A).[22] Syconia maturing progressively fewer seeds plus agaonids contained progressively more vacant ovaries. In *F. aurea*, the correlation between seed and agaonid production is weaker, but still positive (Figure 8B); (Bronstein, unpublished data). Although many more studies must be done before conclusions can be drawn (but see Baijnath and Ramcharun[6] and Herre[87]), these data at least indicate that a tradeoff between seed and wasp maturation is not an intrinsic property of this interaction.

It is useful to specify exactly how such a tradeoff would be produced, to begin to understand why it might not exist in this system. We can envision a scenario (pictured in Figure 9A) in which, when individual pollinators laid more eggs or increasing numbers of pollinators entered the syconium, more total offspring would mature, until a limit set by the number of accessible ovaries was reached. Seed production might be relatively high in underutilized syconia if some pollen commonly landed on accessible but unused ovaries, but with increasing wasp oviposition, it would be forced down to a plateau set by the number of inaccessible ovaries. As Figure 9B shows, this scenario would result in a negative correlation between seed and agaonid maturation.

Some data are available in the literature to look at the predictions of this scenario, and they clearly show it to be inadequate. In different species pairs, the relationship between pollinators per syconium and total offspring they produce (Figure 10) ranges from positive but peaking at intermediate pollinator numbers, to nonsignificant, and even to strongly negative; in no case yet studied does it reflect the pattern shown in Figure 9A. The more limited data on pollinator number and seed production are no less confusing (Figure 11). This correlation ranges from positive and peaking at intermediate pollinator numbers, to negative and linear, to negative and plateauing, as predicted in Figure 9A. Clearly there is a great deal that we do not know about the control of seed and wasp production within syconia.

3. Resource Constraints on Maturation

Genetic tradeoffs in investment in the male and female reproductive functions in plants are often hidden in nature by phenotypically *positive* correlations, caused by resource limitation that affects the expression of both functions in the same direction.[51,114,193] In the fig/pollinator interaction, it would not be surprising if resource availability simultaneously affected the maturation success of both seeds and wasp larvae, particularly since the latter feed on the former. If this were the case, real tradeoffs between the number of seeds and agaonids that began to develop within syconia might be hidden by correlated patterns of mortality during development. Although this hypothesis has not yet been directly tested, it is essential to consider it in studies of fig/pollinator conflicts. We might find, for example, that the reproductive success of wasps actually decreases with decreasing seed maturation. If maturation success were linked in this

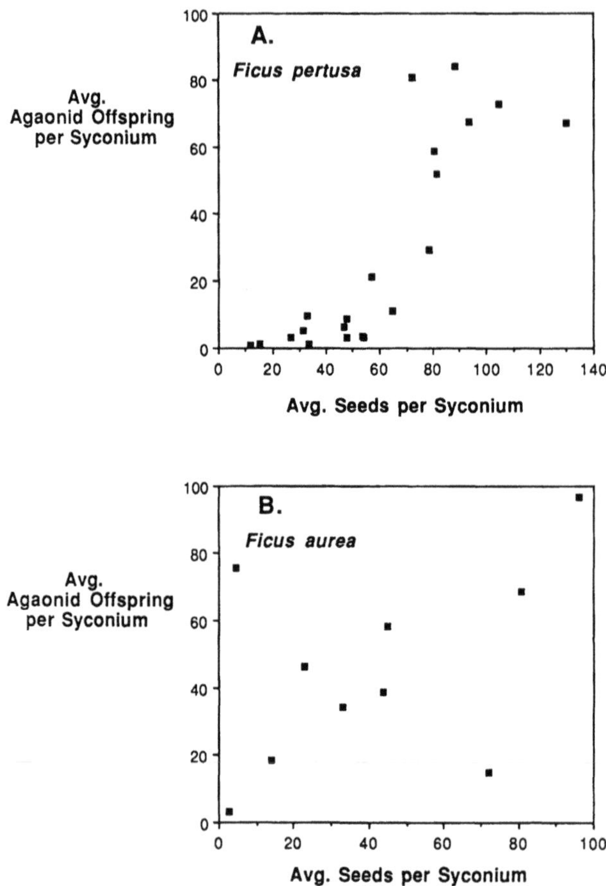

FIGURE 8. Positive correlations between the numbers of seeds and agaonid wasps maturing per syconium in (A). *Ficus pertusa* from Costa Rica, and (B). *Ficus aurea* from Florida. Each point is the average value for one crop of one tree.

manner, then it might actually be in the wasps' evolutionary interests to *guarantee* that some seeds develop — quite a different scenario than originally posed in this section.

One indirect test of the resource-limitation hypothesis is to examine the contents of syconia that are entered by identical numbers of pollinators but that have access to different quantities of resources. We would expect that the number of seeds and wasps in these syconia would increase, and the number of ovaries vacant at maturity would decrease, with increasing resource levels. These predictions proved to be accurate in an experiment I conducted on *F. pertusa* in Costa Rica.[22] In this species, pollinators arrive at different trees when the syconia have expanded to different points relative to their ultimate attainable size; there is a twofold difference in diameter and a ninefold difference in volume between the smallest and largest pollinated syconia.[24] (It is reasonable to suppose that syconium size is a reflection of resources available for maturation in fig species. In *F. carica*, reserve materials, including starch, begin to accumulate in the parenchyma cells of the syconium wall immediately after pollination, and are gradually translocated within.[49,115] Larger syconia should therefore be more effective storage units.) In nature I had observed that small, early-pollinated syconia matured very few seeds or wasps relative to larger, later-pollinated ones,[22] but interpretation of this pattern was confounded by the fact that early-pollinated syconia were also

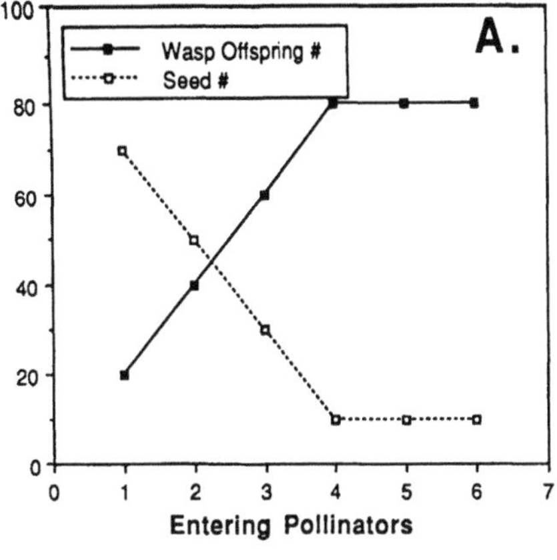

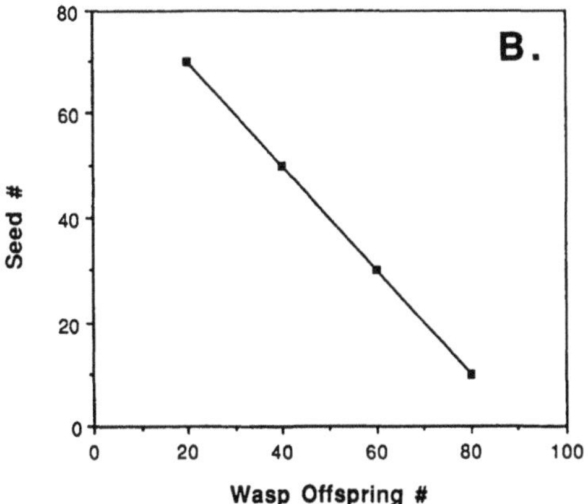

FIGURE 9. Hypothetical mechanism generating a tradeoff between seed and agaonid wasp maturation within a 90-ovary monoecious syconium. (A). As more pollinators enter the syconium, wasps oviposit into more accessible ovaries, until a limit is reached that is set by the number of flowers with short styles relative to wasp ovipositor length (= 80 flowers). The number of seeds initially exceeds the number of relatively long-styled flowers (= 10), but decreases and plateaus at this value with increasing occupancy of short-styled flowers. (B). The negative correlation between seed and agaonid wasp maturation success generated from the data in Figure 9A.

entered by significantly more pollinators.[25] In an attempt to tease apart whether seed and wasp maturation were primarily influenced by pollinator numbers per syconium or by syconium size, I allowed single pollinators to enter syconia on one tree that had grown to different proportions of their ultimate size. Seed and wasp numbers both increased linearly with syconium size (regressions: seeds, $r^2 = 0.349$, $P < 0.01$, $n = 21$; agaonids, $r^2 = 0.796$, $P < .01$, $n = 13$), implying a direct role for syconium size or a close correlate of it.

There is other evidence scattered in the fig literature suggestive of high mortality within developing syconia. As previously mentioned, vacant ovaries are extremely abundant, and often appear swollen as if development of a seed or wasp had begun but then ceased.[40,69] Total offspring maturation is often low when pollinator entries are high (Figure 10), possibly indicating mortality due to resource competition or to double oviposition within individual ovaries. Finally, in *F. pertusa* the seeds and wasps developing in small syconia are smaller as well as fewer.[22] The resource limitation hypothesis still needs to be tested directly, however. Ideally, one would determine both the numbers

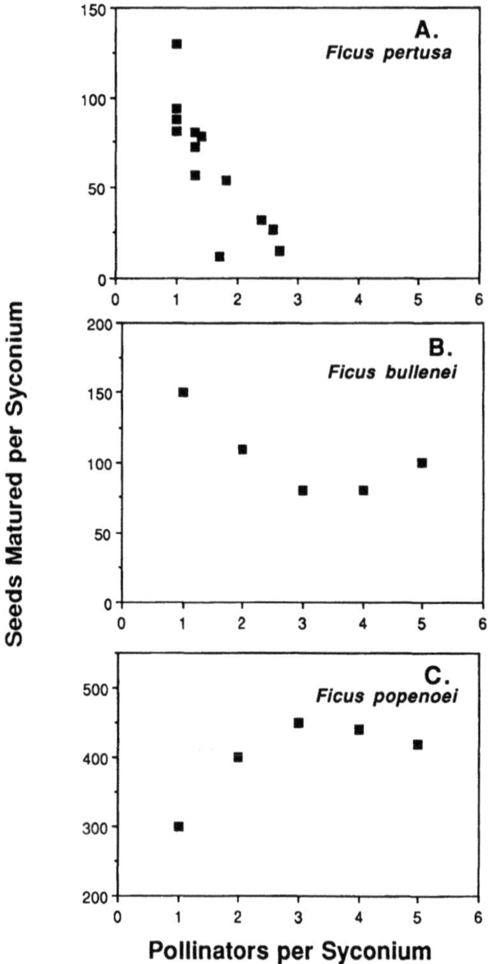

FIGURE 10. Correlations between the number of entering pollinators and the number of successfully maturing pollinators in five New World fig species; compare with the prediction in Figure 9A. (A). *Ficus pertusa*, Costa Rica.[22] (B). *Ficus popenoei*, Panama; data from Herre.[87] (C). *Ficus bullenei*, Panama; data from Herre.[87] (D). *Ficus aurea*, Florida; data from Frank.[58] (E). *Ficus citrifolia*, data from Frank.[58] In A, points represent the average agaonid maturation in different crops with known average pollinators/syconium; in B–E, points represent the average maturation in experiments in which known numbers of pollinators were introduced directly into receptive syconia.

of seeds and wasps that begin to develop within a syonium and the number of each that ultimately mature, at different levels of available resources.

4. Consequences of Resource Limitation for Fig/Pollinator Conflicts

Agaonids may be selected for increasing exploitation of figs, even if resource availability does stringently limit maturation success. Wasps with longer ovipositors may still leave *relatively* more offspring, even at very high rates of larval mortality, because they are able to lay a few more eggs. On the other hand, resource-limited maturation success could provide a simple explanation for why fig seeds continue to be produced despite the wasps' evident evolutionary interests in occupying all oviposition sites. In *F. carica*, either the seeds or ovary walls act as the source of growth hormones that induce mobilization of metabolites into the syconium.[47,50] If the physiology of other figs is

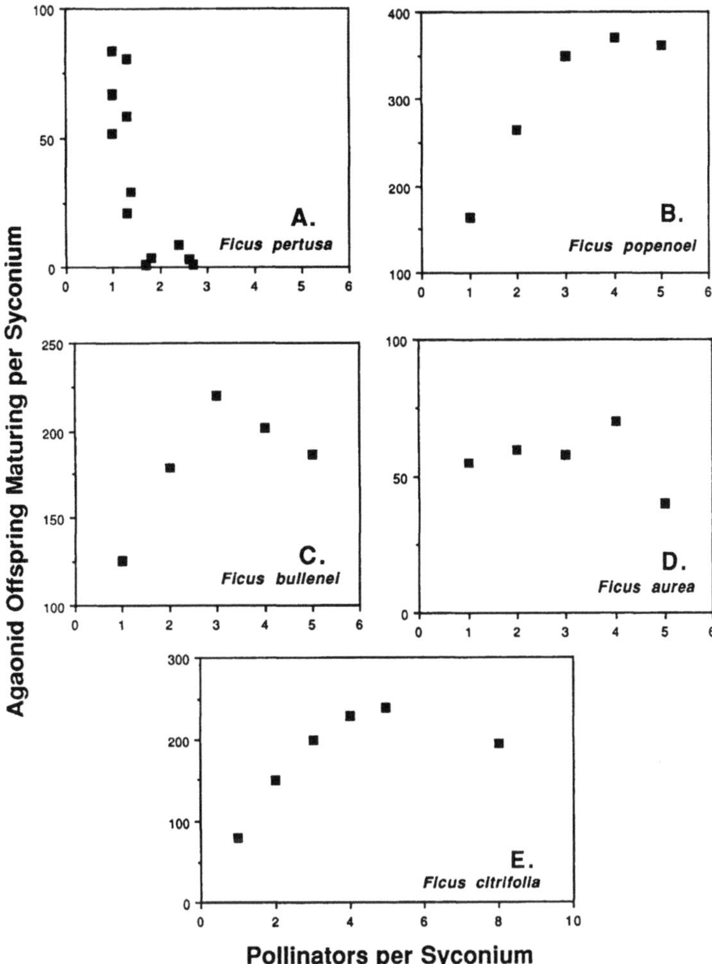

FIGURE 11. Correlations between the number of entering pollinators and the number of successfully maturing seeds in three New World fig species; compare with the prediction in Figure 9A. (A). *Ficus pertusa,* Costa Rica.22 (B). *Ficus bullenei,* Panama; redrawn from Herre.[87] (C). *Ficus popenoei,* Panama; redrawn from Herre.[87] In A, points represent the average agaonid maturation in different crops with known average pollinators/syconium; in B–C, points represent the average maturation in experiments in which known numbers of pollinators were introduced directly into receptive syconia.

similar, then wasp maturation should always be lower in fewer seeded syconia, since these syconia would be the poorest competitors for resources. Wasp *maturation* success might thus be lowest when *oviposition* success was highest. If this is the case, there may even be selection for agaonids to purposefully pollinate some flowers in which they do not subsequently oviposit. This could account for why, in a variety of fig species, so many seeds develop within the "accessible" class of fig ovaries.[24,40,68,130]

In conclusion, while it is logical to suppose that strong conflicts must exist between the evolutionary interests of monoecious figs and their pollinators, the simple scenario of conflict portrayed in Figure 3 has not been supported over the last two decades of fig research. Although wasps with longer ovipositors can reach more oviposition sites, this access is an inadequate predictor of the number of seeds and agaonids that ultimately mature, at least in part due to high mortality during development. This mortality seems to destroy any initial tradeoff between seed and wasp occupancy of fig ovaries, and often produces a phenotypically *positive* correlation between them. There is as yet no evidence that a fig able to restrict wasp "overexploitation" would produce more seeds. It is probable that it would simply experience lower total reproductive output; if so, the frequently discussed fig adaptations restricting wasp success are unlikely to evolve. Whether correlated patterns of seed and wasp mortality are sufficient in themselves to explain the persistence of seed production in figs, counter to the wasps' *apparent* evolutionary interests, remains to be verified by experiments in which resource levels are manipulated and mortality is quantified rather than inferred. Other possible factors restricting wasp success, such as interference competition during pollination or multiple ovipositions into individual fig ovaries at high pollinator densities, have not yet been studied extensively in the field.

C. EVOLUTIONARY CONFLICTS IN GYNODIOECIOUS FIGS

The conflict over fig ovary use that I have discussed so far is relevant only to monoecious fig species. In gynodioecious figs, pollinators have access to every ovary in hermaphroditic syconia and to no ovary in female syconia (Figure 2A). Thus the situation reflects the pattern shown in Figure 4B, in which bimodality of style lengths precludes any advantage of a marginally longer wasp ovipositor.

However, gynodioecious figs do face a distinct conflict with their pollinators, one that has only recently been identified. Agaonids associated with gynodioecious figs can only reproduce if they enter a syconium on a hermaphroditic tree; entering a syconium on a female tree is evolutionarily lethal. As Kjellberg et al.[108] point out, there should therefore be intense selection in favor of wasps able to discriminate between hermaphroditic and female trees, and avoid to the females. If a discriminating genotype did evolve, the interaction would quickly become destabilized, since seed production would stop; the mutualist would have become an antagonist, and the fig would quickly be driven to extinction. What, then, can account for the persistence of over 300 species of gynodioecious figs and their pollinators?

Kjellberg et al.[108] argue that for *F. carica*, the answer lies in its complex population-level flowering phenology. Syconia are not receptive to pollinator (*Blastophaga psenes)* entry on hermaphroditic and female trees simultaneously. In southern France, wasps emerging from hermaphroditic trees in May can only enter syconia on hermaphroditic trees; in July, the offspring of those individuals have access only to syconia on female trees, and in August their offspring can once again only find syconia on hermaphroditic trees.[172] Thus, *B. psenes* may never have had a choice between hermaphroditic and female syconia, and have never evolved the ability to disciminate between fig sexes.

This explanation for the persistence of the *F. carica/B. psenes* mutualism will not hold for all gynodioecious figs. *F. carica*'s mechanism clearly depends on tight within-sex flowering synchrony; it is possible that the environmental cues necessary to synchronize trees exist only in distinctly seasonal environments, such as that in which *F. carica* is found. The few data we have on flowering phenology in gynodioecious figs in more tropical habitats show much less within-sex synchrony, particularly for hermaphrodites, as well as the simultaneous availability of receptive hermaphroditic and female syconia.[44,90,128] Field studies of these species are essential to resolve the mystery of how their interactions with pollinators can persist evolutionarily. But, as for monoecious figs, tests of the fundamental assumptions underlying apparent conflicts between the mutualists are even more necessary at this point than tests of hypothetical mechanisms that would mitigate those conflicts. For instance, no one has yet done choice tests to see if syconia of both sexes are equally attractive to agaonids.

IV. POPULATION-LEVEL CONSEQUENCES OF FIG/POLLINATOR CONFLICTS

To summarize so far, it is evident that the evolutionary interests of figs and their pollinators do not completely coincide, although it is much less clear whether the wasps have the potential to respond to selection to become more efficient exploiters of figs. Selection for increased overexploitation, as well as the complete dependence of agaonids on figs, have been assumed to make pollinator overabundance the rule for figs.[56] However, evidence is accumulating that figs are in fact often reproductively limited by an *underabundance* of pollinators. How can we reconcile ideas of limitation and overexploitation by the same mutualist? Here it is helpful to return to basic natural history and to ask simply: What proximate factors determine whether or not a fig tree will be pollinated? In answering this question, it will become clear that pollen-limited fig reproductive success and high probabilities of local wasp extinction should be inherent features of this interaction.

A. FLOWERING PHENOLOGY AND THE AVAILABILITY OF POLLINATORS

There are three possible outcomes of a given reproductive cycle for a monoecious fig tree (Figure 12). A tree can successfully attract pollinators during its brief female phase, and several weeks later the departing offspring of those wasps can successfully deliver pollen to another tree. That tree is therefore fertile in both its male and female phases. Alternatively, the tree can successfully attract pollinators, but when the pollen-carrying offspring of those wasps depart weeks later, there may not be another tree flowering close enough in space and time for them to reach. That tree is therefore fertile in its female phase but sterile as a pollen donor. Finally, the tree might flower at a time or place where no searching pollinators exist; it will be sterile both as a female and as a male.

The frequency of these three outcomes depends on the flowering pattern at the level of the fig population. Flowering within a hypothetical four-tree population is illustrated in Figure 13. Two critical points emerge from this figure. First, population-level flowering asynchrony is clearly essential for wasps to be able to move successfully between trees, and thus for trees and wasps both to reproduce. Moreover, such asynchrony must extend *year-round*; if there is a sufficiently long gap in the flowering sequence, the local wasp population will go extinct, and no tree will reproduce again until wasps recolonize.

The second critical point is that reproductive failure should be very common for figs, even when flowering is asynchronous enough for pollinators to persist. Of the nine complete reproductive cycles shown in Figure 13, three fail entirely because the female phase fell at a time when no searching pollinators were available. One other cycle is successful only in the female phase, because the pollen carriers departing from it could reach no flowering neighbor. Furthermore, success is relative; for example, in the simple case presented here, pollen carriers can arrive from either one or two neighbors.

The general validity of this simple phenological model can be tested by asking whether reproductive success is, in fact, highly variable within monoecious fig populations. Quantitative data are available for very few species, and then only for the female component of reproductive success.[22,87,119,121] These data do, however, support the model. For example, I found that the proportion of syconia entered by pollinators (and consequently the proportion matured) ranged from 1 to 100% among 21 *F. pertusa* crops studied over 2 years in Costa Rica and averaged only 65%.[25] Furthermore, differences among crops in the number of pollinators per entered syconium were highly

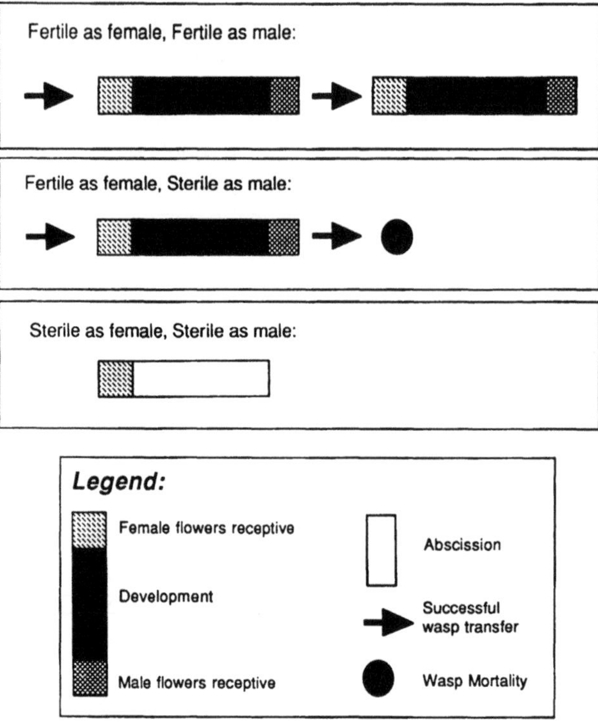

FIGURE 12. Three outcomes of a reproductive cycle in monoecious figs. Depending on the timing of receptivity and wasp release in other trees, an individual may be fertile in both the male and female reproductive phases, fertile only in the female phase, or sterile.

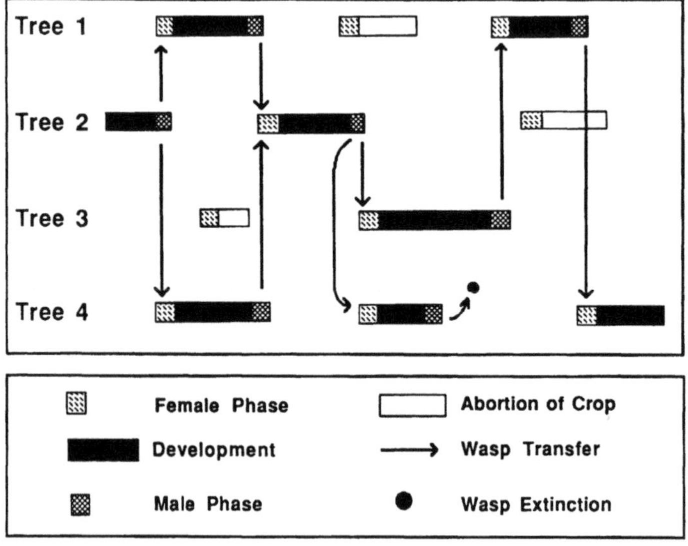

FIGURE 13. Hypothetical flowering sequence of a four-tree monoecious fig population. Note that (1) if any one of these trees were absent, the pollinator population would become locally extinct during the time period shown. (2) Although pollinators can persist in this example, there is variability in the success of individual crops: of the complete crops shown, four are successful in both the male and female phase, one is successful only in the male phase, and three are sterile.

significant.[22] The number of arriving pollinators caught on sticky traps placed in some trees reflected this great variation.[23]

1. Frequency-Dependent Reproductive Success

According to the model of flowering phenology in Figure 13, the answer to the question of what determines whether a fig tree will attract pollinators is the probability that wasps are departing from a nearby tree that day. But what determines *that* probability? In part, it is fig population size. In fact, both the success of individual trees and the ability of a pollinator population to locally persist should be frequency dependent. The larger the population, the higher the probability that pollen-carrying wasps will be departing from at least one plant near in space and time to others bearing receptive flowers. Beyond some critical population size, there will be no gaps in the flowering sequence, allowing a local pool of pollinators to be maintained. For example, in Figure 13 the pollinator population can persist if successive generations transfer from Tree 2 to either Trees 1 or 4, to Tree 2, to Tree 3, to Tree 1, to Tree 4. But in the absence of any one of these trees, local wasp extinction would occur during the time period illustrated.

In order to determine how many trees were needed in a more realistic fig population to prevent wasp extinction, Bronstein et al.[33] devised a simulation model of the interaction between one common African fig species, *F. natalensis*, and its pollinator. Georges Michaloud has accumulated a very complete 7-year record of flowering phenology in a large population of *F. natalensis* in Gabon.[119] From these data, we calculated the mean and variance in the duration of reproductive episodes, and the mean and variance in the intervals between episodes. Using these distributions, we then generated a 4-year flowering sequence for one tree, by randomly selecting an episode length, an interval length, an episode length, and so on. Sequences were generated similarly for other trees. After each new tree was added, we asked: if wasps arrived at time = 0, could the wasp population persist for the full 4-year period? (We initially used the assumption that wasps arrived and departed from the trees during 1-week periods.) The simulation ended after the first tree was added that permitted the wasp population to survive. Notice that this simple model includes no spatial structuring: we defined a fig population as the number of trees within an area in which wasp transfer is possible, provided that wasps are available at the right time.

In 100 simulations, we found that a median of 95 *F. natalensis* trees was necessary to allow pollinators to persist for 4 years. (We referred to this as the *critical population size.*) The range, however, was very large, from 65 to 294 trees. Furthermore, median critical population size would have been larger if we had required the pollinators to persist for longer periods. These results imply that the long-term persistence of pollinator populations is problematic, even at large fig-population sizes.

The model also supported another predicted consequence of fig flowering phenology, namely, that reproductive success within fig populations should be highly variable, even when there are enough trees to assure pollinator persistence. In the average reproductive cycle, only 55% of trees within the simulated population were fertile in both the male and female reproductive phases; 10% were female-fertile only, and the rest were sterile. (In nature, of course, we would also expect great variability in relative success among trees "equivalently" successful in our simple model.) Only in simulated populations two to three times larger than the critical population size did every tree reproduce successfully in both sexual functions.

There are as yet no field data available to confirm that the likelihood of maintaining pollinators is in fact a function of fig population size. An ideal situation would be to study this prediction on islands: the likelihood of successful establishment of colonizing fig wasps should depend on the number of fig trees present. Some suggestive information is offered from studies of the recolonization of Krakatau after the volcanic eruptions of 1896 to 1897 and of Anak Krakatau after the 1952 eruption.[41] In both cases, figs were among the first plants to recolonize the islands, probably due to bat dispersal of seeds from the mainland, and within several years were noted to be quite abundant. However, in both cases, agaonid wasps were first collected 20 to 40 years later. Provided that wasps are commonly blown to the islands from the nearby mainland, their successful establishment may have had to await sufficiently large fig population sizes.

2. The Importance of Phenological Traits

It can be seen in Figure 13 that the proximate factors producing an asynchronous flowering sequence in a fig population are the somewhat variable durations of flowering episodes, and the highly variable intervals between successive episodes. We therefore decided to use the simulation model[33] to test whether critical population size was sensitive to episode and interval length, as well as whether average reproductive success shifted with changes in these phenological traits. In general, the simulations showed that for pollinators to persist within a fig population, a minimum number of crops had to be receptive to pollinators each week. This number depended directly on the function (population size/[episode duration + interval between episodes]). Consequently, increasing either mean duration or mean interval increased the critical population size, which, in effect, is the population size necessary to maintain the same minimum number of crops receptive each week. Increasing the mean duration of a reproductive episode (but not the interval between episodes) also caused progressively greater reproductive failure within fig populations able to maintain pollinators.

One other phenological parameter has a major effect on fig and pollinator success. In our model, we initially assumed that trees were receptive to pollinators for 1 week, and that pollinator offspring departed from their natal tree for 1 week. Lengthening the sexual phases greatly increased the chance of a "match" between female-phase trees and pollen-carrying wasps, thus reducing critical population sizes and elevating individual reproductive success. Most of the very few species whose phenologies have been quantified to date do in fact attract and release pollen-carriers for very brief periods (e.g., see Wharton et al.[181]), but it may be premature to conclude that this is typical for figs. A few species have been reported to initiate syconia asynchronously within trees, a phenomenon that prolongs both sexual phases. As I discuss in the following section, this behavior has been thought to be an adaptation specifically for reducing wasp mortality in seasonal conditions. However, given the pollen-limited reproductive success and pollinator extinction that seem to be inherent to this interaction, prolonged sexual phases may be more generally advantageous, and more common than previously assumed.

Results from the model, therefore, indicate that certain fig flowering traits influence both individual reproductive success and the likelihood that a population of pollinators can persist locally. This is especially true at fairly small fig population sizes. Interestingly, it is evident from recent studies[5,6,27,33,144,195] that there is great interspecific variation for the phenological traits in question. These observations suggest that life-history strategies may differ among monoecious fig species. For example, given equivalent dispersal rates and ages at first reproduction, fig species with long sexual phases and short intervals between successive reproductive episodes may be the most successful colonizers of new habitats, since they can maintain a pollinator pool at the smallest population size.

B. PERSISTENCE OF THE MUTUALISM IN SEASONAL ENVIRONMENTS

One reasonable place to look for unusual flowering traits in figs is in seasonal environments. *Ficus* and its pollinators evolved and are most abundant in the equatorial tropics, but representatives are present over a very broad latitudinal range (45° N to 35° S).[11, 45] In seasonal sites, fig species show a marked decline in flowering frequency during the coldest or driest months.[5,6,32,66,90,123,181,195] These conditions may be simultaneously more stressful for the wasps, which, even under optimal conditions, seem to survive only a few days at most between trees.[110] These phenomena greatly increase the likelihood that an unbridgeable gap in the population flowering sequence will arise. Might fig flowering traits have evolved that decrease the likelihood either of a winter flowering gap or of the local wasp extinction that we would expect to follow from it?

Janzen[95] hypothesized that seasonality should select for a breakdown in within-tree flowering synchrony in figs. To understand his reasoning, first consider the supposed advantage of the characteristically tight synchrony of syconium initiation and development. Pollen-carrying agaonids probably experience high mortality during the dispersal phase. From the wasp's perspective, it would therefore be advantageous to stay and oviposit at the natal tree if female-phase syconia were present there. However, this is another instance in which the evolutionary interests of the fig and wasp do not coincide: from the plant's perspective, it is more advantageous if the wasps depart in search of another tree, because this is the only way for outcrossing to occur. The observation that sexual phases are tightly synchronized within individual trees in most fig species has been taken as evidence that figs have evolved to force the wasps to depart with pollen.

However, in seasonal environments, wasps departing in winter would seem very unlikely to successfully find another tree, which would be disadvantageous for both the mutualists. Janzen,[95] therefore, suggested that in winter it would benefit individual trees to allow some overlap between sexual phases, rather than to force all wasps to depart simultaneously with almost no chance of success: although self-pollination would result, a new generation of pollen carriers might be able to mature successfully. If trees then regained tight sexual synchrony when better conditions returned, the new generation of pollen carriers would be forced to depart. Janzen's hypothesis predicts, then, that the cost of within-tree asynchrony to figs (self-pollination) is lower than its benefit (wasp persistence) only at times when outcrossing is impossible; thus, asynchrony should be seen only among species in seasonal environments, and then only in winter.

So far, studies have failed to support Janzen's seasonality hypothesis. As I summarize elsewhere,[27] some but not all fig species in seasonal environments show within-tree flowering asynchrony, and some that do are equally asynchronous in more tropical parts of their ranges. Within-tree asynchrony may be a species-specific, rather than site-specific, trait; furthermore, there is some evidence that related fig species exhibit similar levels of within-tree asynchrony.[12] Additionally, the temporal pattern of asynchrony within a seasonal site does not reflect Janzen's[95] prediction in the few species studied to date. A 2-year study of *F. aurea* at its northern range limit[32] has shown that crops of individual trees are most, not least, synchronous during cold, dry months; in this species, maximum asynchrony occurs in seasons that are probably *most* favorable for wasp transit between trees. Flowering asynchrony in *F. aurea* is generated by the initiation of multiple cohorts of syconia a few days or weeks, rather than a few months, apart. It is not surprising that this does not occur during winter, the most resource-stressed season. (Interestingly, consecutive cohorts are spaced in time in a way that limits overlap of the male and female phases within trees, so that self-pollination is not a significant cost of flowering asynchrony in *F. aurea*.[32]) *F. aurea* phenology reflects patterns seen in the only other asynchronously flowering figs studied to date. At the same study site, *F.*

citrifolia shows cycles of within-tree synchrony roughly parallel to those of *F. aurea*.[27] Similarly, Baijnath and Ramcharun[5,6] have shown that the South African species *F. sur* and *F. burtt-davyi* exhibit maximum within-tree asynchrony during the warm, wet season, when the largest proportion of trees are in a reproductive phase, rather than in winter.

How the mutualism can persist in relatively seasonal conditions is still an open question. Pollination success for *F. aurea* trees flowering in early spring is extremely low (Bronstein, unpublished data), suggesting that the wasps do in fact undergo a population bottleneck during winter. Relative to equatorial habitats, much larger fig population sizes may be necessary in seasonal environments to allow a pollinator population to persist year-round.

Kjellberg and Maurice[111] argue that these difficulties of wasp persistence in seasonal environments may have been critical selective forces in the evolution of gynodioecy from monoecy in *Ficus*. They point out that in seasonal habitats, the periods of the year most favorable for production of seeds and pollen-carrying wasps may differ. For instance, the presence of seed dispersers or good germination conditions should partially determine the optimal period for seed production, but have no consequences for wasp production. If certain trees within the population flower at the best time for seed production they would become relatively superior females. Other trees would then be selected to flower *before* them, at a time at which they could act as their pollen donors; these would become relatively superior males. Kjellberg and Maurice discuss how strong seasonality might thus lead either to wasp extinction and the breakdown of the mutualism, or to the evolution of gynodioecy. Some gynodioecious fig species are, in fact, successful under far more seasonal conditions than any monoecious species can tolerate (e.g., in France[108,172]).

C. CONSEQUENCES FOR FIG/POLLINATOR CONFLICTS

Close examination of population-level flowering patterns in monoecious fig populations strongly suggests that pollinator populations could never become stable and predictably overabundant for figs. Wasp abundance should fluctuate dramatically with the number of trees that flower in a given month; due to the stochasticity of flowering time necessary to produce a year-round flowering sequence, that number of trees is never constant. At low fig population sizes, gaps in the flowering sequence should arise that lead to the local extinction of the fragile, short-lived wasps, in which case no local tree could reproduce again until wasps recolonized the area. But even in larger populations, fig reproductive success should be highly variable. Furthermore, when a tree is in fact well visited, most of the pollen that arrives should be from a very few donors, since so few trees are likely to be in the male phase in the same week that it is receptive to pollination (cf. Figure 12).

One interesting conclusion emerges that is directly relevant to the issue of fig/pollinator conflicts over ovary use. It would appear that by decreasing the fecundity of wasp genotypes that can leave more offspring, e.g., by aborting overexploited syconia, a fig would only lose. It would lose in the short term, because its own success as a pollen donor would decrease in a system in which pollen availability commonly limits the reproductive success of its neighbors. Perhaps less importantly, it would also lose in the long term, because it would decrease the chance that any local pollinator population would be left the next time it flowered itself. However, the magnitude of these costs of overexploitation should themselves shift over time, as fig population size changes.

I wish to emphasize, however, that these conclusions are based primarily on computer models and extrapolations from fig natural history. As I have stressed in discussing evolutionary conflicts between the mutualists, field studies are urgently needed, not only to test these theories, but to test the central assumptions that underlie them.

V. COMMUNITY-LEVEL CONSEQUENCES OF THE INTERACTION

No two-species interaction, even such a highly specific one as the fig/pollinator mutualism, behaves in isolation from its community context. The outcome of competitive interactions can be altered by the presence of a predator,[134] parasite,[139] or mutualist.[19] Similarly, predator/prey interactions are often influenced by organisms at a third trophic level.[138,139] Mutualisms, in turn, are affected by at least three classes of organisms: predators, species that exploit available rewards but do not benefit either mutualist partner, and species involved in other mutualistic relationships with one of the partners. Interactions with these organisms range from facultative and generalized[92] to highly specialized.[94] One could envision networks of direct and indirect interactions becoming built up around a mutualism, such that it became a "keystone" of community structure; that is, loss of the mutualism would result in widening circles of extinction.[77,161] The fig/pollinator relationship has recently been proposed to be one of the most important keystone interactions in tropical forests.[117,161] In this section I review the ecology of the broader group of species that interact with, and influence the outcome of, the fig/pollinator relationship, and consider evidence for the role this mutualism may play in structuring tropical communities.

A. "INTERLOPERS" IN THE FIG/POLLINATOR INTERACTION

Mutualisms are almost ubiquitously exploited by species that can obtain available rewards without providing any service in return. The effect of such "cheating" on a mutualism is potentially great. The reward-providing species loses its investment in the reward, which may not be renewable. Its partner loses a necessary resource (generally, food or shelter), which may not be obtainable elsewhere. The ecology and evolution of cheating within mutualisms has received scant attention (but see Soberon and Martinez del Rio[157]). Major unanswered questions include: How costly is the presence of cheaters to mutualists? Do cheaters evolve from mutualists, or do they invade mutualisms from outside the interaction? Are there mechanisms by which mutualists can effectively exclude cheaters from their interaction? Is cheating a facultative or obligate strategy, and how diverse is the range of mutualisms with which a cheater interacts?

The fig/fig wasp relationship probably cannot be fully understood without considering certain "interlopers" that develop within syconia but do not transport pollen. (These have been referred to as parasites[79,80,90,168] or parasitoids,[95] but since it is almost never known what a particular species eats, a term that refers to their exploitation of the mutualism seems more appropriate.) These interlopers can often outnumber the legitimate mutualists (Bronstein, unpublished data). In terms of diversity, up to 30 species have been reared from a single fig species (S. Compton, unpublished data; see host records in Wiebes[186] and Hill[90]). Many appear to be as species specific as the pollinators themselves.[80,168] Remarkably little is known about the biology of these insects, their impact upon the mutualism, or the history of their association with it. Below, I briefly review the current status of our knowledge; further information is provided by Wiebes,[188] Hamilton,[81] Bouček al.,[16] Ulenberg,[168] and Bronstein.[28] I will not attempt to review here the fascinating mating behavior of these wasps, which is largely independent of their association with the mutualists. Those interested are referred to Hamilton,[81] Murray,[127] Frank,[60] and Godfray.[79]

1. Classification of Interlopers

The great majority of interlopers in the fig/fig wasp interaction are chalcidoid wasps (however, see Roskam and Nadel's[150] description of a cecidomyid fly that develops within *F. citrifolia* syconia). It is not possible yet to offer a clear picture of the taxonomic position of fig-associated wasps; most of these species (and many common genera) are still undescribed, and the higher level systematics within the Chalcidoidea are still debated.[15] Some are currently placed with the pollinators in the Agaonidae,[187] while others have been put into various subfamilies of the Torymidae,[80, 168] Pteromalidae,[37] and Eurytomidae.[16] Only two genera, *Idarnes* and *Apocrypta* (Torymidae), have received major taxonomic treatment;[80,168] the entire suite of chalcidoids cohabiting one fig has been described in a few instances.[16,90] Classification is greatly complicated by many striking polymorphisms, both between sexes and among males, involving traits such as size, wingedness, and modifications for fighting.

2. Interactions Between Interlopers and Mutualists

Depending on the species, interloper females either enter the syconium to oviposit (like the pollinator) or oviposit through the syconium wall. Like the pollinators, they deposit eggs directly into the ovaries of the fig, and each larva feeds strictly within a single ovary. The critical difference is that interloper larvae do *not* feed on seeds initiated by pollen carried by their mother into the syconium.

Larval diets, which vary among species, determine the impact a particular interloper will have on the mutualists. Condit[43] proposed a dichotomy between interloper species based on diet. "Primary sycophiles" (e.g., *Sycophaga sycomori*[72]) deposit their eggs into ovaries unoccupied by agaonid larvae, then induce swelling of the nucellus within each ovary. The larvae feed on this gall-like sterile tissue. These wasps may have little effect on the mutualists, since they kill neither developing seeds nor agaonids, although it is possible that they drain fig resources stored for current or future reproduction. In contrast, "secondary sycophiles" have an unambigously negative impact. These species (e.g., *Philotrypesis caricae*[112]) oviposit into fig ovaries already occupied by agaonid larvae; the agaonid is killed and the seed that started to develop within that ovary is fed upon. Other modes of feeding may exist among the interlopers as well. There is some evidence that certain interlopers are parasitoids of others.[79] No fig associate has yet been shown to be a parasitoid of agaonid larvae, or to feed on seeds in fig ovaries that were not also occupied by an agaonid, although it should be emphasized that the feeding ecology of very few interlopers has yet been studied. It would be fascinating to know if the many interloper species sharing a single fig species have divergent diets; the strikingly different ovipositor lengths within the fauna of a single fig[28,81] suggests that they may at least partition ovaries by their depth within the syconium.[40]

Interlopers obviously depend on the existence of the mutualism for a site to develop and feed, regardless of their exact diet. They depend on the mutualists in three other ways as well. First, there is evidence that many species follow the chemical attractant that the pollinators use to locate a flowering fig tree.[23,28] Second, even if the interlopers do not feed on seeds or agaonids, they must oviposit within *pollinated* syconia if their offspring are to mature successfully, because most fig species abort unpollinated syconia after several days of receptivity. Certain externally ovipositing torymids investigate syconia for a prolonged period before inserting their ovipositors,[80,169] a behavior that may be directed in part to selecting pollinated ones, and I have shown[28] that one *Idarnes* species does in fact oviposit significantly more frequently in syconia that have recently been entered by a pollinator. Finally, most interlopers depend on the agaonid males to chew the exit hole through the syconium wall; the mandibles of the interlopers are either small and weak, or scythe-like, as if modified for fighting.[81] Some interlopers, however, do seem relatively independent of the mutualism. *Sycophaga sycomori*, a primary sycophile associated with *F. sycomorus*, inhibits abscission of unpollinated syconia when it feeds within them; not surprisingly, males show morphological and behavioral adaptations for constructing an exit passage.[65,69,72] In other fig species, the ostiolar bracts loosen at maturity to form a natural exit for pollinators,[144] allowing the interlopers to escape unaided.

It is difficult to identify fig traits that could effectively exclude interlopers from the syconium. I have suggested three explanations to account for the interlopers' great evolutionary success.[28] (1) *Coevolutionary race hypothesis:* Due to the great inequity of generation times and selective pressures, interlopers may be able to rapidly counter figs' adaptations to exclude them. In particular, while the thick and tough syconium wall may have evolved to deter externally ovipositing interlopers, the often extremely long ovipositors suggest that they have ready access to the fig ovaries anyway. (2) *Exploiter/mutualist mimicry hypothesis*: Figs may not be able to distinguish between larvae of potential pollen carriers and larvae of interloper species, and to selectively eliminate the interlopers. (3) *Commensalism hypothesis*: Interlopers, particularly the primary sycophiles, may be tolerated because their presence inflicts no fitness costs on the fig. Problems with each of these hypotheses and possible tests of them are discussed elsewhere.[28]

3. Evolution of Interloper/Pollinator/Fig Interactions

It seems obvious that these interlopers are highly specialized to exploit the fig/fig wasp interaction; every fig species has a suite of them, none of which, to my knowledge, has been reported to reproduce anywhere except within syconia. However, their existence probably will not help us to understand the still-obscure origins of the fig/pollinator mutualism. It seems clear, even at this early stage of phylogenetic understanding, that most interlopers are relatively distantly related to the agaonid pollinators. There are exceptions, however: some fig-inhabiting agaonids superficially resemble pollinators but, when studied in greater depth, have been found not to transfer pollen. In the best-documented example, Galil and Eisikowitch[65] found two agaonid species, *Ceratosolen arabicus* and *C. galili,* to be frequently associated within *F. sycomorus* syconia in East Africa. They demonstrated that syconia entered by *C. galili* alone never set seeds.[68] Further studies have shown that while this species does have pollen pockets, it exhibits neither pollen-collecting nor pollen-unloading behaviors.[42, 68] It seems reasonable, in this case, that cheating is the derived condition, since *C. galili* possesses the morphological modifications, but not the behaviors, for pollen transport. Morphological evidence indicates that *C. galili* is not the sister species of *C. arabicus,*[190] and therefore cannot be derived from it. It is more likely that the ancestor of *C. galili* originally pollinated a different fig species, and subsequently colonized *F. sycomorus* after it was already being pollinated by *C. arabicus.*[42]

There are suggestions that many non-agaonid interlopers have radiated in parallel with figs and their pollinators. Members of a single genus tend to be spread across related fig species, while the suite associated with a particular fig/pollinator pair seems, in general, not to be closely related, although they often converge in traits related to exploitation of that mutualism (e.g., body size, developmental time, and oviposition behavior[28]). Gordh,[80] for example, described a unique *Idarnes* species from each of 10 Mexican and Central American fig species, and Ulenberg and van Pelt[170] reared a unique *Apocrypta* species from each of 24 Indo-Australian fig species. Furthermore, levels of species specificity are often as high among interlopers as among pollinators. Gordh[80] and Ulenberg and van Pelt[170] found no cases of *Idarnes* and *Apocrypta*, respectively, using more than one fig species (although some figs examined lacked an *Idarnes* or *Apocrypta* associate).

There is a real possibility, however, that some of these apparent patterns may be artifacts of unresolved taxonomic relationships and incomplete host records. Some fig species, especially those with a large interloper fauna, are in fact exploited by more than one interloper species per genus (e.g., *F. thonningii;*[16] see also Hill[90]). Seven of 15 Hong Kong figs share one or more interloper species.[90] In half of these instances, Hill found the shared wasp to be equally common in both figs, suggesting that these were not rare instances of straying among usually species-specific wasps. It should be evident that answers to most of the really exciting questions about the evolution of interlopers on the fig/fig wasp interaction must await further phylogenetic studies on these organisms. Their impressive diversity, compounded by increasing endangerment of their tropical habitats, makes this a daunting task.

However, ecological studies should help us understand how different selective pressures have produced contrasting levels of species specificity in different interloper groups. Intuitively, we would expect interlopers to exploit many fig/pollinator pairs. Because interlopers do not transfer pollen between trees, they will not experience the intense selection for specificity that the agaonids must; furthermore, a broader host range would greatly increase the number of oviposition sites available to them. On the other hand, requirements for coordination of developmental time with the mutualists may enforce specificity anyway. Agaonid maturation times, and thus the time between pollination and formation of the exit hole, vary greatly among fig/pollinator pairs.[33,90,144] Consequently, if interlopers closed well before the pollinators, adults would be trapped within sealed syconia, whereas if they developed more slowly, larvae would still be present when syconia ripened and became attractive to frugivorous animals. I have made two predictions regarding fig species specificity of the interlopers.[28] (1) Those species requiring the aid of agaonid males to escape the syconium (i.e., those without morphological modifications for chewing and digging) will be found to be the most species specific. (2) Sympatric fig species whose syconia (a) open naturally[144] or (b) have similar developmental times should be those most likely to share interlopers. Additionally, if developmental times of interlopers are flexible, an uninvestigated possibility, then many more than expected should be nonspecific.

B. INTERACTIONS WITH PREDATORS

Predation on one mutualist can ultimately affect its partner as well, when the partner's fitness is a function of the mutualist's abundance. In the fig/fig wasp interaction, predators range from insects highly modified to feed on wasps within syconia to birds that destroy the seeds of figs and many other plants. Even generalized predators, however, may be highly dependent upon figs and fig wasps as food, since they are often among the few relatively reliable and abundant food resources in an unpredictable world.

1. Fig Wasp Predation

Predators of fig wasps can be divided into those that feed on individuals developing within syconia and those that feed on free-living adults. For predators capable of penetrating and feeding within syconia, fig wasp larvae form an abundant food source for several weeks. I described the larvae of an unidentified curculionid beetle and pyralid moth that mature within up to 20% of the syconia of *F. pertusa* in Monteverde, Costa Rica.[26] These larvae feed indiscriminately on developing wasps, seeds, and anthers, ultimately destroying the entire contents of the syconium. The presence of these predators somehow inhibits the normal fig ripening process; unripe fruits fall to the ground, at which point the larvae bore through the syconium wall and probably enter the soil to pupate. *F. pertusa* is also commonly exploited by a staphylinid beetle, *Charoxus bicolor. C. bicolor* adults enter syconia soon after the exit hole is formed, but before all the wasps have exited. Upon entry, they pack frass back into the exit hole, effectively trapping the wasps, and then prey upon them. The staphylinids then apparently remove the plug and move on to other syconia.[26] These *F. pertusa* associates may be relatively specialized for exploiting fig/fig wasp interactions. The timing of their arrival relative to the developmental stage of the syconia is crucial for them, and all exhibit behaviors and morphological modifications associated with exploiting the interaction. The genus *Charoxus* has in fact been described exclusively in association with figs.[106] Other organisms feeding within syconia may be much less specialized; for example, many ants enter through the exit hole of ripening syconia to scavenge on the few dead wasps found within.

Fig trees are a concentrated source of small insect prey during the brief periods when adult female fig wasps arrive and depart. Janzen[95] observed hundreds of large dragonflies feeding on fig wasps at a Costa Rican fig species. A pseudomyrmicine ant patrols branches and syconia of *F. pertusa* at these stages, and at least 11 species of birds glean the wasps as well.[26] Hespenheide[89] reported that at least 12% of the diet of two Central American swift species was made up of fig wasps.

The loss of agaonids to predation can have the indirect effect of lowering the fig's success, both as a pollen recipient and a pollen donor. However, it is possible that in many cases predation is heavier on the interloper wasps than on the pollinators themselves. Whereas pollinators oviposit, develop, and mate within shelter of the syconium, many interloper species only develop there; during oviposition and mating, they are active on the tree and are exposed to predators. In cases where the interlopers are species that kill pollinator larvae (i.e., secondary sycophiles), their loss to predators can benefit the mutualists. Compton and Robertson[39] have shown by exclusion experiments that ant predation on ovipositing *Apocrypta guineensis*, a secondary sycophile of *F. capensis*, significantly reduces its offspring numbers and, as a consequence, raises the maturation success of the pollinator *Ceratosolen capensis*.

2. Fig Seed Predation

A series of predators attacks fig seeds from the time they are initiated to the time they germinate. The greatest loss, perhaps 50% on average, is to the developing pollinator larvae.[87,96] Other fig seed predators are only known anecdotally. Insects such as weevil and pyralid larvae[26,95] and fruit-piercing lygaeid bugs[155] destroy seeds developing within syconia. A large variety of insects destroy figs in commercial situations;[95] in addition to the ripe pulp, some probably feed on seeds as well. Some vertebrates that swallow ripe figs destroy seeds rather than dispersing them intact; for example, parakeets crack fig seeds in their beaks, rendering them ungerminable.[97] Postdispersal predation on fig seeds can also be intense, particularly in the vicinity of the parent tree. Slater[155] has described a large suite of lygaeid bugs that destroys fig seeds at this stage; some appear to be fig specialists. Slater calculates that in South Africa, lygaeids are so abundant as to be capable of destroying 100% of the seed crop that falls beneath the parent.

C. DISPERSERS OF FIG SEEDS

As in most other plant species, seeds of figs must escape from the vicinity of the parent plant in order to elude seed predators such as lygaeids that converge under adult trees, as well as to reach vacant sites for establishment. Thus, fig species are involved in a second set of mutualistic interactions, with frugivorous, seed-dispersing animals. Pollination and seed dispersal mutualisms involving figs form a particularly interesting contrast because of their extreme differences in specificity. Whereas the fig/pollinator interaction is one of the most species-specific mutualisms known, fig seed dispersal is extremely generalized. "Who eats figs? Everybody."[95]

The attractiveness of fig fruits does not seem to be a product of their nutritional value. Figs are generally considered to be a good source of readily utilizable carbohydrates, but are watery and low in protein relative to other wild tropical fruits.[84,101,122] Yet Janzen[95] states that figs constitute a large part of the diet for more animals than any other genus of wild tropical perennial fruit, and that in the typical mainland tropical forest all terrestrial species of frugivorous vertebrates eat some species of figs at some time during the year. For instance, figs sustain 40% of the biomass of frugivorous monkeys in Cocha Cashu, Peru,[160] constitute 70% of the total diet of one Panamanian fruit bat,[124] and make up nearly the entire diet of some Bornean birds of paradise.[9]

Rather, it is the pattern of temporal availability of fig fruits that makes them so critical in the diets of tropical vertebrates. One direct consequence of year-round flowering asynchrony in figs (Figure 13) is that fruits are also present year-round somewhere in the fig population. Hence figs are a relatively reliable food source in tropical forests.[123,160,195] In Panama, for instance, figs are one of few fruit resources available during the late rainy season;[57] even in the equatorial tropics, most other species fruit distinctly seasonally.[76,118] As a consequence, the numbers and diversity of frugivores at fig trees are often extremely high at times of year when few other species are in fruit, whereas fig individuals that fruit when richer foods are available can go virtually unvisited and disperse few seeds.[30]

The popularity of figs does *not* imply that each fig species is visited sooner or later by every local frugivore species. Fruits of different fig species contrast wildly in traits such as size, shape, color, and position on trees. For example, the Indo-Malaysian fig flora includes *F. ramentacea*, an epiphyte that bears its tiny (0.3 cm diameter) orange-ripe fruits in small clusters at the tips of branches; *F. callicarpa*, a creeping shrub whose fruits are large (2 cm), oblong, yellowish when ripe, and borne singly in leaf axils; *F. roxburghii*, whose large (2 cm) urn-shaped, brown-ripe fruits are borne in clusters at the base of the trunk; and *F. geocarpa*, which bears small (1 cm), round, spine-covered figs on long runners on or even underground (data from King[105]). These contrasting traits must have consequences for frugivores, and in fact have probably been shaped at least in part by selection for seed dispersal. At this point, however, the disperser assemblage of very few fig species has been described, and then usually from a single tree;[20,21,30,101,116,153] this limits our ability to attribute particular fruit traits to adaptations for seed dispersal. However, predictions can readily be made regarding the dispersers of some of the more unusual figs. For instance, dull-colored figs borne on the trunk or roots are probably most attractive to ground-dwelling mammals that primarily use odor rather than vision to locate food. Furthermore, anecdotal observations suggest that some suites of fig fruit traits can be categorized into well-known "dispersal syndromes".[173] Birds are the primary visitors to figs that ripen during the day to a red or black color and juicy texture,[20,30,101,153] whereas bat-dispersed figs tend to ripen nocturnally with no color change.[2,14,124] The seeds of a few species attract ant dispersers via odors and food bodies, once the fruit is removed from the tree, broken apart, and dropped by a vertebrate frugivore.[52,103,149,154] It would be very enlightening to study the dispersal biology of a group of sympatric fig species with contrasting fruit traits, to obtain a more complete picture of the diversity that exists within these interactions and the selective pressures that may have generated it.

What consequences might fig/seed disperser interactions have for the ecology and evolution of fig/pollinator interactions? As yet, there are no data or theories pertaining to this intriguing issue, but I can at least suggest some of the important potential effects. (1) Different frugivores generate "seed shadows" of different size and shape;[93] for instance, bats tend to deposit large numbers of seeds under their feeding roosts, well away from the fruiting tree, whereas birds drop many more seeds close to the tree, where seed predators may congregate.[162] The abundance and spatial distribution of fig trees thus established should feed back to partially regulate the host-finding success of searching pollinators, and thus both the likelihood of local pollinator persistence and fig reproductive success. (2) Syconium size, and hence ovary numbers per syconium, may be shaped in part by selection for seed dispersal; different-sized frugivores take fruits of different sizes[184] and differ in their effectiveness as seed dispersers.[30] But, as discussed earlier, Herre[87] has found that ovary number per syconium also influences the outcome of interactions with pollinators: species with relatively small syconia are very efficient producers of pollinator offspring, but mature relatively few viable seeds. (3) Pulp characteristics of syconia should influence the abilities of both frugivores and the externally ovipositing interlopers to exploit the fruit. In particular, thick-walled syconia may effectively exclude many interlopers, but also restrict the suite of potential seed dispersers to those that can penetrate the rind.

These speculations suggest that compromises may exist between syconium traits that maximize seed and pollen-vector *production* and those that maximize seed *dispersal*. This is a rich topic for further research.

D. FIGS AND POLLINATORS: A KEYSTONE INTERACTION?

Two observations emerge regarding the community consequences of the fig/pollinator interaction. First, many animals are remarkably specialized to exploit its products (seeds, wasp larvae, and ripe fruit). Second, many animals with fairly generalized diets rely on the products of the interaction for food as well, since these products are relatively abundant and are present at times of year when no other food resources are available. Terborgh,[161] therefore, concluded that fig trees act as a "keystone resource" in tropical forests: if they were removed, large numbers of species could starve. He urged that figs be a focus of tropical conservation efforts. McKey[117] refined this point by specifying that it is the fig/pollinator *interaction*, rather than figs per se, that produces the essential food resources, and thus management efforts must focus on preserving enough trees to allow local pollinator populations to persist. He suggested that Bronstein et al.'s[33] critical population size model could be used to predict "minimum viable population sizes"[158] for figs in tropical reserves.

Are figs really all that important, though, for preserving tropical diversity? For certain groups of invertebrates, such as the interlopers, certainly: extinction of a particular fig/pollinator interaction will lead directly to the loss of these dependent species. But would vertebrates (the usual focus of conservation efforts) really starve if figs did not fruit, or could they instead switch to alternative, although possibly less preferred, fruit? The problem with the concept of keystone interactions or resources, of course, is that it is not morally defensible to perform the critical test of removing the resource in question to see if local extinctions would follow. In the case of figs, no weaker test, for instance, comparing vertebrate diversity in nearby reserves with different numbers of fruiting fig trees, has yet been performed either. Clearly, figs are not *always* a critical food source. Gautier-Hion and Michaloud[75] have questioned the notion of figs as a keystone resource for vertebrates in Gabon, on the basis that at this site figs are too rare to sustain significant biomass, even when few other plants are in fruit. But Terborgh,[161] for instance, has clearly documented the dominance of figs in the diets of monkeys and birds at Cocha Cashu, Peru. Perhaps an assessment of the proportion of figs in the diets of large vertebrates should be made at a site of conservation interest, before an effort is focused specifically on preserving the appropriate number of fig trees. Since fig trees tend to be widely dispersed in tropical forests,[120,165] preserving enough land to include a viable population of figs will probably also save many alternative, and possibly even more important, food resources for vertebrates.

VI. CONCLUSIONS

The interaction between figs and their pollinators may be the most extreme example of plant/animal coevolution that has yet been studied in depth. In spite of its widespread treatment as a "special case", its unique aspects create an ideal system in which to study the processes of coevolution and the patterns it produces. Some of the advantages of the fig/pollinator interaction for studying coevolution, and the unexpected messages it has so far delivered, are as follows.

(1) *Selective effects persist in space and time.* In species-specific interactions, variation in traits involved with exploitation of the partner can have persistent, long-term effects on both partners. This is useful if we wish to assess the likelihood and the potential rate of evolutionary change within interactions. Spatial and temporal variability in the identity and abundance of different partners is increasingly cited as the critical factor preventing coevolution, particularly in mutualisms;[56,88,91,183] this can be taken to imply that in more specific interactions, there are few constraints to reciprocal change. However, studies of this obligate mutualism are revealing that in many ways, figs and pollinators have *not* responded coevolutionarily to certain strong and persistent selective pressures. Wasps do not succeed in exploiting every fig ovary available to them; on occasions when figs are overexploited, they demonstrate no obvious response to decrease wasp success. These results imply that there are forces, as yet largely unidentified, that can prohibit further coevolution within even the most highly specific interactions.

(2) *Traits affecting the success of each partner can be identified and studied.* In facultative, nonspecific associations (including fig/seed disperser interactions), it is frequently unclear which traits are of pivotal importance within the interaction, or, if these can be identified, whether they evolved in the context of that relationship.[82,99,166,184] In contrast, in species-specific interactions, it can be much more straightforward to identify the most critical morphological and behavioral traits, to measure and even to manipulate them in order to study if and how they coevolve (e.g., see Nilsson[131]). In the case of the fig/pollinator interaction, wasp ovipositor length has long been recognized as a critical trait, and more recently, as one expected to be subject to strong directional selection. The potential value of an ovipositor of a given length is easy to quantify, using assumptions basic to our understanding of this interaction. As I have discussed, however, recent tests of the actual role of ovipositor length have shown some of these assumptions to be deeply flawed. Direct studies of *apparently* pivotal traits can thus fundamentally change our understanding of the processes of coevolution that occur within interactions.

(3) *Success of both partners can be measured simultaneously.* While coevolution is by definition a reciprocal process, most empirical studies of coevolved systems are *unilateral*: the ecological consequences are quantified for only one of the partners. While such studies are simpler to perform, they do not provide more than inferential information on the ecological or evolutionary dynamics of the interaction in question. Coevolutionary scenarios have been proposed, and commonly defended, in many of these

studies anyway. In the fig pollination mutualism, some of these scenarios (e.g., the one illustrated in Figure 3) are actually amenable to direct tests, due to the ease of measuring plant and insect reproductive success with the same currency (fig ovary use). These tests have resoundingly rejected some of the most accepted causes of coevolution within this interaction. In particular, there is as yet no evidence from nature for any simple tradeoff between seed and wasp maturation, a phenomenon that had been expected to generate fig/pollinator conflicts and coevolutionary responses to them. The way these results are altering our view of this interaction argues strongly for the more widespread use of the reciprocal approach (see e.g., Morse,[125] Wilson and Knollenberg,[194] Schmid-Hempel and Schmid-Hempel,[152] and Pierce et al.[137]).

The fig/pollinator interaction is particularly useful as a model system for understanding mutualism. It has become a truism that mutualisms are the least studied and most poorly understood interspecific interaction;[29,34,104,148] many common mutualisms, including some of the most spectacular and frequently cited cases of coevolution, remain known only anecdotally. The fig pollination mutualism, however, has been approached from an exceptional diversity of angles (taxonomic, physiological, anatomical, ecological, and evolutionary), and a concerted attempt has been made to propose and test theoretical scenarios for how the interaction functions. Particularly important are the insights that have emerged regarding the origins and consequences of conflicts between coevolved mutualists. Reciprocal benefits emerge within mutualisms as a result of reciprocal exploitation. If selection on one partner leads to increased exploitation of its mutualist, and counterselection does not or cannot act on that species, the interaction *should* be driven away from mutualism towards greater and greater unilateral exploitation.[63] Testing this theory demands simultaneous measures of the success of both partners, as well as the ability to identify conflicts and traits controlling them. The advantages of the fig pollination interaction mentioned above — its species specificity, the measurability of traits central to the conflict, and the ability to quantify the reproductive output of the mutualists — create a system in which some tests of these theories are rather simple. Interestingly, though, these tests have produced results that are far from straightforward. In particular, it seems possible that fig seed maturation actually benefits an agaonid wasp in the short term; seed production may continue for this reason, not because figs have evolved responses to overexploitation. Much more work needs to be done, however, to test existing evolutionary scenarios, as well as to propose and test more accurate ones.

Perhaps the most evident conclusion about coevolution within the fig/pollinator interaction is that it is far messier than anyone would have guessed. Here is a highly species-specific, obligate interaction with an exceptionally long history of coevolution, in which the interests of the partners clearly conflict; but no simple scenario of *ongoing coevolutionary change* has yet proven to be a good description of the system. The patterns that we *have* detected point to the importance of regulatory factors extrinsic to the interaction itself. Simultaneous resource limitation of seed and wasp maturation success, for example, may account for positive correlations between them that contradict predictions based exclusively on conflicts between the partners. Wasp population dynamics and, consequently, fig reproductive success seem more commonly limited by climate, fig population size, and phenological traits than by traits that act directly at the stage of oviposition/pollination. Furthermore, the community context of the interaction cannot be ignored. Fig and wasp reproductive success may be directly and indirectly limited by "interloper" species in ways yet unknown; we may find that fig traits central to interactions with pollinators (e.g., syconium size) are subject to conflicting selection pressures from pollinators, seed dispersers, predators, and interlopers. Heithaus et al.[83]

and others have argued that population growth of mutualists is constrained by factors extrinsic to the mutualism itself. Studies of fig/pollinator interactions are suggesting that coevolutionary processes may be constrained similarly.

It is important to recognize that a good understanding of coevolutionary processes within any interaction rests on detailed information about the natural history and phylogenies of the organisms in question. There are still major gaps in our knowledge about the fig pollination mutualism, which, when filled, may well change our impression about how coevolution can proceed within it. For instance, what are the evolutionary origins of the interaction? Have figs and pollinators cospeciated, and what accounts for their explosive evolutionary radiation? In monoecious figs, how can the interaction persist in seasonal environments, and in gynodioecious figs, how can the interaction persist at all? What happens to agaonid females between the time they leave their natal fig and when they arrive at a receptive tree? What accounts for the large numbers of vacant ovaries within a mature syconium? How do the ubiquitous nonpollinating wasps affect the mutualism? How much diversity is there among fig/pollinator species pairs in the answers to these questions, and what are the origins and significance of that diversity? As attested to by the recent dates of many of the articles cited in this review, studies are rapidly proliferating that address these questions, as well as the broader evolutionary issues to which they relate.

REFERENCES

1. **Addicott, J. F., Bronstein, J., and Kjellberg, F.,** Evolution of mutualistic life cycles: yucca moths and fig wasps, in *Genetics, Evolution, and Coordination of Insect Life Cycles,* Gilbert, F., Ed., Springer-Verlag, London, 1990, 143.
2. **August, P. V.,** Fig (*Ficus trigonata*) fruit consumption and seed dispersal by *Artibeus jamaicensis* in the llanos of Venezuela, *Biotropica,* 13, 70, 1981.
3. **Axelrod, R.,** *The Evolution of Cooperation,* Basic Books, New York, 1984.
4. **Axelrod, R. and Hamilton, W. D.,** The evolution of cooperation, *Science,* 211, 1390, 1981.
5. **Baijnath, H. and Ramcharun, S.,** Aspects of pollination and floral development in *Ficus capensis* Thunb. (Moraceae), *Bothalia,* 14, 883, 1983.
6. **Baijnath, H. and Ramcharun, S.,** Reproductive biology and chalcid symbiosis in *Ficus burtt-davyi* (Moraceae), *Monogr. Syst. Bot. Missouri Bot. Gard.,* 25, 227, 1988.
7. **Barker, N. P.,** Evidence of a volatile attractant in *Ficus ingens* (Moraceae), *Bothalia,* 15, 607, 1985.
8. **Beattie, A. J.,** *The Evolutionary Ecology of Ant-Plant Mutualisms,* Cambridge University Press, Cambridge, 1985.
9. **Beehler, B.,** Frugivory and polygamy in birds of paradise, *Auk,* 100, 1, 1983.
10. **Berg, C. C.,** Floral differentiation and dioecism in *Ficus* (Moraceae), in *Minisymposium, Figs and Fig Insects,* CNRS, Montpellier, France, 1984, 15.
11. **Berg, C. C.,** Classification and distribution of *Ficus, Experientia,* 45, 605, 1989.
12. **Berg, C. C.,** Reproduction and evolution in *Ficus* (Moraceae): traits connected with the adequate rearing of pollinators, *Mem. N.Y. Bot. Gard.,* 55, 169, 1990.
13. **Bernays, E. and Graham, M.,** On the evolution of host specificity in phytophagous arthropods, *Ecology,* 69, 886, 1988.
14. **Bonaccorso, F.,** Foraging and reproductive ecology in a Panamanian bat community, *Bull. Fla. St. Mus. Biol. Sci.,* 24, 359, 1979.
15. **Bouček, Z.,** Australian Chalcidoidea (Hymenoptera), in *CAB International,* Wallingford, United Kingdom, 1988.
16. **Bouček, Z., Watsham, A., and Wiebes, J.T.,** The fig wasp fauna of the receptacles of *Ficus thonningii* (Hymenoptera, Chalcidoidea), *Tijd. Ent.,* 124, 149, 1981.
17. **Boucher, D. H.,** The idea of mutualism, past and future, in *The Biology of Mutualism,* Boucher, D.H., Ed., Croom Helm, London, 1985, 1.

18. **Boucher, D. H., Ed.,** *The Biology of Mutualism: Ecology and Evolution,* Croom Helm, London, 1985.
19. **Boucher, D. H.,** Mutualism in agriculture, in *The Biology of Mutualism,* Boucher, D. H., Ed., Croom Helm, London, 1985, 375.
20. **Breitwisch, R.,** Frugivores at a fruiting *Ficus* vine in a southern Cameroon tropical wet forest, *Biotropica,* 15, 125,1983.
21. **Brockelman, W. Y.,** Observations of animals feeding in a strangler fig, *Ficus drupacea* in southeast Thailand, *Nat. Hist. Bull. Siam Soc.,* 30, 33, 1982.
22. **Bronstein, J. L.,** Coevolution and Constraints in a Neotropical Fig-Pollinator Wasp Mutualism, Dissertation, University of Michigan, Ann Arbor, Michigan, 1986.
23. **Bronstein, J. L.,** Maintenance of species-specificity in a Neotropical fig-pollinator wasp mutualism, *Oikos,* 48, 39, 1987.
24. **Bronstein, J. L.,** Mutualism, antagonism, and the fig-pollinator interaction, *Ecology,* 69, 1298, 1988.
25. **Bronstein, J. L.,** Fruit production in a monoecious fig: consequences of an obligate mutualism, *Ecology,* 69, 207, 1988.
26. **Bronstein, J. L.,** Predators of fig wasps, *Biotropica,* 20, 215, 1988.
27. **Bronstein, J. L.,** A mutualism at the edge of its range, *Experientia,* 45, 622, 1989.
28. **Bronstein, J. L.,** The nonpollinating wasp fauna of *Ficus pertusa*: exploitation of a mutualism?, *Oikos,* 61, 175, 1991.
29. **Bronstein, J. L.,** Mutualism studies and the study of mutualism, *Bull. Ecol. Soc. Am.,* 72, 6, 1991.
30. **Bronstein, J. L. and Hoffmann, K.,** Spatial and temporal variation in frugivory at a neotropical fig, *Ficus pertusa, Oikos,* 49, 261, 1987.
31. **Bronstein, J. L. and McKey, D.,** The fig/pollinator mutualism: a model system for comparative biology, *Experientia,* 45, 601, 1989.
32. **Bronstein, J. L. and Patel, A.,** Causes and consequences of within-tree phenological patterns in the Florida strangling fig, *Ficus aurea* (Moraceae), *Am. J. Bot.,* 79, 41, 1992.
33. **Bronstein, J. L., Gouyon, P.H., Gliddon, C., Kjellberg, F., and Michaloud, G.,** Ecological consequences of flowering asynchrony in monoecious figs: a simulation study, *Ecology,* 71, 2145, 1990.
34. **Cherif, A. H.,** Mutualism — the forgotten concept in teaching science, *Am. Biol. Teacher,* 52, 206, 1990.
35. **Chopra, R. N. and Kaur, H.,** Pollination and fertilization in some *Ficus* species, *Beitr. Biol. Pflanzen,* 45, 441, 1969.
36. **Clay, K.,** Clavicipitaceous fungal endophytes of grasses: coevolution and the change from parasitism to mutualism, in *Coevolution of Fungi with Plants and Animals,* Pirozynski, K. A. and Hawksworth, D. L., Eds., Academic Press, New York, 1988, 79.
37. **Compton, S.,** The fig wasp, *Odontofroggatia galili* (Hymenoptera: Pteromalidae), in the Greek Isles, *Ent. Gaz.,* 40, 183, 1989.
38. **Compton, S.,** A collapse of host specificity in some African fig wasps, *S. Afr. J. Sci.,* 86, 39, 1990.
39. **Compton, S. G. and Robertson, H.G.,** Complex interactions between mutualisms: ants tending homopterans protect fig seeds and pollinators, *Ecology,* 69, 1302, 1988.
40. **Compton, S. G. and Nefdt, R. J. C.,** The figs and fig wasps of *Ficus burtt-davyii, Mitt. Inst. Allg. Bot. Hamburg,* 23a, 441, 1990.
41. **Compton, S. G., Thornton, I. W. B., New, T. R., and Underhill, L.,** The colonization of the Krakatau Islands by fig wasps and other chalcids (Hymenoptera, Chalcidoidea), *Phil. Trans. R. Soc. Lond. B.,* 322, 459, 1988.
42. **Compton, S. G., Holton, K. C., Rashbrook, V. K., van Noort, S., Vincent, S. L., and Ware, A. B,** Studies of *Ceratosolen galili,* a non-pollinating agaonid fig wasp, *Biotropica,* 23, 188, 1991.
43. **Condit, I. J.,** *The Fig,* Chronica Botanica, Waltham, 1947.
44. **Corlett, R. T.,** The phenology of *Ficus fistulosa* in Singapore, *Biotropica,* 19, 122, 1987.
45. **Corner, E. J. H.,** An introduction to the distribution of *Ficus, Reinwardtia,* 4, 357, 1958.
46. **Corner, E. J. H.,** *Ficus* (Moraceae) and Hymenoptera (Chalcidoidea): figs and their pollinators, *Biol. J. Linn. Soc.,* 25, 187, 1985.
47. **Crane, J. C.,** The chemical induction of parthenocarpy in the Calimyrna fig and its physiological significance, *Plant Physiol.,* 40, 606, 1965.
48. **Crane, J. C.,** Fig, in *CRC Handbook of Fruit Set and Development,* Monselise, S. P., Ed., CRC Press, Boca Raton, FL, 1986, 153.
49. **Crane, J. C. and Blondeau, R.,** Hormone-induced parthenocarpy in the Calimyrna fig and a comparison of parthenocarpic and caprified syconia, *Plant Physiol.,* 26, 136, 1951.

50. **Crane, J. C. and van Overbeek, J.,** Kinin-induced parthenocarpy in the fig, *Ficus carica* L., *Science,* 147, 1468, 1965.
51. **Cruden, R. W. and Lyon, D. L.,** Patterns of biomass allocation to male and female functions in plants with different mating systems, *Oecologia,* 66, 299, 1985.
52. **Davidson, D. W.,** Ecological studies of neotropical ant gardens, *Ecology,* 69, 1138, 1988.
53. **Dawkins, R. and Krebs, J. R.,** Arms races between and within species, *Proc. R. Soc. Lond. B.,* 205, 489, 1979.
54. **Ehrlich, P. R. and Raven, P. H.,** Butterflies and plants: a study in coevolution, *Evolution,* 18, 586, 1964.
55. **Ewald, P. W.,** Transmission modes and evolution of the parasitism-mutualism continuum, *Ann. N.Y. Acad. Sci.,* 503, 295, 1987.
56. **Feinsinger, P.,** Coevolution and pollination, in *Coevolution,* Futuyma, D. J. and Slatkin, M., Eds., Sinauer, Sunderland, 1983, 282.
57. **Foster, R. B.,** The seasonal rhythm of fruitfall on Barro Colorado Island, in *The Ecology of a Tropical Forest,* Leigh, E. G., Rand, A. S., and Windsor, D. W., Eds., Smithsonian Institution Press, Washington, D.C., 1982, 151.
58. **Frank, S. A.,** Theoretical and Empirical Studies of Sex Ratios, Mainly in Fig Wasps, M.S. Thesis, University of Florida, Gainesville, 1983.
59. **Frank, S. A.** The behavior and morphology of the fig wasps *Pegoscapus assuetus* and *P. jimenezi*: descriptions and suggested behavioral characters for phylogenetic studies, *Psyche,* 91, 289, 1984.
60. **Frank, S. A.,** Weapons and fighting in fig wasps, *Trends Ecol. Evol.,* 2, 259, 1987.
61. **Futuyma, D. J. and Slatkin, M.,** Epilogue: the study of coevolution, in *Coevolution,* Futuyma, D. J. and Slatkin, M., Eds., Sinauer, Sunderland, 1983, 459.
62. **Futuyma, D. J. and Slatkin, M., Eds.,** *Coevolution,* Sinauer, Sunderland, 1983.
63. **Galil, J.,** Topocentric and ethodynamic pollination, in *Pollination and Dispersal,* Brantjes, N. B. M., and Linskens, H. F., Eds., Univ. of Nijmegen Press, Nijmegen, 1973, 85.
64. **Galil, J.,** Pollination in dioecious figs: pollination of *F. fistulosa* by *Ceratosolen hewitti, Gard. Bull. (Singapore),* 26, 303, 1973.
65. **Galil, J. and Eisikowitch, D.,** On the pollination ecology of *Ficus sycomorus* in East Africa, *Ecology,* 49, 259, 1968.
66. **Galil, J. and Eisikowitch, D.,** Flowering cycles and fruit types in *Ficus sycomorus* in Israel, *New Phytol.,* 67, 745, 1968.
67. **Galil, J. and Eisikowitch, D.,** On the pollination ecology of *Ficus religiosa* in Israel, *Phytomorph.,* 18, 356, 1968.
68. **Galil, J. and Eisikowitch, D.,** Further studies on the pollination ecology of *Ficus sycomorus* L. (Hymenoptera, Chalcidoidea, Agaonidae), *Tijd. Ent.,* 112, 1, 1969.
69. **Galil, J. and Eisikowitch, D.,** Studies on mutualistic symbiosis between syconia and sycophilous wasps in monoecious figs, *New Phytol.,* 70, 773, 1971.
70. **Galil, J. and Eisikowitch, D.,** Further studies on the pollination ecology of *Ficus sycomorus.* II. Pollen filling and emptying by *Ceratosolen arabicus* Mayr, *New Phytol.,* 73, 551, 1974.
71. **Galil, J. and Meiri, L.,** Number and structure of anthers in fig syconia in relation to behaviour of the pollen vectors, *New Phytol.,* 88, 83, 1981.
72. **Galil, J., Dulberger, R., and Rosen, D.,** The effects of *Sycophaga sycomori* L. on the structure and development of the syconia in *Ficus sycomorus* L., *New Phytol.,* 69, 103, 1970.
73. **Galil, J., Ramirez B. W. and Eisikowitch, D.,** Pollination of *Ficus costaricana* and *F. hemsleyana* by *Blastophaga estherae* and *B. tonduzii* in Costa Rica (Hymenoptera: Chalcidoidea, Agaonidae), *Tijd. Ent.,* 116, 175, 1973.
74. **Galil, J., Stein, M., and Horovitz, A.,** On the origins of the sycamore fig (*Ficus sycomorus* L.) in the Middle East, *Gard. Bull. (Singapore),* 29, 191, 1976.
75. **Gautier-Hion, A. and Michaloud, G.,** Figs: are they keystone resources for frugivorous vertebrates throughout the tropics? A test in Gabon, *Ecology,* 70, 1826, 1989.
76. **Gautier-Hion, A., Duplantier, J.-M., Quris, R., Feer, F., Sourd, C., Decoux, J.-P., Dubost, G., Emmons, L., Erard, C., Hecketsweiler, P., Moungazi, A., Roussilhon, C., and Thiollay, J.-M.,** Fruit characters as a basis of fruit choice and seed dispersal in a tropical forest vertebrate community, *Oecologia,* 65, 324, 1985.
77. **Gilbert, L. E.,** *Food* web organization and the conservation of neotropical diversity, in *Conservation Biology,* Soulé, M. E. and Wilcox, B. A., Eds., Sinauer, Sunderland, 1980, 11.
78. **Gilbert, L. E. and Raven, P. H., Eds.,** *Coevolution of Animals and Plants,* Univ. of Texas Press, Austin, 1975.
79. **Godfray, H. C. J.,** Virginity in haplodiploid populations: a study on fig wasps, *Ecol. Entom.,* 13, 283, 1988.

80. **Gordh, G.,** The comparative external morphology and systematics of the neotropical parasitic fig wasp genus *Idarnes* (Hymenoptera: Torymidae), *U. Kansas Sci. Bull.,* 50, 389, 1975.
81. **Hamilton, W. D.,** Wingless and fighting males in fig wasps and other insects, in *Sexual Selection and Reproductive Competition in Insects,* Blum, M. S. and Blum, N. A., Eds., Academic Press, London, 1979, 167.
82. **Heads, P. A. and Lawton, J. H.,** Bracken ants and extrafloral nectaries. II. The effect of ants on the insect herbivores of bracken, *J. Anim. Ecol.,* 53, 1015, 1984.
83. **Heithaus, E. R., Culver, D. C., and Beattie, A. J.,** Models of some ant-plant mutualisms, *Am. Nat.,* 116, 347, 1980.
84. **Herbst, L. H.,** The role of nitrogen from fruit pulp in the nutrition of the frugivorous bat *Carollia perspicillata, Biotropica,* 18, 39, 1986.
85. **Herre, E. A.,** Sex ratio adjustment in fig wasps, *Science,* 228, 896, 1985.
86. **Herre, E. A.,** Optimality, plasticity, and selective regime in fig wasp sex ratios, *Nature,* 329, 627, 1987.
87. **Herre, E. A.,** Coevolution of reproductive characteristics in twelve species of New World figs and their pollinator wasps, *Experientia,* 45, 637, 1989.
88. **Herrera, C. M.,** Determinants of plant-animal coevolution: the case of mutualistic dispersal of seeds by vertebrates, *Oikos,* 44, 132, 1985.
89. **Hespenheide, H. A.,** Selective predation by two swifts and a swallow in Central America, *Ibis,* 117, 82, 1975.
90. **Hill, D. S.,** *Figs of Hong Kong,* Hong Kong University Press, Hong Kong, 1967.
91. **Howe, H. F.,** Constraints on the evolution of mutualisms, *Am. Nat.,* 123, 764, 1984.
92. **Inouye, D. W.,** The ecology of nectar robbing, in *The Biology of Nectaries,* Bentley, B. and Elias, T., Eds., Columbia University Press, New York, 1983, 153.
93. **Janzen, D. H.,** Herbivores and the number of tree species in tropical forests, *Am. Nat.,* 104, 501, 1970.
94. **Janzen, D. H.,** *Pseudomyrmex pilosa*: a parasite of a mutualism, *Science,* 188, 936, 1975.
95. **Janzen, D. H.,** How to be a fig, *Ann. Rev. Ecol. Syst.,* 10, 13, 1979.
96. **Janzen, D. H.,** How many babies do figs pay for babies?, *Biotropica,* 11, 48, 1979.
97. **Janzen, D. H.,** *Ficus ovalis* seed predation by an orange-chinned parakeet (*Brotogeris jugularis)* in Costa Rica, *Auk,* 98, 841, 1981.
98. **Janzen, D. H.,** Dispersal of seeds by vertebrate guts, in *Coevolution,* Futuyma, D. J. and Slatkin, M., Eds., Sinauer, Sunderland, 1983, 232.
99. **Janzen, D. H. and Martin, P.,** Neotropical anachronisms: what the gomphotheres ate, *Science,* 215, 19, 1982.
100. **Johri, B. M. and Konar, R. N.,** The floral morphology and embryology of *Ficus religiosa* Linn., *Phytomorph.,* 6, 97, 1956.
101. **Jordano, P.,** Fig-seed predation and dispersal by birds, *Biotropica,* 15, 38, 1983.
102. **Joseph, K. J.,** Recherches sur les chalcidiens, *Blastophaga psenes* L. and *Philotrypesis* L. du figuier (*Ficus carica* L.), *Ann. Sci. Nat. Zool.,* 2e série 20, 13, 187, 1958.
103. **Kauffmann, S., McKey, D. B., Hossaert-McKey, M., and Horvitz, C. C.,** Adaptations for a two-phase seed dispersal system involving vertebrates and ants in a hemiepiphytic fig (*Ficus microcarpa:* Moraceae), *Am. J. Bot.,* 78, 971, 1991.
104. **Keddy, P.,** Is mutualism really irrelevant to ecology?, *Bull. Ecol. Soc. Am.,* 71, 101, 1990.
105. **King, G.,** The species of *Ficus* of the Indo-Malayan and Chinese countries, *Ann. R. Bot. Gard. Calcutta,* 1, 1, 1888.
106. **Kistner, D. H.,** The reclassification of the genus *Charoxus* Sharp with the description of new species (Coleoptera: Staphylinidae), *J. Kans. Entomol. Soc.,* 54, 841, 1981.
107. **Kjellberg, F.,** La Stratégie Reproductive du Figuier (*Ficus carica* L.) et de son Pollinisateur (*Blastophaga psenes* L.): Un Exemple de Coévolution, Dissertation, Institut Nationale Agronomique, Paris, 1983.
108. **Kjellberg, F., Gouyon, P. H., Ibrahim, M., Raymond, M., and Valdeyron, G.,** The stability of the symbiosis between dioecious figs and their pollinators: a study of *Ficus carica* L. and *Blastophaga psenes* L., *Evolution,* 41, 693, 1987.
109. **Kjellberg, F., Michaloud, G., and Valdeyron, G.,** The *Ficus-Ficus* pollinator mutualism: how can it be evolutionarily stable?, in *Insects-Plants,* Labeyrie, V., Fabres, G., and Lachaise, D., Eds., W. Junk Publ., Dordrecht, 1987, 335.
110. **Kjellberg, F., Doumesche, B., and Bronstein, J. L.,** Longevity of a fig wasp (*Blastophaga psenes*), *Proc. K. Ned. Akad. Wet. Ser. C.,* 91, 117, 1988.
111. **Kjellberg, F. and Maurice, S.,** Seasonality in the reproductive phenology of *Ficus*: its evolution and consequences, *Experientia,* 45, 653, 1989.

112. **Kuttamathiathu, J. J.**, The biology of *Philotrypesis caricae* (L.), parasite of *Blastophaga psenes* (L.) (Chalcidoidea: parasitic Hymenoptera), *XV Int. Congr. Zool., Sec. VIII*, 1, 1955.
113. **Lee, T. D.**, Patterns of fruit and seed production, in *Plant Reproductive Ecology*, Lovett Doust, J. and Lovett Doust, L., Eds., Oxford University Press, New York, 1988, 179.
114. **Lovett Doust, J. N. and Harper, J. L.**, The resource costs of gender and maternal support in an andromonoecious Umbellifer, *Smyrnium olusatrum* L., *New Phytol.*, 85, 251, 1980.
115. **Maxie, E. C. and Crane, J. C.**, Effect of ethylene on growth and maturation of the fig (*Ficus carica* L.) fruit, *Proc. Am. Soc. Hort. Sci.*, 92, 255, 1968.
116. **McClure, H. E.**, Flowering, fruiting, and animals in the tropical rain forest, *Malay For.*, 29, 182, 1966.
117. **McKey, D.**, Population biology of figs: applications for conservation, *Experientia*, 45, 661, 1989.
118. **Medway, L.**, Phenology of a tropical rain forest in Malaya, *Biol. J. Linn. Soc.*, 4, 117, 1972.
119. **Michaloud, G.**, Aspects de la Reproduction des Figuiers Monoïques en Forêt Equatoriale Africaine, Dissertation, Université des Sciences et Techniques du Languedoc, Montpellier, France, 1988.
120. **Michaloud, G. and Michaloud-Pelletier, S.**, *Ficus* hemi-epiphytes (Moraceae) et arbres supports, *Biotropica*, 19, 125, 1987.
121. **Michaloud, G., Michaloud-Pelletier, S., Wiebes, J.T., and Berg, C. C.**, The cooccurrence of two pollinating species of fig wasp and one species of fig, *Proc. K. Ned. Akad. Wet. Ser. C*, 88, 93, 1985.
122. **Milton, K. and Dintzis, F. R.**, Nitrogen-to-protein conversion factors for tropical plant samples, *Biotropica* 13, 177, 1981.
123. **Milton, K., Windsor, D. M., Morrison, D. W., and Estribi, M. A**, Fruiting phenologies of two neotropical *Ficus* species, *Ecology*, 63, 752, 1982.
124. **Morrison, D. W.**, Foraging ecology and energetics of the frugivorous bat *Artibeus jamaicensis*, *Ecology*, 59, 716, 1978.
125. **Morse, D. H.**, Costs in a milkweed-bumblebee mutualism, *Am. Nat.*, 125, 903, 1985.
126. **Murray, M. G.**, Figs (*Ficus* spp.) and fig wasps (Chalcidoidea, Agaonidae): hypotheses for an ancient symbiosis, *Biol. J. Linn. Soc.*, 26, 69, 1985.
127. **Murray, M. G.**, The closed environment of the fig receptacle and its influence on male conflict in the Old World fig wasp, *Philotrypesis pilosa*, *Anim. Behav.*, 35, 488, 1987.
128. **Nair, P. B. and Abdurahiman, U. C.**, Population dynamics of the fig wasp *Kradibia gestroi* (Grandi) (Hymenoptera Chalcidoidea, Agaonidae) from *Ficus exasperata* Vahl, *Proc. K. Ned. Akad. Wet. Ser. C*, 87, 365, 1984.
129. **Neeman, G. and Galil, J.**, Seed set in the male syconia of the common fig, *Ficus carica* (caprificus), *New Phytol.*, 81, 375, 1978.
130. **Newton, L. E. and Lomo, A.**, The pollination of *Ficus vogelii* in Ghana, *Bot. J. Linn. Soc.*, 78, 21, 1979.
131. **Nilsson, L. A.**, The evolution of flowers with deep corolla tubes, *Nature*, 334, 147, 1988.
132. **Nitecki, M. H., Ed.**, *Coevolution*, Univ. of Chicago Press, Chicago, 1983.
133. **Okamoto, M. and Tashiro, M.**, Mechanism of pollen transfer and pollination in *Ficus erecta* by *Blastophaga nipponica*, *Bull. Osaka Mus. Nat. Hist.*, 34, 7, 1981.
134. **Paine, R. T.**, Food web complexity and species diversity, *Am. Nat.*, 100, 65, 1966.
135. **Pemberton, C. E.**, Fig wasps established on Kauai, *Proc. Hawaiian Ent. Soc.*, 8, 399, 1934.
136. **Pellmyr, O.**, The cost of mutualism: interactions between *Trollius europaeus* and its pollinating parasites, *Oecologia*, 78, 53, 1989.
137. **Pierce, N. E., Kitching, R. L., Buckley, R. C., Taylor, M. F. J., and Benbow, K. F.**, The costs and benefits of cooperation between the Australian lycaenid butterfly, *Jalmenus evagoras*, and its attendant ants, *Behav. Ecol. Sociobiol.*, 21, 237, 1987.
138. **Price, P. W., Bouton, C. E., Gross, P., McPheron, B., Thompson, J., and Weis, A. E.**, Interactions among three trophic levels: influence of plants on interactions between insect herbivores and natural enemies, *Ann. Rev. Ecol. Syst.*, 11, 41, 1980.
139. **Price, P. W., Westoby, M., Rice, B., Atsatt, P. R., Fritz, R. S., Thompson, J. N., and Mobley, K.**, Parasite mediation in ecological interactions, *Ann. Rev. Ecol. Syst.*, 17, 487, 1986.
140. **Ramcharun, S., Baijnath, H., and van Greuning, J. V.**, Some aspects of the reproductive biology of the *Ficus natalensis* complex in southern Africa, *Mitt. Inst. Allg. Bot. Hamburg*, 23a, 451, 1990.
141. **Ramirez, B. W.**, Fig wasps: mechanism of pollen transfer, *Science*, 163, 580, 1969.
142. **Ramirez, B. W.**, Host specificity of fig wasps (Agaonidae), *Evolution*, 24, 681, 1970.
143. **Ramirez, B. W.**, Taxonomic and biological studies of neotropical fig wasps, *U. Kansas Sci. Bull.*, 49, 1, 1970.

144. **Ramirez, B. W.,** Coevolution of *Ficus* and Agaonidae, *Ann. Miss. Bot. Gard.,* 61, 770, 1974.
145. **Ramirez, B. W.,** A new classification of *Ficus, Ann. Miss. Bot. Gard.,* 64, 296, 1977.
146. **Ramirez, B. W.,** Evolution of mechanisms to carry pollen in Agaonidae (Hymenoptera, Chalcidoidea), *Tijd. Ent.,* 121, 279, 1978.
147. **Ramirez, B. W.,** Evolution of the monoecious and dioecious habit in *Ficus* (Moraceae), *Brenesia,* 18, 207, 1980.
148. **Risch, S. and Boucher, D.,** What ecologists look for, *Bull. Ecol. Soc. Am.,* 57, 8, 1978.
149. **Roberts, J. T. and Heithaus, E. R.,** Ants rearrange the vertebrate-generated seed shadow of a neotropical fig tree, *Ecology,* 67, 1046, 1986.
150. **Roskam, J. C. and Nadel, H.,** Redescription and immature stages of *Ficiomyia perarticulata* (Diptera: Cecidomyiidae), a gall midge inhabiting syconia of *Ficus citrifolia, Proc. Ent. Soc. Wash.,* 92, 778, 1990.
151. **Schemske, D. W.,** Limits to specialization and coevolution in plant-animal mutualisms, in *Coevolution,* Nitecki, M. H., Ed., Univ. of Chicago Press, Chicago, 1983, 67.
152. **Schmid-Hempel, P. and Schmid-Hempel, R.,** Efficient nectar-collecting by honeybees II. Response to factors determining nectar availability, *J. Anim. Ecol.,* 56, 219, 1987.
153. **Scott, P. E. and Martin, R. F.,** Avian dispersers of *Bursera, Ficus,* and *Ehretia* fruit in Yucatán, *Biotropica,* 16, 319, 1984.
154. **Seidel, J. L., Epstein, W. W., and Davidson, D. W.,** Neotropical ant gardens. I. Chemical constituents, *J. Chem. Ecol.,* 16, 1791, 1990.
155. **Slater, J. A.,** Lygaeid bugs (Hemiptera: Lygaeidae) as seed predators of figs, *Biotropica,* 4, 145, 1972.
156. **Smith, D. C. and Douglas, A. E.,** *The Biology of Symbiosis,* Edward Arnold, London, 1987.
157. **Soberon Mainero, J. and Martinez del Rio, C.,** Cheating and taking advantage in mutualistic associations, in *The Biology of Mutualism,* Boucher, D. H., Ed., Croom Helm, London, 1985, 192.
158. **Soulé, M.,** *Viable Populations for Conservation,* Cambridge University Press, Cambridge, 1987.
159. **Stephenson, A. G. and Bertin, R. I.,** Male competition, female choice, and sexual selection in plants, in *Pollination Biology,* Real, L., Ed., Academic Press, New York, 1983, 109.
160. **Terborgh, J.,** *Five New World Primates: a Study in Comparative Ecology,* Princeton University Press, Princeton, NJ,1983.
161. **Terborgh, J.,** Keystone plant resources in the tropical forest, in *Conservation Biology: the Science of Scarcity and Diversity,* Soulé, M. E., Ed., Sinauer, Sunderland, 1986, 330.
162. **Thomas, D. W., Cloutier, D., Provencher, M., and Houle, C.,** The shape of bird- and bat-generated seed shadows around a tropical fruiting tree, *Biotropica,* 20, 347, 1988.
163. **Thompson, J. N.,** *Interaction and Coevolution,* Wiley, New York, 1982.
164. **Thompson, J. N.,** Concepts of coevolution, *Trends Ecol. Evol.,* 4, 179, 1989.
165. **Todzia, C.,** Growth habits, host tree species, and density of hemiepiphytes on Barro Colorado Island, Panama, *Biotropica,* 18, 22, 1986.
166. **Tomback, D. F.,** Nutcrackers and pines: coevolution or coadaptation?, in *Coevolution,* Nitecki, M. H., Ed., Univ. of Chicago Press, Chicago, 1983, 179.
167. **Turner, J. R. G., Kearney, E. P., and Exton, L. S.,** Mimicry and the Monte Carlo predator: the palatability spectrum and the origins of mimicry, *Biol. J. Linn. Soc.,* 23, 247, 1984.
168. **Ulenberg, S. A.,** The phylogeny of the genus *Apocrypta* Coquerel in relation to its hosts, *Ceratosolen* Mayr (Agaonidae) and *Ficus* L., in *The Systematics of the Fig Wasp Parasites of the Genus Apocrypta Coquerel,* Ulenberg, S. A., Ed., North-Holland Publishing Co., Amsterdam, 1985, 149.
169. **Ulenberg, S. A. and Nübel, B. K.,** Oviposition of *Apocrypta* fig wasp parasites (Hymenoptera: Chalcidoidea: Torymidae), *Proc. K. Ned. Akad. Wet. Ser. C,* 85, 607, 1982.
170. **Ulenberg, S. A. and Van Pelt, W.,** A revision of the species of *Apocrypta* Coquerel: a taxonomic study incorporating uni- and multivariate data analyses, in *The Systematics of the Fig Wasp Parasites of the Genus Apocrypta Coquerel,* Ulenberg, S. A., Ed., North-Holland Publishing Co., Amsterdam, 1985, 41.
171. **Valdeyron, G.,** Sur le système génétique du figuier *Ficus carica* L. Essai d'interprétation évolutive, *Ann. Inst. Nat. Agr. Paris,* 5, 1, 1967.
172. **Valdeyron, G. and Lloyd, D. G.,** Sex differences and flowering phenology in the common fig, *Ficus carica* L., *Evolution,* 33, 673, 1979.
173. **van der Pijl, L.,** *Principles of Dispersal in Higher Plants,* Springer-Verlag, Berlin, 1982.

174. **vander Wall, S. B.,** *Food Hoarding in Animals,* Univ. of Chicago Press, Chicago, 1990.
175. **Van Noort, S., Ware, A. B., and Compton, S. G.,** Pollinator-specific volatile attractants released from the figs of *Ficus burtt-davyi, Suid-Afr. Tyd. Wet.,* 85, 323, 1989.
176. **Verkerke, W.,** Syconial anatomy of *Ficus asperifolia* (Moraceae), a gynodioecious tropical fig, *Proc. K. Ned. Akad. Wet. C,* 90, 461, 1987.
177. **Verkerke, W.,** Flower development in *Ficus sur* Forsskål (Moraceae), *Proc. K. Ned. Akad Wet. C,* 91,175, 1988.
178. **Verkerke, W.,** Sycone morphology and its influence on the flower structure of *Ficus sur* (Moraceae), *Proc. K. Ned. Akad Wet. C,* 91, 319, 1988.
179. **Verkerke, W.,** Structure and function of the fig, *Experientia,* 45, 612, 1989.
180. **Verkerke, W.,** Fig anatomy and reproductive biology of African *Ficus* species (Moraceae), *Mitt. Inst. Allg. Bot. Hamburg,* 23a, 427, 1990.
181. **Wharton, R. A., Tilson, J. W., and Tilson, R. L.,** Asynchrony in a wild population of *Ficus sycomorus, S. Afr. J. Sci.,* 76, 478, 1980.
182. **Wheelwright, N. T.,** Fruit size, gape width, and the diets of fruit-eating birds, *Ecology,* 66, 808, 1985.
183. **Wheelwright, N. T. and Orians, G. H.,** Seed dispersal by animals: constraints with pollen dispersal, problems of terminology and constraints on coevolution, *Am. Nat.,* 119, 402, 1982.
184. **Wheelwright, N. T. and Janson, C. H.,** Colors of fruit displays of bird-dispersed plants in two tropical forests, *Am. Nat.,* 126, 777, 1985.
185. **White, M. J. D.,** *Modes of Speciation,* Freeman, San Francisco, 1978.
186. **Wiebes, J. T.,** Provisional host catalog of fig wasps (Hymenoptera: Chalcidoidea), *Zool. Verh.,* 83, 1, 1966.
187. **Wiebes, J. T.,** Fig wasps from Israeli *Ficus sycomorus* and related East African species (Hymenoptera, Chalcidoidea). 2. Agaonidae (concluded) and Sycophagini, *Zool. Meded. Leiden,* 42, 307, 1968.
188. **Wiebes, J. T.,** A short history of fig wasp research, *Gard. Bull. (Singapore),* 29, 207, 1977.
189. **Wiebes, J. T.,** The phylogeny of the Agaonidae (Hymenoptera, Chalcidoidea), *Neth. J. Zool.,* 32, 395, 1982.
190. **Wiebes, J. T.,** Agaonidae (Hymenoptera Chalcidoidea) and *Ficus* (Moraceae): fig wasps and their figs, IV (African *Ceratosolen*), *Proc. K. Ned. Akad. Wet. C,* 92, 251, 1989.
191. **Wiebes, J. T. and Compton, S. G.,** Agaonidae (Hymenoptera Chalcidoidea) and *Ficus* (Moraceae): fig wasps and their figs, VI (Africa concluded), *Proc. K. Ned. Akad. Wet. C,* 93, 203, 1990.
192. **Willson, M. F. and Burley, N.,** *Mate Choice in Plants: Tactics, Mechanisms and Consequences,* Princeton University Press, Princeton, NJ, 1983.
193. **Willson, M. F. and Ruppel, K. P.,** Resource allocation and floral sex ratios in *Zizania aquatica, Can. J. Bot.,* 62, 799, 984.
194. **Wilson, D. S. and Knollenberg, W. G.,** Adaptive indirect effects: the fitness of burying beetles with and without their phoretic mites, *Evol. Ecol.,* 1, 139, 1987.
195. **Windsor, D. M., Morrison, D. W., Estribi, M.A., and de Leon, B.,** Phenology of fruit and leaf production by "strangler" figs on Barro Colorado Island, Panama, *Experientia,* 45, 647, 1989.

2 Plant Sterols and Host-Plant Affiliations of Herbivores

E. A. Bernays
Department of Entomology and Center for Insect Science
University of Arizona
Tucson, Arizona

TABLE OF CONTENTS

I. INTRODUCTION

Nutritionally, the phylum Arthropoda is most clearly distinguished from the vertebrates by an inability to biosynthesize sterols, despite analogous physiological needs for sterols in lipid biostructures and as steroid hormone precursors. Insects have been shown to be unable to condense farnesyl pyrophosphate to squalene and to lack squalene epoxide cyclase, the enzyme needed to catalyze the cyclization of squalene to lanosterol.[19]

Initially, a general division was made between insects feeding on animal tissue, mostly using cholesterol, which is abundant in their food, and those feeding on plants and using phytosterols, either directly — at least for the bulk of the structural requirements — or converting them to cholesterol. In recent years, a considerable diversity has been found in insect sterol metabolism, with some strikingly different abilities to use particular sterols among insects with very specialized diets and clear limitations in some insect groups with respect to the variety of plant sterols they are able to use.[17,52]

Given that phytosterols are not all equally utilizable by all phytophagous insects, it becomes particularly important to investigate the qualitative and quantitative aspects of sterol composition of plant tissue, and whether or not sterol nutrition is commonly one of the determinants of host selection and use by arthropod herbivores in general. Over 20 years ago, the singular dependence of *Drosophila pachea* on the unusual sterols of its host plant *Lophocereus schottii* (senita cactus) was worked out by Heed and Kircher,[26] yet little progress has been made in the study of other systems, much less in the determination of whether there may be any generalizations to be made on the relative importance of specific phytosterols in insect herbivore nutrition. Since plants have a diversity of sterols,[3,35] and the profiles of sterol composition are being shown to vary more than was formerly thought,[40] it is becoming more important to investigate what the significance may be for the arthropods that feed upon them. This short review highlights the inadequacy of our knowledge and suggests that the importance of sterols has been underestimated.

II. THE STEROL STRUCTURES

The sterol structures mentioned in this review are shown in Figure 1. The major departures from the structure of cholesterol involve the presence or absence of double bonds: Δ 7 added, Δ 22 added, and C5 saturated; or C24 ethyl or methyl groups added to produce C28 or C29 compounds. The latter require dealkylation to produce cholesterol. Inspection of the two-dimensional representations in Figure 1 shows what steps are required to convert the various different sterol species to cholesterol. The pathways of phytosterol metabolism in insects are reviewed by Svoboda and Thompson.[52]

Introduction of unsaturation at either the 7 or the 22 position has a marked effect on the shape of the sterol molecule, and hence the potential to greatly alter their availability as enzyme substrates or their potential to bind with specific receptors.

III. THE INSECT GROUPS

A. ORTHOPTERA

The phytophagous group most studied to date is the Acrididae. Here, the most detailed study is that of Dadd,[15] who examined the differential use of sterols in the locusts *Schistocerca gregaria* (Cyrtacanthacridinae) and *Locusta migratoria* (Oedipodinae). Although dealkylation apparently occurred, he showed that, among

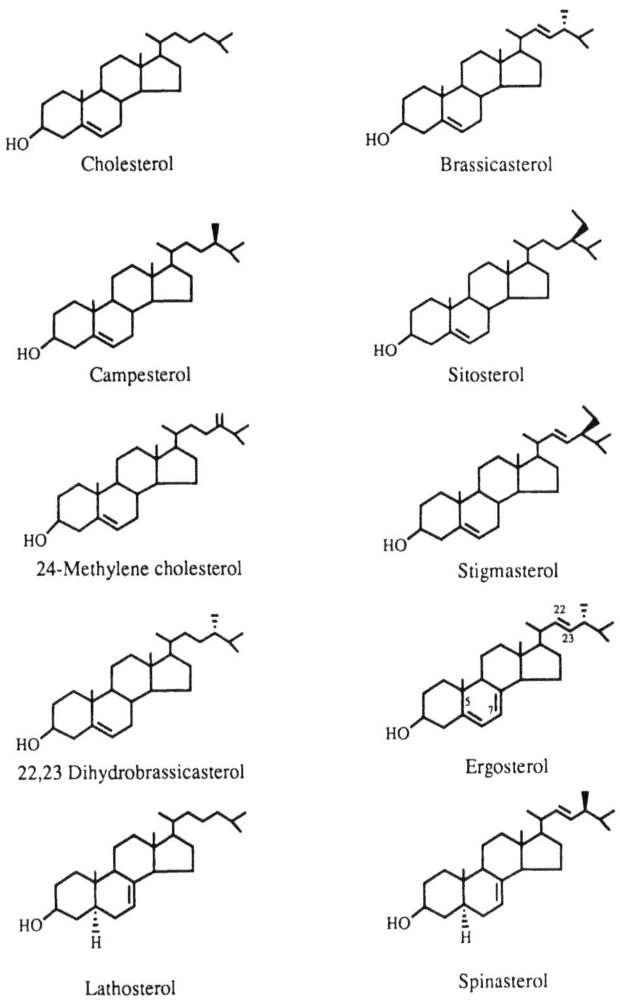

Cholesterol

Brassicasterol

Campesterol

Sitosterol

24-Methylene cholesterol

Stigmasterol

22,23 Dihydrobrassicasterol

Ergosterol

Lathosterol

Spinasterol

FIGURE 1. Some common sterol structures.

phytosterols, utilization was prevented by the presence of a double bond at either position Δ 7 or Δ 22. This contrasts with the situation in a number of lepidopteran species that appear to have a broader capability,[57] and he noted that apparently quite minor structural characters determine the sterols that are utilizable in these insects. It was suggested that the specificity in sterol requirements may be due to highly specific systems governing uptake from the gut. It does seem likely that the limitations occur at the level of the gut, since sterols that are produced metabolically during the synthesis of ecdysone, such as 7-dehydrocholesterol, are of no value to locusts when provided as the sole source of dietary sterol. Further, Dadd[15] found that in the two locust species he studied, sitosterol was slightly superior to cholesterol as a dietary sterol, even though cholesterol provides the essential metabolic building block.

Three acridid species were selected for a recent study by the author: *Melanoplus differentialis* (Melanoplinae), *Taeniopoda eques* (Romaleidae), and *Cibolacris parviceps* (Gomphocerinae). All three occur in Arizona and have very wide host ranges, although the recorded hosts are mostly different for all three. Individuals were fed the artificial diet developed by Dadd,[16] but in some treatments cholesterol was replaced by either stigmasterol or sitosterol. The results obtained were different for all three species. Cholesterol was best for the survivorship of *T. eques*, which seems logical, since it is normally suitable for insects, and the steroid hormones are usually produced via this sterol, whatever the normal diet. However, there is something of a mystery in that cholesterol is not commonly found in plants, except in trace amounts. One possibility is that this species obtains a significant proportion of its dietary sterol from pollen in which 24-methylene-cholesterol is extremely common,[2,49] and recent studies have demonstrated that flowers form one of the major dietary items for *T. eques* in the field.[42] The very common phytosterol, sitosterol, gave moderately good survival, but with Δ 22 stigmasterol survivorship was no better than in the absence of sterols.

Sitosterol was the only one of the three sterols that supported development in *M. differentialis*. It is interesting that cholesterol is of no value, even though it is the major tissue sterol. This is perhaps suggestive of insects with a long history of foliage feeding, with cholesterol a rare dietary component and sitosterol the most common phytosterol.

Although work is still in progress the pattern with *C. parviceps is* apparently similar to that found for *S. gregaria* and *L. migratoria.* The fact that differences in sterol utilization patterns occur among Acrididae is interesting, but in nature, when individuals feed on foliage from one or more plant species, the differences in their dietary sterol intake may be less obvious. This is because there usually are mixtures of sterols in plants, and even cholesterol may be present in small amounts. Therefore, needs for structural purposes and for ecdysteroid synthesis may be met by different dietary sterols. For example, plants containing mainly Δ 7 or Δ 22 sterols, which cannot be converted to cholesterol by certain species, can often be used in membrane structures, while smaller quantities of Δ 5 sterols are absorbed and dealkylated for the production of cholesterol and thence ecdysone.

A further interesting question is whether an increasing ability to use sitosterol rather than cholesterol, demonstrated by the sequence — *T. eques, S. gregaria, M. differentialis* — represents a longer evolutionary history of herbivory.

B. HEMIPTERA

The first species studied in detail was the milkweed bug, *Oncopeltus fasciatus,* by Svoboda and his co-workers.[54] The tissue sterol composition was found to be a direct reflection of the proportions of sitosterol, campesterol, and stigmasterol in its food, and no dealkylation could be demonstrated. In addition, it appeared to use campesterol as a precursor for makisterone A, the C28 molting hormone, instead of the usual C27 ecdysones, which in other insects are derived from cholesterol. More recently, there have been similar findings in other heteropterans. For example, *Dysdercus fasciatus* is unable to dealkylate phytosterols but uses mainly sitosterol as the tissue sterol, while makisterone A is the important ecdysteroid.[25] Similar patterns are found in *Nezara viridula, Podisus maculiventris*, *Dysdercus cingulatus,*[58] and *Megalotomus quinquespinosus.*[22]

C. HYMENOPTERA

The Hymenoptera have not had extensive study, but it is known that some sawflies at least are able to convert dietary sitosterol to cholesterol.[47] Honeybees, on the other hand, apparently engage in very little metabolism of their dietary sterols and contain a mixture of plant sterols, dominated by 24-methylenecholesterol, an abundant pollen sterol. When given artificial diets with different phytosterols, 24-methylene-cholesterol proved to be the best for rearing brood; this could be replaced by cholesterol, but not sitosterol or stigmasterol. In fact, honeybees apparently cannot dealkylate the C28 and C29 phytosterols to produce cholesterol,[55] and the molting hormone is makisterone A.[21]

Leaf-cutting ants are among the most severe defoliators of plants, but the food consists mainly of the fungi that they cultivate. The main dietary sterol is the fungal Δ 7 sterol, ergosterol. Ritter et al.[46] found in analyzing these ants that they contained only Δ 7 sterols and that there appeared to be a major structural requirement for 7-dehydro-24-methylenecholesterol specifically. The indication is that their sterol physiology may have evolved specifically for the use of Δ 7 compounds.

D. LEPIDOPTERA

A number of lepidopteran species have been examined, and it appears to be common that a wide variety of phytosterols is utilizable by caterpillars. Not only have dietary studies been undertaken, but in many cases the enzymic pathways for their conversion to cholesterol are known.[52] In their review, Svoboda and Thompson provide a table listing six species of Lepidoptera able to convert sitosterol to cholesterol as shown by biochemical studies. The prime example is *Manduca sexta*, a species feeding on plants in the family Solanaceae, in which many different sterols occur together. Most of these sterols appear to be used by this species. Appropriate saturation of double bonds is widespread, as is dealkylation.

Among lepidopterans that can survive on plants containing either Δ 5 or Δ 7 phytosterols, the former proved to be superior for *Heliothis zea*. The caterpillars of this species were unable to metabolize the B ring of Δ 7 sterols, and thus accumulated and partially used Δ 7 sterols without change.[34] Clearly, uptake of the sterols from the gut was adequate, but enzymes involved in conversion to cholesterol were lacking. It is perhaps interesting that this extremely polyphagous species is apparently less versatile that the more host-restricted *M. sexta*.

Sitosterol is better than cholesterol for the silkworm *Bombyx mori*,[31] while the grass-feeder *Crambus trisectus* is able to survive with sitosterol in the diet but not cholesterol.[20] In these two species, the restricted diets have sterol profiles dominated by sitosterol. However, *Homona coffearea*, the tea tortrix, requires the Δ 7 sterol ergosterol, and it too is unable to utilize dietary cholesterol.[48] Its host plant contains exclusively Δ 7 sterols.[35] *Pectinophora gossypiella* grew better with stigmasterol in the diet than with cholesterol.[61]

E. COLEOPTERA

Relatively few beetles have been studied in detail, but it is apparent that great diversity exists. The conifer weevil, *Hylobius pales*, appears to be similar to *Melanoplus* among the grasshoppers and *Crambus* among caterpillars, in that it is able to use dietary sitosterol but not cholesterol, although the latter is its tissue sterol.[45] However, the stored-products pest *Trogoderma granarium* cannot dealkylate phytosterols but uses them unchanged to a large extent, though it can best use cholesterol, as does its animal-feeding relatives.[53] Ishii[30] found that *Callosobruchus chinensis* grew better with sitosterol than with cholesterol in the diet, while McDonald and co-workers[34] showed that the alfalfa weevil, *Hypera postica*, which feeds only on alfalfa (a plant rich in Δ 7 sterols), did not convert absorbed Δ 7 sterols to cholesterol but apparently used the derivative lathosterol. A similar pattern of ability to use Δ 7 sterols was found in the mexican bean beetle, *Epilachna varivestis*, though its close relatives, which are the predatory coccinellids, use mainly cholesterol.[51] Another example is known of a dietary need for Δ 7 sterols, matching the normal availability of food components: the ambrosia beetle, *Xyleborus ferrugineus*, which feeds on a symbiotic fungus in wood, has been shown by Norris and his co-workers to require ergosterol.[38]

F. DIPTERA

Least studied and most difficult are the Diptera. This is because many of the phytophagous flies feed on vegetation that is at least partly rotted and contains many microorganisms, which may themselves contribute to or alter the food. *Drosophila melanogaster*, for example, feeds on many plant-derived substrates. In artificial diet it grows well if cholesterol is the sterol provided, but it grows better if this is replaced with stigmasterol or ergosterol.[13] Interestingly, the authors tested two different strains of the fly and found significant differences: one was able to use sitosterol and one could not. Such intraspecific variation gives a first indication of the genetic variation that may occur and the potential for evolutionary change, and may also explain the more recent contradictory evidence that *D. melanogaster* cannot dealkylate phytosterols.[56] However, this species does have a rather versatile endocrine ability, as demonstrated in vitro, to use cholesterol, on the one hand (and to produce 20-hydroxyecdysone as the molting hormone), but to be able to switch to using phytosterols without dealkylating them (and to produce makisterone A as the molting hormone).[43]

Most is known about the now famous *Drosophila pachea*, which feeds on senita cactus and requires the unusual Δ 7 sterols occurring in its host.[26] However, this is not a major step from a specialization for ergosterol, a common fungal sterol that may be in many substrates utilized by other *Drosophila* spp. Perhaps the cabbage fly, which also feeds on rotting vegetation, also depends on such sterols; this could explain the finding that cholesterol is detrimental,[18] since sterols can compete for uptake in the gut (see below).

G. CONCLUSIONS ON INSECT DIVERSITY

Even from the few studies mentioned above and summarized in Table 1, it is obvious that there is considerable diversity in the abilities of phytophagous insects to utilize sterols. These abilities, demonstrated by use of artificial diets or by biochemical studies, indicate considerable adaptation of sterol uptake mechanisms, biochemical pathways, and the actual sterol type used in the tissues to suit the available sterols in the diet. In one case, intraspecific variation has been demonstrated, and the data, taken together, indicate considerable evolutionary versatility in the insects, just as in other aspects of their biology.

Many questions remain. Is it possible that we will find some meaningful patterns in the differences among polyphagous acridid species when we know which are the most abundant sterols in the particular host plants selected? Why should the plant family specialist *Manduca sexta* be more versatile than the generalist *Heliothis zea*? Do flower-feeding herbivores require high levels or specific types of sterols?

Since in the majority of insects the steroid hormones are produced from cholesterol, it is interesting that the ability to take up cholesterol from the gut appears to have been lost in some herbivores. This, together with the demonstrated limitations in the use of available phytosterols in many species, suggests that maintaining a broad capacity may have its cost. Perhaps the specificity of the gut sterol receptors also prevents the uptake of deleterious steroidal compounds, which are abundant and varied in plants. Alternatively, if the postabsorption enzymic pathways have limited substrate specificity, a narrow specificity in the uptake mechanism may be needed. Certainly, relatively high dietary levels of sterols suboptimal for the performance of particular insects have been shown to be detrimental to *Locusta*,[16] *Taeniopoda* (Bernays, unpublished), and *Hylemya*.[18]

Table 1
DIFFERENTIAL USE OF STEROLS BY SOME HERBIVOROUS INSECTS[a]

	1	2	3	4	5	6	7	8	9	10
Orthoptera										
L. migratoria	++	.	.	.	.	+++	-	-	-	.
S. gregaria	++	.	.	.	.	+++	-	-	-	.
M. differentialis	-	.	.	.	.	+++	-	.	.	.
T. eques	+++	.	.	.	.	+	-	.	.	.
C. parviceps	++	.	.	.	.	+++	+	.	.	.
Hemiptera										
O. fasciatus	.	++	.	.	.	+	+	-	-	-
D. fasciatus	.	++	.	.	.	++	.	.	.	.
Hymenoptera										
N. pratti	.	.	.	.	.	+++	.	.	.	.
A. mellifera	++	+	+++	.	.	+	+	.	.	.
A. cephalotes	.	.	.	.	.	.	.	.	++	+
Lepidoptera										
M. sexta	++	++	++	++	++	++	++	++	.	.
H. zea	++	++	.	.	+	++	++	-	+	+
B. mori	+	.	.	.	.	+++	.	.	.	.
C. trisectus	-	.	.	.	.	+++	.	.	.	.
H. coffearea	-	.	.	.	.	-	.	.	+++	.
P. gossypiella	+	.	.	.	.	+	++	.	.	.
Coleoptera										
H. pales	-	.	.	.	.	+++	.	.	.	.
H. postica	.	.	.	.	.	.	.	++	++	++
E. varivestis	+	++	.	.	.	++	++	++	.	.
X. ferrugineus	-	.	.	.	.	-	.	.	+++	.
C. chinensis	+	.	.	.	.	++	.	.	.	.
T. granarium	+++	.	.	.	.	+	+	.	.	.
Diptera										
D. melanogaster	++	.	.	.	.	-	+++	.	+++	.
D. pachea	lophenol and schottenol only									

1 = cholesterol; 2 = campesterol; 3 = 24-methylene cholesterol; 4 = dihydrobrassicasterol; 5 = brassicasterol; 6 = sitosterol; 7 = stigmasterol; 8 = spinasterol; 9 = ergosterol; 10 = lathosterol.

[a] See text for source references.

An intriguing possibility is that sterols may be detrimental at high concentration because they can inhibit the activity of polysubstrate monooxygenases. These important detoxifying enzymes can be produced at high levels when caterpillars are fed on certain plant secondary compounds, yet the induction was prevented in *Spodoptera frugiperda* by both sitosterol and stigmasterol.[60]

IV. COMPLICATIONS IN DETERMINING DIETARY NEEDS FOR STEROL

As mentioned in Section I above, sterols have two major and very different roles in insects. First, there is the requirement for structural purposes, for which the bulk of sterols are needed. Second, there is steroid hormone synthesis. Since the hormones are usually synthesized via cholesterol, conversion of some phytosterols to cholesterol is required, and in many cases the pathways are known. The structural role may be partly met in many cases by different sterol structures. This is well reviewed by Svoboda and Thompson.[52]

Another complication in determining the precise value of a particular foliage for a particular insect is that sterols may themselves interact. Phytosterols are known to compete for absorption in vertebrates, and such an effect can be shown for phytosterols in the zoophagous *Dermestes maculatus*.[32] When various nonusable phytosterols were added to the cholesterol-containing diet of this insect, cholesterol uptake was halved, as was growth rate. There are, in addition, several examples in the Lepidoptera where cholesterol in the diet alleviates the detrimental effects of steroidal alkaloids,[5,8] suggesting that the significance of dietary sterols may go beyond the need for membrane structures and for steroid hormones.

There are probably several different limiting factors for sterol utilization. Firstly, there may be effects in the gut lumen. For example, Heinemann et al.[27] indicated that in humans the inhibition of cholesterol absorption by certain phytosterols could have been due to their preferential affinity for mixed micelles (with bile salts, glycerides, fatty acids, and lysolecithin) formed in the lumen prior to uptake. Secondly, there may be differential uptake, due to the specificity of one or more receptor mechanisms. There is a suggestion that esterification in midgut cells varies and that esterases aid in uptake when concentrations are low.[52] Thirdly, there are clear differences in the complement of specific enzymes capable of transforming the variety of different phytosterols into cholesterol.[44]

Other dietary components may influence the rates of uptake of sterols from the gut. For example, dietary lipids influence the efficiency of sterol uptake in caterpillars,[20] though the mechanisms are unknown. Cholesterol absorption in mammals is inhibited by a number of drugs, including neomycin,[33] which brings up the possibility that secondary metabolites of food plants, fungi, and bacteria may have the potential for influencing the availability of dietary sterols. It is known at least that alfalfa saponins and sterols interact in some way in the gut lumen, since added sterols reduce the detrimental effects of ingested saponin in beetles.[29]

A final complication, and one that will not be dealt with here, is that a number of insects from different orders contain steroidogenic symbiotes. In some cases this is well established, while in others, such as aphids, there is no general agreement.[7]

V. AVAILABILITY OF STEROLS IN PLANTS

Different plant species have different sterol profiles. However, as a starting point, Nes[36] suggests that there are many species with a generally similar range of compounds. He says that "a common sterol mixture without regard for its evolutionary position would be dominated by sitosterol, often together with stigmasterol. Campesterol is usually next most abundant, accompanied by perhaps half as much 22-dihydrobrassicasterol. Lesser but significant amounts of cholesterol, and such sterols as 22-dihydrospinasterol, are also frequently present." However, from the point of view of phytophagous insects, this is too simplistic, and the variations are probably very important. For example, the Cucurbitaceae and Theaceae contain almost exclusively Δ 7 sterols, while the Poaceae and Asteraceae have almost exclusively Δ 5 sterols. The

Amaranthaceae and Chenopodiaceae vary from species exclusively with Δ7 to species exclusively with Δ 5 sterols.[1] An increasingly large proportion of plant species are being found to contain mainly Δ 7 sterols,[1,40] and this may prove important with respect to differential plant use in a given habitat. Hence the fact that the polyphagous insect species have very different abilities to use different sterols may relate to the differences in plant species they tend to feed upon.

The leaves of plants contain total sterol concentrations varying from 0.01% dry weight to approximately 0.3% dry weight, with occasional values of 1% dry weight in meristematic tissues.[35] Seeds are often higher, with values generally 0.2 to 0.5% dry weight, while pollen is usually also relatively high, 0.13 to 0.5%. Profiles also vary in the different tissues. Thus, if sterols are in any way limiting for insects, the actual plant species or tissue value could be extremely important for growth and reproduction of the herbivore. In insect artificial diets, the required overall sterol concentrations vary from 0.01 to 0.2%, with around 0.1% dry weight being most common.[52] There is thus potential for sterols to be limiting. This is particularly true for insects with a restricted ability with respect to sterol type, for whom the effective sterol concentration in any plant may be considerably less than optimal.

Even within a plant species, big changes can occur with development. For example, seedlings of some *Brassica* species produce Δ 7 and Δ 22 phytosterols, but mature plants produce Δ 5 sterols.[37] Ontogenetic changes in overall sterol profile have also been shown to be dramatic in several other dicot species.[3,23] Large differences can also occur between different parts of the plant: Garg and co-workers[24] showed that shoot apices of flowering plants had cholesterol concentrations of up to 1% of the dry weight, which is at least two orders of magnitude greater than the concentrations found in leaves.

Other sources of considerable variation in sterol profile and/or quantity in a plant include infection with fungi, levels of light, and quantities and type of exogenous chemicals (see Bean, 1973 for a short review).[3] Even pollutants produced by human activity can have significant effects on plant sterol metabolism.[39] Recent studies on fungicides have demonstrated that they may totally alter the plant sterol type,[41] with extreme consequences for the insect herbivores. For example, wheat treated with fungicides produced, instead of the usual Δ 5 sterols, a suite of Δ 0 sterols unusable by the test insect *Locusta migratoria*.[14] It would be interesting to know whether endophytic fungi, which are now known to be very widespread,[9] influence the sterol profiles of their hosts, and thus alter the nutritional quality of plants for insect herbivores.

Phytecdysteroids are commonly considered to be defenses of plants against insect herbivores, although the evidence is by no means clear cut, due to the rather unrealistic bioassays in many cases. However, about 70 different steroids in plants have insect molting hormone activity when tested in various ways, and currently well over 100 plant families are known to contain species with these types of compounds.[4] Concentrations usually vary between 0.001 and 0.1% of the dry weight of the plant,[28] although the flowers of *Serratula inernis* (Asteraceae) contain 2% of 20-hydroxyecdysone,[59] the most important ecdysteroid in insects. Mulberry leaves, the food of the silkworm, *Bombyx mori*, have high levels of ecdysteroids, and it appears that this insect grows better on an artificial diet when these compounds are added.[12] It would be surprising if some herbivorous insects did not make use of phytecdysteroids, especially if sterols in their foods are limiting, but there is no ecologically relevant experimentation on any other species.

VI. HOST-PLANT SELECTION

Food-plant selection in relation to the presence or absence of utilizable sterols has not been studied, though there is some evidence that sterols are tasted by Lepidoptera.[11] This is probably not the case in grasshoppers, however. On the other hand, food acceptability can alter with experience in grasshoppers, depending on the sterols present.[10] In a study with *Schistocerca americana*, learned avoidance of a plant dominated by Δ 7 and Δ 22 phytosterols was demonstrated, but this did not occur if cholesterol or sitosterol was added to the leaves. Therefore, in nature it is possible that even with polyphagous species, phytosterol profile is important in food-plant selection. It may be important, for example, in driving switching between food items when the insect concerned uses only certain phytosterols that are not particularly abundant. The grasshopper *Taeniopoda eques* switches very frequently between apparently acceptable foods in the field, and may be habitually short of its required cholesterol or related sterols.[42]

In the case of more host-specific insects with restricted abilities in the use of phytosterols, it is assumed that other plant features determine selection of the food. For example, it is likely that *Crambus*, which is restricted to the Poaceae and requires sitosterol, uses quite different cues in selecting hosts, but since all grasses are quite rich in sitosterol, the sterol needs will automatically be catered for. However, the sterol profile is likely to provide an important constraint in the ability of oligophagous species to use alternative food plants or to switch to a new plant host. An area totally uninvestigated is the possibility of induction of particular enzymes required to convert different phytosterols to cholesterol. This may be worth examining in individuals of polyphagous Lepidoptera, since they are usually not in a position to choose a new host plant when they hatch, and will almost certainly experience different plant sterol profiles.

Specialization on host-plant part may prove to have implications with respect to dietary sterols. For example, species that feed preferentially on meristematic tissue, such as some of the Diptera, will encounter relatively high concentrations of sterols with a greater proportion than elsewhere of cholesterol. Feeding on pollen, which is a very widespread habit,[6] will generally provide insects with relatively high levels of sterols with a preponderance of 24-methylene-cholesterol.[50] Are such insects constrained in their feeding habits by a specificity or special need with respect to sterols?

It is possible, but still unproved, that dietary sterols are limiting nutrients for many herbivores, especially those that feed mainly on older foliage or stems. Svoboda and Thompson[52] mention that insect requirements for dietary concentrations of sterol vary between 0.01 and 0.2% dry weight of diet, with 0.1% being a common value. The lower values are mainly for carnivorous species, and this may reflect the fact that sterols needs may be related to protein concentration in the diet. However, plant concentrations can be well below 0.1% dry weight of sterols (see Section V), so the implication would be that sterols may often be in short supply. When one considers the case of a grasshopper such as *Taeniopoda*, where a proportion of the available plant sterols are unusable, the likelihood of sterol shortage becomes even greater. In fact, in this species flower feeding is common, which may be useful for supplementing sterols in the diet. Many plant-feeding insects occasionally or often eat animal matter, and for those able to use cholesterol this may be an important supplementary source of sterols. It would be interesting to find out if the ability to use cholesterol is related to a shorter evolutionary history of herbivory, which in turn is associated with the degree to which animal matter is included in the diet. Are inveterate cannibals, such as *Heliothis virescens*, always ready to supplement their diets with sterol-rich foods? How often does variation in host-plant quality depend on the quality and quantity of utilizable sterols for the insect concerned?

VII. CONCLUSIONS

The relationship of plant sterols and plant-feeding insects is a subject that needs considerably more work. Might sterols be one of the keys to some of the combinations of plants or plant parts selected by certain insects? What is the role of sterols in host range generally? Protein availability has been the focus of much work on plant quality for insects but further interesting questions arise. Is young, protein-rich tissue also the best from the point of view of sterols? Only with further study will we be able to say, but it is probable that sterols have been underemphasized relative to protein as quantitative nutrients that may be limiting.

ACKNOWLEDGMENTS

Thanks to D. Champagne for getting me interested in sterols and for helpful discussions of the questions addressed here. R. F. Chapman and R. Feyereisen provided helpful criticism. Betty Estesen carried out experiments with grasshoppers. Partial support came from NSF grant BSR 8705014.

REFERENCES

1. **Adler, J. H. and Salt, T. A.**, Phytosterol structure and composition in the chemosystematics of the Caryophyllales, in *The Metabolism, Structure, and Function of Plant Lipids*, Stumpf, P. K., Mudd, J. B., and Nes, W. D., Eds., Plenum Press, New York, 1986, 119.
2. **Barbier, M., Hugel, M. F., and Lederer, E.**, Isolation of 24-methylenecholesterol from the pollen of different plants, *Soc. Chem. Bull.*, 42, 91, 1960.
3. **Bean, G. A.**, Phytosterols, *Adv. Lipid Res.*, 11, 193, 1973.
4. **Bergamasco, R. and Horn, D. H. S.**, Insect hormones in plants, in *Endocrinology of Insects*, Downer, R. G. H. and Laufer, H., Eds., Alan R. Liss, New York, 1983, 627.
5. **Bloem, K. A., Kelley, K. C., and Duffey, S. S.**, Differential effect of tomatine and its alleviation by cholesterol on larval growth and efficiency of food utilization in *Heliothis zea* and *Spodoptera exigua*, *J. Chem. Ecol.*, 15, 387, 1989.
6. **Burgess, K. H.**, Florivory: The Ecology of Flower-Feeding Insects and their Host Plants, Ph.D. Thesis, Harvard University, Cambridge, MA, 1991, 224.
7. **Campbell, B. C.**, On the role of microbial symbiotes in herbivorous insects, in *Insect-Plant Interactions Vol 1*, Bernays, E. A., Ed., CRC Press, Boca Raton, FL, 1989, 1.
8. **Campbell, B. C. and Duffey, S. S.**, Alleviation of alpha-tomatine induced toxicity to the parasitoid *Hyposoter exiguae*, by phytosterols in the diet of the host, *Heliothis zea*, *J. Chem. Ecol.*, 7, 927, 1981.
9. **Carroll, G.**, Fungal endophytes in stems and leaves: from latent pathogen to mutualistic symbiont, *Ecology*, 1988, 69, 2.
10. **Champagne, D. and Bernays, E. A.**, Phytosterol unsuitability as a basis of food aversion learning in grasshoppers, *Physiol. Entomol.*, 16, 391, 1991.
11. **Chippendale, G. M.**, Insect metabolism of dietary sterols and essential fatty acids, in *Insect and Mite Nutrition*, Rodriguez, J. G., Ed., Elsevier, London, 1972, 423.
12. **Chou, W.-S. and Lu, H.-S.**, Growth regulation and silk production in *Bombyx mori* L. from phytogenous ecdysteroids, in *Progress in Ecdysone Research*, Hoffman, J. A., Ed., Elsevier, North Holland, 1980, 281.

13. **Cooke, J. and Sange, J. H.**, Utilization of sterols by larvae of *Drosophila melanogaster, J. Insect Physiol.*, 16, 801, 1970.
14. **Corio-Costet, M. F., Charlet, M., Benveniste, P., and Hoffman, J.**, Metabolism of dietary 8 sterols and 9beta, 19-cyclopropyl sterols by *Locusta migratoria, Arch. Insect Biochem. Physiol.*, 11, 47, 1989.
15. **Dadd, R. H.**, The nutritional requirements of locusts — II Utilization of sterols, *J. Insect Physiol.*, 5, 161, 1960.
16. **Dadd, R. H.**, The nutritional requirements of locusts — I Development of synthetic diets and lipid requirements, *J. Insect Physiol.*, 4, 319, 1960.
17. **Dadd, R. H.**, Nutrition: Organisms, in *Comprehensive Insect Physiology Biochemistry and Pharmacology, Vol 4*, Kerkut, G. A. and Gilbert, L. I., Eds., Pergamon Press, Oxford, 1985, 313.
18. **Dambrae-Raes, H.**, The effect of dietary cholesterol on the development of *Hylemya brassicae, J. Insect Physiol.*, 22, 1287, 1976.
19. **Downer, R. G. H.**, Functional role of lipids in insects, in *Biochemistry of Insects*, Rockstein, M., Ed., Academic Press, New York, 1978, 57.
20. **Dupnik, T. D. and Kamm, J. A.**, Development of an artificial diet for *Crambus trisectus, J. Econ. Entomol.*, 63, 1578, 1970.
21. **Feldlaufer, M. F., Herbert, E. W., Svoboda, J. A., Thompson, M. J., Lusby, W. R., and Makisterone, A.**, The major ecdysteroid from the pupa of the honey bee, *Apis mellifera, Insect Biochem.*, 15, 597, 1985.
22. **Feldlaufer, M. F., Svoboda, J. A., Aldrich, J. R., and Lusby, W. R.**, The neutral sterols of *Megalotomus quinquespinosus* (Say) (Hemiptera: Alydidae) and identification of makisterone A as the major free ecdysteroid, *Arch. Insect Biochem. Physiol.*, 3, 432, 1986.
23. **Garg, V. K. and Nes, W. R.**, Changes in sterol biosynthesis from [2. ^{14}C] mevalonic acid during development of *Cucurbita maxima* seedlings, in *The Metabolism, Structure and Function of Plant Lipids*, Stumpf, P. K., Mudd, J. B., and Nes, W. D., Eds., Plenum, New York, 1986, 87.
24. **Garg, V. K., Douglas, T. J., and Paleg, L. G.**, Presence of unusually high levels of cholesterol in the shoot-apices of flowering plants, in *The Metabolism, Structure and Function of Plant Lipids*, Stumpf, P. K., Mudd, J. B., and Nes, W. D., Eds., Plenum, New York, 1986, 83.
25. **Gibson, J. M., Majumder, M. S. I., Mendis, A. H. W., and Rees, H. H.**, Absence of phytosterol dealkylation and identification of the major ecdysteroid as makisterone A in *Dysdercus fasciatus* (Heteroptera, Pyrrhocoridae), *Arch. Insect Biochem. Physiol.*, 105, 1983.
26. **Heed, W. and Kircher H. W.**, Unique sterol in the ecology and nutrition of *Drosophila pachea, Science*, 149, 758, 1965.
27. **Heinemann, T., Kullak-Ublick, G.-A., Pietruck, B., and von Bergmann, K.**, Mechanisms of action of plant sterols on inhibition of cholesterol absorption, *Eur. J. Clin. Pharmacol.*, 40, S59, 1991.
28. **Hikino, H. and Hikino, Y.**, Arthropod molting hormones, *Progr. Chem. Org. Nat. Products*, 28, 256, 1970.
29. **Horber, E.**, Alfalfa saponins significant in resistance to insects, in *Insect and Mite Nutrition*, Rodriguez, J. G., Ed., Elsevier, London, 1972, 611.
30. **Ishii, S.**, Studies on the host-plants of the cowpea weevil (*Callosobruchus chinensis* L.) IX. On the relation between the chemical constituents of sterols and the development of larvae, *Botyu-Kagaku*, 16, 83, 1951 (in Japanese).
31. **Ito, T. and Horie, Y.**, Utilization of sterols and related compounds by the silkworm, *Bombyx mori* L., *Annot. Zool. Jpn.*, 39, 1, 1966.
32. **Katz, M., Budowski, P., and Bondi, A.**, The effect of phytosterols on the growth and sterol composition of *Dermestes maculatus, J. Insect Physiol.*, 17, 1295, 1971.
33. **Kesaniemi Y. A. and Miettinen, T. A.**, Inhibition of cholesterol absorption by neomycin, benzodiasepine derivatives and ketoconazole, *Eur. J. Clin. Pharmacol.*, 40 (Suppl. 1), S65, 1991.
34. **McDonald, D. L., Nham, D. N., Cochran W. K., and Ritter, K. S.**, Differences in the sterol composition of *Heliothis zea* fed *Zea mays* versus *Medicago sativa, Insect Biochem.*, 20, 437, 1990.
35. **Nes, W. R.**, The biochemistry of plant sterols, *Adv. Lipid Res.*, 15, 233, 1977.
36. **Nes, W. R.**, Multiple roles for plant sterols, in *The Metabolism, Structure, and Function of Plant Lipids*, Stumpf, P. K., Mudd, J. B., and Nes, W. D., Plenum Press, New York, 1986, 3.
37. **Nes, W. R. and McKean, M. L.**, *Biochemistry of Steroids and Other Isopentenoids*, University Park Press, Baltimore, 1977.

38. **Norris, D. M. and Moore, C. L.,** Lack of dietary Δ 7 sterol markedly shortens the periods of locomotory vigor, reproduction and longevity of adult female *Xyleborus ferrugineus*) (Coleoptera, Scolytidae), *Exp. Geront.*, 15, 359, 1980.
39. **Ormrod, E. P.,** *Pollution in Horticulture. Fundamental Aspects of Pollution Control and Environmental Science 4,* Elsevier, Amsterdam, 1978.
40. **Patterson, G. W. and Xu, S.**, Sterol composition in five families of the order Caryophyllales, *Phytochemistry*, 29, 3539, 1990.
41. **Rahier, A., Schmitt, P., Huss, B., Benveniste, P., and Pommer E. H.,** Chemical structure-activity relationships of the inhibition of sterol biosynthesis by N-substituted morpholines in higher plants, *Pestic. Biochem. Physiol.*, 25, 112, 1986.
42. **Raubenheimer, D. and Bernays, E. A.,** Food selection by *Taeniopoda eques*: a field study, *Anim. Behav.,* in press, 1992.
43. **Redfern, C. P. F.,** Changes in patterns of ecdysteroid secretion by the ring gland of *Drosophila* in relation to the sterol composition of the diet, *Experientia*, 42, 307, 1986.
44. **Rees, H. H.,** Biosynthesis of ecdysone, in *Comprehensive Insect Physiology Biochemistry and Pharmacology, Vol 7,* Kerkut, G. A. and Gilbert, L. I., Eds., Pergamon Press, Oxford, 1985, 249.
45. **Richmond, J. A. and Thomas, H. A.,** *Hylobius pales*: effect of dietary sterols on development and on sterol content of somatic tissues, *Ann. Ent. Soc. Am.*, 68, 329, 1975.
46. **Ritter, K. S., Weiss, B. A., Norrbom, A. L. and Nes, W. R.,** Identification of 5,7-24methylene- and methylsterols in the brain and whole body of *Atta cephalotes isthmicola*, *Comp. Biochem. Physiol. B.*, 1982.
47. **Schaefer, C. H., Kaplanis, J. N., and Robbins, W. E.,** The relationship of the sterols of the Virginia pine sawfly, *Neodiprion pratti* Dyar, to those of two host plants, *Pinus virginiana* Mill and *Pinus rigida* Mill, *J. Insect Physiol.*, 11, 1013, 1965.
48. **Sivapalan, P. and Gnanapragasam, N. C.,** The influence of linoleic and linolenic acid on adult moth emergence of *Homona coffearia* from meridic diets in vitro, *J. Insect Physiol.*, 25, 393, 1979.
49. **Standifer, L. N., Davys, M., and Barbier, M.,** Pollen sterols — a mass spectrographic survey, *Phytochemistry*, 7, 1361, 1958.
50. **Stanley, R. G. and Linskens, H. F.,** *Pollen: Biology, Biochemistry and Management,* Springer-Verlag, New York, 1974, 307.
51. **Svoboda, J. A. and Robbins, W. E.,** Comparison of sterols from a phytophagous and predacious species of the family Coccinellidae, *Experientia*, 35, 186, 1979.
52. **Svoboda, J. A. and Thompson, M. J.,** Steroids, in *Comprehensive Insect Physiology Biochemistry and Pharmacology, Vol. 10,* Kerkut, G. A. and Gilbert, L. I., Eds., Pergamon Press, Oxford, 1985, 137.
53. **Svoboda, J. A., Agarwal, N., Robbins, W. E., and Nair, A. M. G.,** Lack of conversion of C29 phytosterols to cholesterol in the Khapra beetle, *Trogoderma granarium*, *Experientia,* 36, 1029, 1980.
54. **Svoboda, J. A., Dutky, S. R., Robbins, W. E., and Kaplanis, J. N.,** Sterol composition and phytosterol utilization and metabolism in the milkweed bug, *Lipids* 12, 318, 1977.
55. **Svoboda, J. A., Herbert, E. W., and Thompson, M. J.,** Definitive evidence for lack of phytosterol dealkylation in honey bees, *Experientia*, 39, 1120, 1983.
56. **Svoboda, J. A., Imberske, R. B., and Lusby, W. R.,** *Drosophila melanogaster* does not dealkylate sitosterol, *Experientia*, 45, 983, 1989.
57. **Svoboda, J. A., Kaplanis, J. N., Robbins, W. E., and Thompson, M. J.,** Recent developments in insect steroid metabolism, *Ann. Rev. Entomol.*, 20, 205, 1975.
58. **Svoboda, J. A., Lusby, W. R., and Adlrich, J. R.,** Neutral sterols of representatives of two groups of Hemiptera and their correlation to ecdysteroid content, *Arch. Insect Biochem. Physiol.*, 1984, 139, 1984.
59. **Yatsyuk, J. and Segel, G. M.,** On the isolation of ecdysterone, *Khim. Prir. Soedin*, 6, 281, 1970.
60. **Yu, S. J.,** Consequences of induction of foreign compound-metabolizing enzymes in insects, in *Molecular Aspects of Insect-Plant Associations*, Brattsten, L. B. and Ahmad, S., Eds., Plenum Press, New York, 1986, 153.
61. **Vanderzant E. S. and Reiser, R.**, Studies of the nutrition of the pink bollworm using purified casein media, *J. Econ. Ent.*, 49, 454, 1956.

3

Sensory Coding of Feeding Deterrents in Phytophagous Insects

L. M. Schoonhoven
Department of Entomology
Agricultural University
Wageningen, The Netherlands

W. M. Blaney
Department of Biology
Birkbeck College
London, England

M. S. J. Simmonds
Jodrell Laboratory
Royal Botanic Gardens
Kew, England

TABLE OF CONTENTS

DEDICATION

This review is dedicated to Dr. Tibor Jermy, who is one of the pioneers in the study of feeding deterrents and their importance for host recognition by herbivorous insects.

I. INTRODUCTION

One of the striking features of insect-plant interactions is the high degree of food specialization in phytophagous insects. Although the ultimate causes underlying this strategy are still not well understood, the mechanism governing host selection is open to analysis. It is contained in the chemosensory code, which carries the information on which the decision to feed or not to feed is based. Initially the attention of sensory physiologists was focused on receptors that when stimulated elicited feeding behavior.[73,77] Jermy,[47,49] however, called attention to the importance of feeding deterrents in host-plant selection. The notion that insects avoid eating most plants because of the presence of distasteful compounds raised the question of how such compounds are perceived. Electrophysiological investigation of taste sensilla led to the discovery of a "deterrent neuron" (or "bitter neuron") in larvae of *Bombyx mori* [46] and *Pieris brassicae* .[58] The concept of a chemosensory system with some neurons tuned to detect phagostimulants and others for feeding deterrents, although attractive in its simplicity, appeared to be of limited validity. The sensory coding of feeding deterrents in phytophages appears to be much more intricate than was first thought, and its complete elucidation is still over the horizon.

There are several reasons why the chemosensory perception of feeding deterrents by phytophagous insects warrants special attention.

1. Feeding deterrents are probably more important in host plant recognition than phagostimulants.[4,48,49]
2. The number of feeding deterrents present in the plant kingdom is orders of magnitude larger than the number of phagostimulants. The same applies to their variations in molecular structure, which are much more diverse in deterrents than in phagostimulants.
3. "Deterrent receptors" in lepidopterous larvae[85] and acridids[14] are fewer in number than receptors for phagostimulants, and some insects, e.g., Colorado potato beetles (*Leptinotarsa decemlineata*), even seem to be devoid of them.[62]
4. Different deterrents may elicit different behavioral reactions, indicating the presence of a differential sensory coding system.[57,62,96]

Recently, evidence has been accumulating for the existence of several sensory mechanisms involved in the mode of action of deterrents. Although their principles were outlined in some earlier reviews,[37,84] the new information available and the importance of the subject for understanding the mechanisms involved in insect-plant relationships have stimulated the writing of this paper. Additionally, interest in wider biological aspects of feeding deterrents comes from applied entomology, because it can be argued that compounds that have served for ages as a chemical umbrella to plants may be useful when protecting crops against insect attack (reviewed by Jermy[51]).

II. CHEMORECEPTORS

The sense of taste is located in sensilla whose cuticular parts take the form of hairs or cones, mainly situated on the mouthparts and tarsi.[108] These sensilla contain the

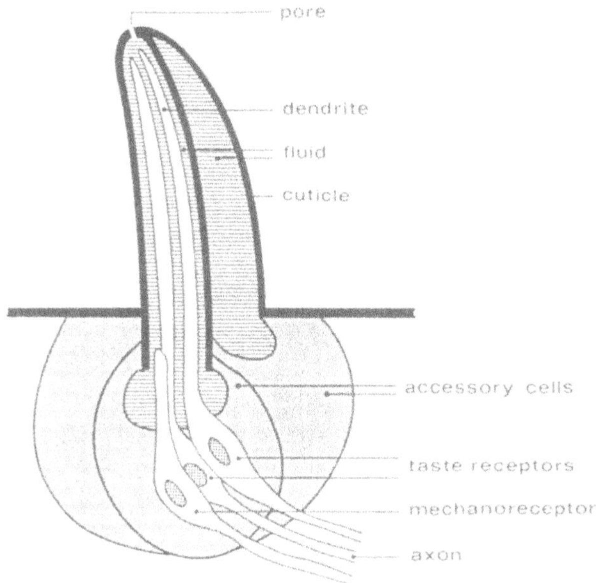

FIGURE 1. Schematic drawing of a longitudinal section of an insect taste hair. The sensillum is innervated by one mechanosensitive and two chemosensitive neurons. (Courtesy of Dr. F. W. Maes, Groningen State University, The Netherlands.)

dendrites of chemosensory neurons, which are connected directly to the central nervous system (CNS) without any intervening synapses, although there is some evidence for the presence of functional interactions between different receptor neurons at the periphery.[8,15,106] Gustatory sensilla often contain three or four neurons, which can be stimulated through a terminal pore in the cuticular structure (Figure 1).

The number of gustatory sensilla varies considerably between different insect groups.[22] In lepidopterous larvae only two sensilla styloconica on each maxilla, and often one pair of epipharyngeal organs, play a dominant role in food recognition,[85] whereas in the adult Colorado potato beetle the palps carry about 15 sensilla basiconica,[63] and the mouthparts of locusts may even accomodate several thousand sensilla.[22]

Electrical activity reflecting neural responses to chemical stimulation can be recorded with electrophysiological techniques. Commonly a single sensillum is stimulated by placing a finely tapered glass capillary, filled with the stimulus solution, over the pore at the tip of the hair. When the capillary contents are connected via a conducting silver wire to an electrical measuring device, action potentials can be recorded relative to an indifferent electrode, which is located somewhere else in the body of the insect. If the sensillum contains only a few chemoreceptors, it is often possible to distinguish neural output from individual neurons on the basis of amplitude, shape, and temporal occurrence of the action potentials recorded.[38] Recently developed computer programs[74,97] can help in the analysis of the recorded data.

III. NEURAL CODING OF FEEDING DETERRENTS: PRINCIPLES

Sensory coding is based upon neural activity in one or more neurons. Three basic types of sensory coding can be distinguished[21]

1. Labeled lines: each neuron conveys a specific message, which can be understood by the CNS without additional information from other neurons.

2. Across-fiber patterns: the message is contained in a neural activity pattern, transmitted by two or more receptors, possessing different stimulus spectra.
3. Temporal patterns: stimulus quality affects nerve-impulse interval patterns and adaptation rates, which may contain additional information.

Several examples can be cited in which one or more of the three coding principles manifests itself.[9,31,88] It should be emphasized that more often than not chemosensory codes in insects are a combination of these three coding systems.

IV. CLASSIFICATION OF NEURAL RESPONSES TO FEEDING DETERRENTS

Feeding deterrents may affect an insect's chemoreceptor system according to different coding principles. Thus they may excite specialized labeled line receptors, and/or distort the integrity of complex neural patterns arising in receptor groups. Additionally, temporal relationships of impulse patterns may be affected in such a way that the CNS does not interpret the incoming message as coming from an acceptable food source. The various ways in which feeding deterrents may alter sensory input are reflected in a classification of neural responses to feeding deterrents, which recognizes five types of neural reactions.[84] These are

- stimulating a neuron specifically tuned to diverse plant compounds that deter feeding
- stimulating certain putative receptor sites on neurons with a broad sensitivity spectrum that includes secondary plant compounds
- inhibiting the responses of neurons that are stimulated by phagostimulants
- changing complex and subtle codes (across-fiber patterns) by stimulating some neurons and inhibiting the activity of others
- evoking irregular impulse patterns, often at high frequency

This classification contains some arbitrariness, and often a given deterrent acts via a combination of two or more of the above mentioned neural reactions. All five modes of action have been confirmed by various studies of perception mechanisms of feeding deterrents.[84] The classification still seems to be a useful one, but present-day knowledge allows the addition of some more details to it. They concern especially temporal aspects, such as latency of the stimulus response and sensory adaptation.

Some recent studies on sensory coding of deterrents will be discussed briefly according to their sensory mechanisms.

A. DETERRENT NEURONS

Feeding deterrent neurons are present in the larvae of several, and possibly all, Lepidoptera (Table 1). Ishikawa,[46] described a "bitter receptor" in the medial maxillary sensillum styloconicum of the silkworm, *Bombyx mori,* which responds to a number of plant alkaloids and phenolics. Similar neurons, although with variations in their reaction spectra, have been found in other lepidopterans.[85] For instance, the tobacco hornworm, *Manduca sexta,* has a neuron in this sensillum that responds to some alkaloids, the triterpene limonin, and to a strong, as yet unidentified, deterrent in *Canna* leaves.[43,72,79] A second deterrent neuron occurs in its lateral sensillum styloconicum. This neuron is stimulated by compounds, such as salicin, caffeine, and aristolochic acid, that act as weak feeding deterrents.[37,72,79] In addition to a deterrent neuron in their medial sen-

Table 1
PRESENCE OF DETERRENT NEURONS IN MEDIAL AND LATERAL TASTE SENSILLA OF LEPIDOPTEROUS LARVAE AS DETERMINED WITH ELECTROPHYSIOLOGICAL METHODS

Insect	Medial sensillum	Lateral sensillum	Ref.
Aglais urticae	+		80
Bombyx mori	+		46
Catocala nupta	+		80
Eldana saccharina	+		101
Episema caeruleocephala	+		80
Lymantria dispar	+		80
Mamestra brassicae	+	+	93, 103
Manduca sexta	+	+	38, 72, 79
Mimas tiliae	+	+	80
Operophtera brumata		+	80
Pieris brassicae	+	+	57, 58
Pieris rapae	+	+	57
Spodoptera exempta	+	+	24, 93, 96
Spodoptera littoralis	+	+	93, 96
Yponomeuta spp.	+	+	34

silla,[58] larvae of *Pieris brassicae* and *P. rapae* have also a deterrent neuron in their lateral sensilla, which responds to phenolic acids, flavonoids, and cardenolides.[57] Thus in many of the species of Lepidoptera studied in some detail, both the lateral and medial maxillary taste sensilla appear to contain a deterrent neuron, although the ranges of chemicals to which they respond are quite different (Table 1).

The well-known deterrent isolated from the neem tree, azadirachtin, stimulates a deterrent neuron in the medial sensilla of *P. brassicae, Lymantria dispar, Hypsipyla grandella,*[82] *Spodoptera exempta, S. littoralis, S. frugiperda, Heliothis virescens, Helicoverpa armigera,* and *Mamestra brassicae.*[93] Although this observation suggests a certain similarity in the sensitivity spectra of this deterrent neuron in the species studied, it should be emphasized that often very closely related species differ in their responsiveness to compounds. This could explain why they exhibit different host preferences[34] (Figure 2).

Deterrent neurons have also been found in tarsal sensilla of adult lepidopterans.[13,74,75,94] In *Papilio polyxenes* these tarsal deterrent neurons are involved in discriminating host plants from nonhosts for oviposition.[74] Neurons that respond to deterrents have also been located in sensilla on the proboscis of adult lepidopterans.[12]

So far there are only a few reports of deterrent neurons in taxa other than the Lepidoptera. In the acridid *Schistocerca americana,* tarsal neurons respond vigorously to stimulation with nicotine, a chemical that is avoided in behavioral experiments.[105] No evidence, however, for the presence of such receptors on the mouthparts of acridids could be obtained (for a review see Blaney and Simmonds[14]).

B. BROAD SPECTRUM NEURONS

Deterrents may stimulate one or more neurons, which also respond to phagostimulants. Stimulating the maxillary sensilla of *Danaus plexippus* with deterrent compounds or saps from unacceptable plants often evoked responses from three or more neurons per sensillum.[28] In *S. littoralis* several flavanones and phenolics stimulate

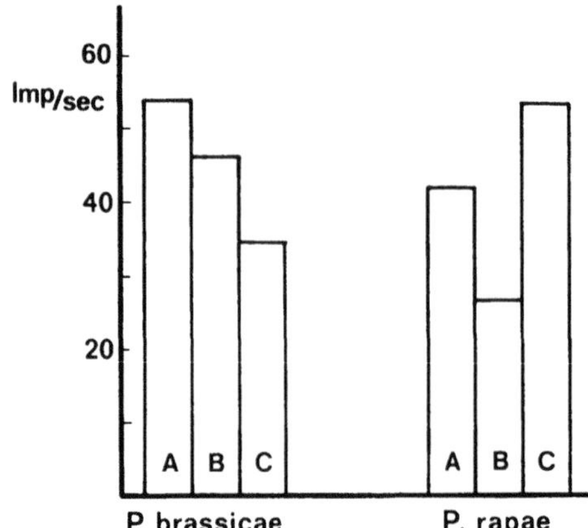

FIGURE 2. Relative sensitivity of the deterrent neuron in the lateral sensillum styloconicum in two related insect species, *P. brassicae* and *P. rapae* , to (A) catechin, (B) protocatechuic acid, and (C) chlorogenic acid. Responses to 1 m*M* concentrations are expresssed in impulses/s. (After data from von Loon, J. J. A., *J. Comp. Physiol.*, A166, 889, 1990. With permission.)

one to three neurons in the medial sensillum. At least one of these neurons responds to both deterrents and phagostimulants.[95,96] In *P. brassicae* some phenolic compounds, besides stimulating the lateral deterrent neuron and, though more weakly, the deterrent neuron located in the medial sensillum, may also elicit some activity in one or two other neurons in the latter sensillum.[57] Likewise, the deterrent azadirachtin stimulates several neurons in both the lateral and medial sensilla of six of the nine species of Lepidoptera studied.[93] Alkaloids, such as sparteine, nicotine, and quinine, are deterrents to *Entomoscelis americana*. However, they stimulate a neuron that is sensitive to glucosinolates, which are phagostimulants to this species.[65]

These examples show that some neurons may respond to a variety of secondary plant compounds, either phagostimulants or deterrents. Although the output from these neurons may be similar when stimulated with either phagostimulants or deterrents, the CNS, however, will be receiving neural input from other neurons at the same time, and thus the total neural input, in the form of an across-fiber pattern, will determine the resulting behavioral output.

C. INHIBITION OF PHAGOSTIMULANT NEURONS

1. Immediate Interactions

The first report on sensory responses to a feeding deterrent in an insect described the inhibition by quinine of a sugar receptor in the fleshfly, *Boettcherisca peregrina*, due to the induction of a hyperpolarization of the sensory neuron[67] (Figure 3). Several examples in phytophagous insects are now known of deterrents diminishing or completely abolishing the responsiveness of phagostimulant receptors, such as sugar cells.[10,30,45,57,62] For example, alkaloids that deter larvae and adults of *Entomoscelis americana* from feeding completely inhibit the activity of the "sugar" neuron.[65] Azadirachtin suppressed the responsiveness of "sugar" neurons in *S. littoralis*. [93] Likewise, quinine inhibits the "sugar" neurons of *Lymantria dispar* and *Malacosoma americana*,[29] cyanin those in *P. brassicae*[57] (Figure 4), and aristolochic acid inhibits the glucose-sensitive neuron in *Manduca sexta*.[37] Interactions between deterrents and phagostimulants could occur at receptor sites[10] or between neurons,[106] but whatever its mechanism, the phenomenon has by now been reported repeatedly.

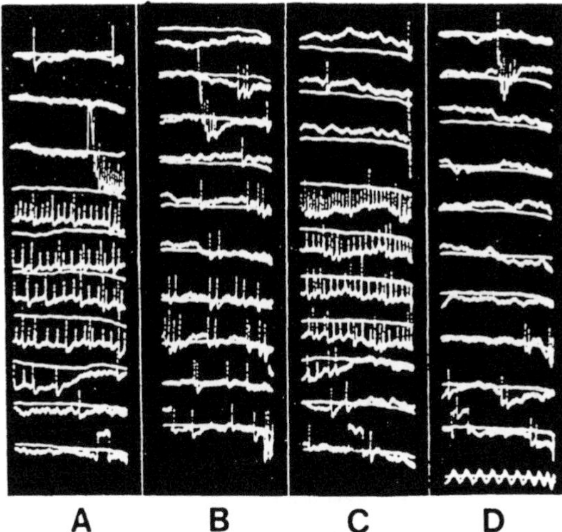

FIGURE 3. Electrophysiological recordings from labellar sensilla of *Calliphora vomitoria* on stimulation with 60 m*M* sucrose (traces A and C); 60 m*M* sucrose + 1.6 m*M* quinine HCl (B); 60 m*M* sucrose + 6 m*M* quinine HCl (D). (From Morita, H., *J. Cell. Comp. Physiol.*, 54, 189, 1959. With permission.)

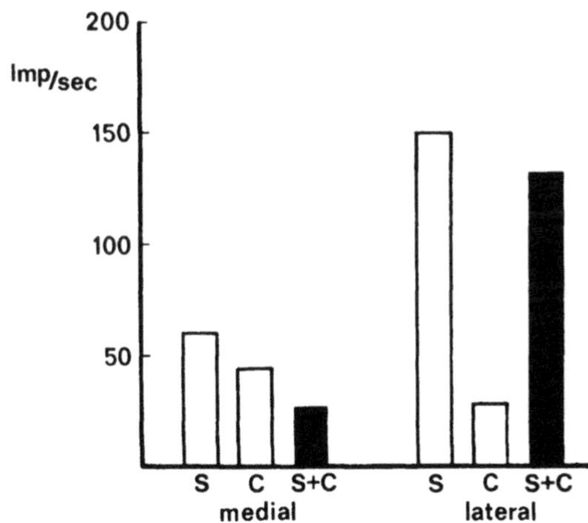

FIGURE 4. Inhibitory effects of cyanine chloride on the sugar responses in the maxillary sensilla styloconica of *Pieris brassicae* . Reactions are presented as impulse frequencies in the medial and lateral sensilla styloconica when stimulating with (S) 15 m*M* sucrose, (C) 2.5 m*M* cyanin chloride, and (S+C), a mixture of these two stimuli. (After data from Van Loon, J. J. A., *J. Comp. Physiol.*, A166, 889, 1990. With permission.)

Interestingly, the reverse reaction, i.e., sucrose inhibiting deterrent receptors, may also occur.[93]

2. Delayed Interactions

Another type of interference with normal chemosensory function was discovered by Ma[60] when studying the effect of warburganal, a sesquiterpene aldehyde, on the sucrose- and inositol-sensitive neurons of *S. exempta.* After prolonged exposure to warburganal, i.e. one to several minutes, depending on concentration, irregular impulse patterns were recorded from sensilla, followed by a reduction in the responses to their

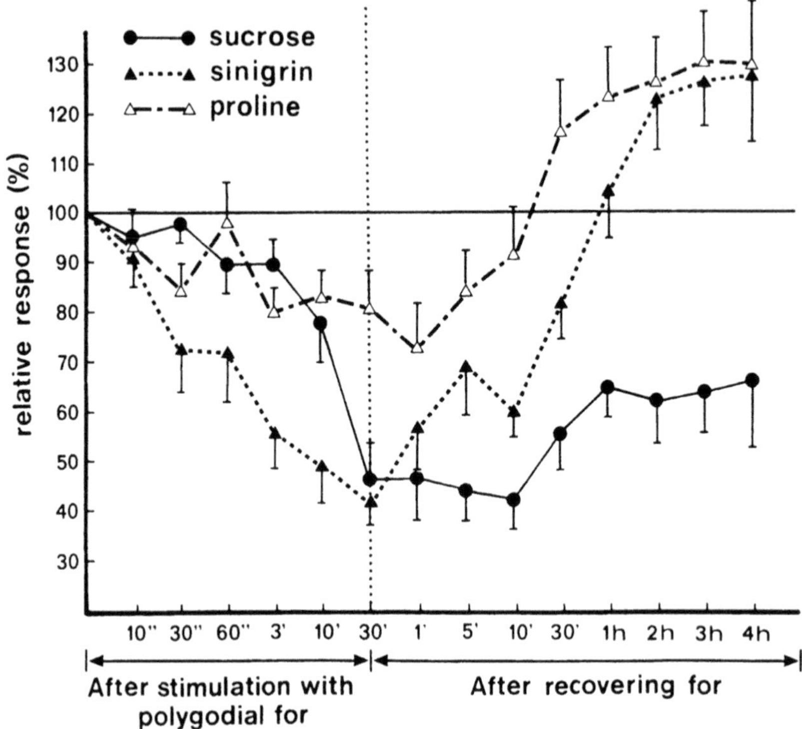

FIGURE 5. Inhibitory effects of 1 m*M* polygodial on sensory responses of three neurons to their adequate stimuli in *Pieris brassicae* after exposure periods of up to 30 min and their recovery profiles. (From Schoonhoven, L. M. and Yan F.-Sh., *J. Insect Physiol.*, 35, 725, 1989. With permission.)

adequate stimuli. Recovery took 0.5 to 1 h. Similarly the glucose- and inositol-sensitive neurons in the maxillary sensilla of *Manduca sexta* were blocked by warburganal[37] and in *Mythimna separata* by toosendanin.[56,92] This delay in the inhibition of the neural response to phagostimulants may be distinguished from the more immediate inhibition discussed in the previous section. We do not know if different mechanisms are involved in the two modes of action.

In *P. brassicae*, warburganal and related drimane compounds were found to elicit action potentials in the medial deterrent neuron. However, after a prolonged stimulation of 1 to 60 min, all four chemoreceptive neurons, including the deterrent neuron, become inhibited, indicating an unspecific blocking effect of these sesquiterpenoids (Figure 5).[90] It is not certain, however, that this blocking is the primary mode of action of these compounds, because in an earlier study Blaney et al.[18] showed that warburganal and some structurally related compounds that deterred a range of caterpillars from feeding all maintained a high level of neural input during a stimulation of 1 to 5 s. However, the exposure to warburganal in this experiment may be too short to induce the drastic effects seen after the longer stimulations undertaken by Ma,[60] and Schoonhoven and Yan.[90] The fact that warburganal has to be applied for relatively long periods, i.e., minutes, to decrease chemosensory responsiveness and that behavioral rejection in *S. exempta* is observed at concentrations one thousandth of that causing sensory inhibition[84] raises doubts about the primary mode of action of this deterrent. Another complication is the fact that the maxillae of caterpillars are constantly being moved during feeding, as in locusts,[11] and their sensilla most likely only contact the substrate intermittently. Therefore, in natural feeding behavior the exposure to warburganal may be too short to induce the striking effects seen after uninterrupted exposure for long periods, as applied in the electrophysiological experiments of Ma[60] and Schoonhoven and Yan.[90] In *Manduca sexta,* on the other hand, feeding activity is interrupted within a few seconds after the mouthparts contact warburganal, indicating that the chemosensory input to the CNS is, to some extent at least, rapidly altered.[39]

Following a 2-min treatment with certain glucose derivatives, the glucose-sensitive neurons in *Manduca sexta* larvae show considerably reduced responses to glucose for periods of at least 15 min. The C-1 fumarate derivative of glucose, which causes a 70% reduction in neuron responsiveness, produces in behavioral tests a total reduction of glucose-stimulated feeding activity.[40] It would be interesting to know whether or not sensory inhibition by these compounds is restricted to glucose sensitive neurons. In other insect species, the fumarate appeared in behavioral tests to be less effective, whereas other glucose analogues did cause strong reductions in food intake. Apparently the differences in the efficacy of various compounds between different insect species reflect functional differences in their taste neurons.[40] The triterpenoid toosendanin, interestingly, resembles in several respects drimane compounds such as warburganal; for instance, by reversibly reducing the responsiveness to sucrose and to inositol of neurons in the maxillary sensilla styloconica of *Mythimna separata* larvae.[92,109]

D. DISTORTION OF SENSORY CODE

Impulse trains, which normally have a certain regularity in their interspike time intervals, may be temporally distorted after contact with a given chemical. This can result in the inhibition of feeding. When in *Leptinotarsa decemlineata* the responses to potato sap were compared with those to saps from nonhosts, it appeared that the response patterns for the nonhost stimuli were considerably less consistent than the patterns evoked by sap from the host plant. Mitchell et al.[66] suggest that the highly variable patterns observed with nonhosts seem to be interpreted by the CNS as "nonsense", with the result that no feeding or only limited feeding occurs. Simmonds and Blaney,[94] using lepidopterous larvae, have found a similar difference in the degree of variability in the sensory patterns evoked by extracts from host and nonhost plants.

Several chemicals, including some heavy metal ions, distort the functioning of chemoreceptors in such a way that, even in the presence of an acceptable plant, the neural "acceptance profile" that the CNS requires to initiate feeding behavior is not evoked.[85,89]

Although many studies make a passing reference to deterrents causing irregular impulse patterns in groups of neurons, attempts to analyze and quantify such responses are few and the reproducibility of "nonsense" or "deterrent patterns" is low.[37] Analysis of temporal relationships of neuron responses, however, will most likely be a fruitful approach, because the temporal element is probably a much more significant feature of sensory codes than has hitherto been assumed.[18,31,33]

E. IRREGULAR IMPULSE PATTERNS

Some deterrents elicit unusually high firing frequencies or "bursting activity" in certain neurons, which soon result in subsequent insensitivity of these cells to their normal stimuli. Alkaloids, which in the literature[64,99] have been considered as feeding deterrents to *Leptinotarsa decemlinetata* (a function, however, which has been questioned[44]), do not stimulate deterrent neurons, but cause irregular firing by several (or perhaps all) neurons in gustatory sensilla on the galeae. After some seconds of stimulation, this firing pattern often develops into "burstlike activity".[64] The glucose receptor in the lateral sensillum of *Manduca sexta* larvae responds vigorously to stimulation by aristolochic acid, but thereafter becomes insensitive to glucose applications.[37] The triterpenoid toosendanin evokes similar reactions in another lepidopterous species.[56]

Irregular impulse patterns or bursting activity in nerve cells are generally considered to reflect injury effects.[70] Since several deterrents immediately or after some delay disrupt the normal functioning of one or more of the stimulated receptors, causing them to respond with irregular bursts of impulses, it is concluded that the CNS interprets such patterns as unacceptable.[16,60,61,90,93] It has also been suggested, however, that bursting activity in chemosensory neurons is a normal mode of sensory coding, rather than an unspecific and fortuitous event.[45,99] This opinion is supported by the observation that the alkaloid DMDP evokes bursting activity in some maxillary receptors of *S. exempta*, but not in those of *S. littoralis*. Apparently there are species-specific differences in sensitivity to this compound, which lead either to bursting activity in some neurons or to inhibition of the sugar receptors.[16] The fact that some glycoalkaloids present in the host plants of *Leptinotarsa decemlineata* and *Manduca sexta* induce neurons in the taste sensilla of these insects to fire in bursts[44,64,72] suggests that, at least in these species, bursting is not an effective code for deterrence. However, these glycoalkaloids were tested in isolation, and it cannot be excluded that the effects would have been different if the taste sensilla had been stimulated with saps from the host plants containing these glycoalkaloids.

Harrison and Mitchell[44] reported that stimulating taste sensilla on the galea and tarsi of the Colorado potato beetle (*Leptinotarsa decemlineata*) with tomatine, solanine, and chaconine modified the subsequent response of the sensilla to stimulation with other chemicals. Earlier experiments had suggested that glycoalkaloids with an unsaturated steroid nucleus were less deterrent to the Colorado potato beetle than those with a saturated nucleus.[91] However, both groups of compounds caused bursting in the beetle's taste sensilla, thus in this case the behavioral response to these compounds is not consistent with the neural activity of these sensilla. It is of interest that all these glycoalkaloids have surfactant properties, and it could be this physical effect on the membrane that is responsible for the observed neural activity. The possibility also remains that the bursting activity is not a normal response and only occurs when the sensilla are stimulated for a prolonged period, as in the experimental protocol adopted by Harrison and Mitchell[44] and Peterson and Hanson.[72]

Clearly the question of whether bursting activity in insect chemoreceptors is a normal physiological phenomenon cannot be answered unequivocally on the basis of present knowledge.

V. CORRELATION BETWEEN CHEMORECEPTOR ACTIVITY AND FEEDING BEHAVIOR

A basic assumption of all studies of chemosensory coding is that there exists a simple and direct relationship between afferent discharge of sensory neurons and feeding behavior. There are only a few studies on the neural coding of deterrents that involve both sensory physiology and behavior. The most detailed investigations have been undertaken in lepidopterous larvae[17,18,20,59,88,96] and lepidopterous adults.[12] In these insects, feeding intensity can be predictably correlated with the neural activity of neurons that respond either to phagostimulants or feeding deterrents. A sensory model developed for *P. brassicae* suggests that one impulse arising in the medial deterrent neuron in this species neutralizes 2.5 impulses from neurons responding to various

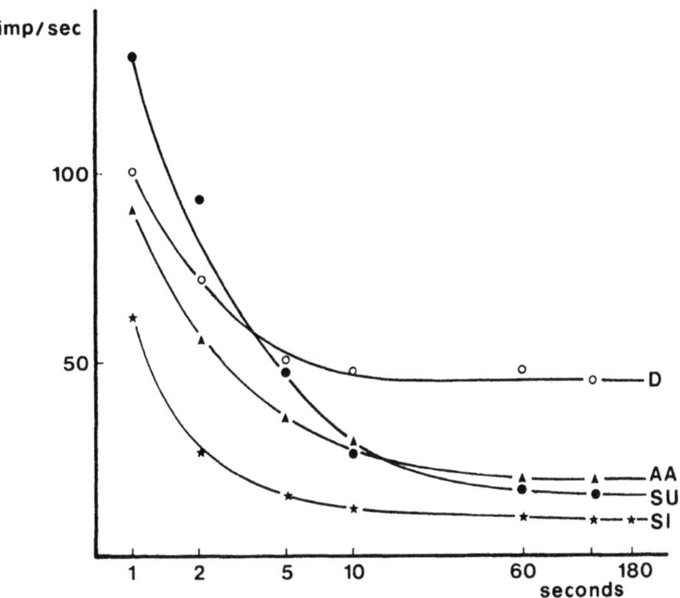

FIGURE 6. Adaptation curves of four chemoreceptory neurons in the maxillary sensilla styloconica of a *Pieris brassicae* larva. (From Schoonhoven, L. M., *Proc. Kon. Ned. Akad. Wetensch. Amsterdam,* C80, 341, 1977. With permission.)

phagostimulants.[88] The role of the lateral deterrent neuron in this model is currently being assessed (van Loon, personal communication).

When insects feed on plants containing deterrents, quantitative aspects of food intake may be modified, such as meal duration,[18] total amount of time spent feeding,[17] amount eaten,[12] and amount of feces produced.[20] Thus many substances that have a marked feeding deterrent effect on larvae of *Spodoptera* spp., when applied to sucrose-treated glass-fiber disks, appear to reduce the duration of the first meal by more than 50% when compared to the duration of the meal taken on a control sucrose-treated disk.[18] Apparently the stimulus situation allows the initiation of a meal, and feeding interruption could be caused by the fact that, in the presence of the deterrent, intake is stopped at a submaximal level of gut filling and/or by a gradual change of neural input during feeding. Evidence for the latter phenomenon is based on the finding that the time characteristics of responses of deterrent receptors differ from those of phagostimulant receptors. Generally deterrent neurons show a latency in their response that is greater than that of the other neurons.[43,57,105] More importantly, their adaptation rate is lower and their phasic-tonic relationships differ from those of the phagostimulant neurons,[34,57,72,74,81,105] as shown, for instance, in *P. brassicae* (Figure 6). As a consequence, the ratio between the number of impulses from the deterrent neurons and those from the phagostimulant neurons changes with time, so that the input from the deterrent neurons gradually becomes more prominent, resulting in a premature end to the meal.

Occasionally it has been reported that certain compounds at low concentration are stimulatory, whereas at higher concentrations they become deterrent. In *S. littoralis* larvae, sinigrin stimulates food intake at a concentration of 1 m*M*, but at higher levels (i.e. greater than 5 m*M*) it acts as an antifeedant.[12] Similar observations have been made in nonphytophagous insects. Thus sodium chloride, stimulating the "salt receptor" in blowfly labellar hairs, at low concentrations enhances drinking activity, whereas above a certain threshold concentration, which varies with the degree of water satiation of the fly, salt solutions evoke aversion behavior.[27,41] A recent study, however, showed that the blowfly salt receptor also possesses some sensitivity to sucrose, which means that it is not a pure and independent deterrent detector, but rather forms an element in a sensory system that functions by producing "across-fiber patterns".[76]

For the sake of simplicity, most electrophysiological studies investigating the mechanism of deterrent perception have employed single compounds. When simple combinations, for instance of azadirachtin and sucrose,[93] were tested, it appeared that deterrents may affect more neurons than expected on the basis of experiments with single chemicals. The "latent spectrum"[85] of phagostimulant receptors includes sensitivity modulations by deterrents. Although there is a paucity of information on sensory responses to mixtures, some studies have shown that combinations of chemicals can have effects not predictable from knowledge of the responses to individual components.[26,52,69,71] Moreover unacceptable plants, in most if not all cases, contain more than one deterrent compound,[50,107] and insect resistance is often based upon combinations of deterrents.[7] Especially interesting, as a near approach to the natural situation, would be to examine chemosensory responses to host plant saps, with and without the addition of deterrent compounds. Unfortunately, few if any such studies have been done. Studies of sensory reactions to stimulations with whole-plant saps from hosts and nonhosts, however, have confirmed the conclusion that complex interactions between plant constituents occur at the receptor level.[31,66,94]

VI. RECEPTOR MECHANISMS

A basic feature of chemoreceptors is that specific chemicals, upon contacting the membrane of chemosensory neurons, elicit some reaction, leading to the formation of action potentials. The specificity of the primary reaction depends on the molecular characteristics of both the stimulus and receptor sites. Although our knowledge about the molecular mechanisms involved in the gustatory transduction process is disappointingly meagre, it is believed that in vertebrates simple ionic compounds, such as salts and acids, may pass through or act directly on membrane ion channels of gustatory cells. More complex chemicals, such as sweet or bitter substances, are presumed to interact with membrane-associated proteins or to influence the conformation of the lipid bilayer.[53,54] Research on the sense of taste in vertebrates has been, and still is, strongly influenced by the classical recognition of four modalities (sweet, sour, salt, bitter), a classification initially also used by students of taste perception in insects (e.g., "bitter receptor" [46]). It was thought that in vertebrates this classification was congruent with the existence of four different types of taste cells. Recent evidence, however, indicates that, as has been inferred earlier to be the case in insects,[32,78] one cell may respond to more than one type of stimulus.[100] This suggests that more than one transduction mechanism is available to each cell.[53,100] It has been hypothesized that in vertebrate taste organs, bitter-tasting compounds interact with specific membrane proteins,[1,100] but alternative modes of action cannot be excluded.[54]

Information available on the transduction process in insect gustatory neurons parallels knowledge of the vertebrate taste system. Thus the dendritic membrane of sugar neurons supposedly contains three,[68] or in blowflies even four, different types of sugar receptor sites: a pyranose site, a furanose site, a site reacting to D-galactose, and one reacting to 4-nitrophenyl-α-glucoside.[104] After binding with a stimulus molecule, a cascade of biochemical events is started, including the production of a second messenger,[2] ultimately leading to changes in the flow of ions through the dendritic membrane. The presence of sugar receptor sites is not limited to the dendrites of sugar neurons,

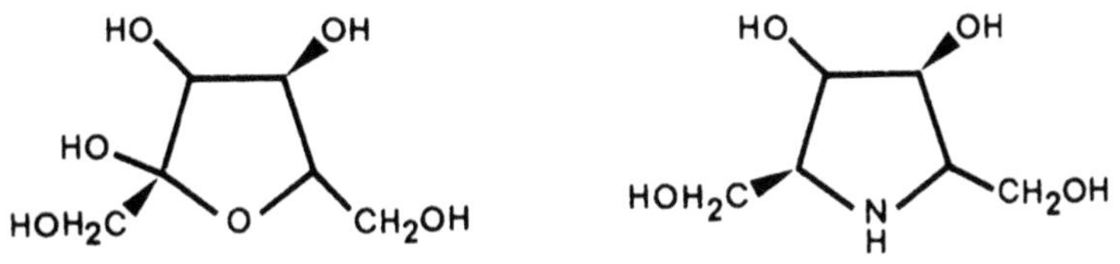

FIGURE 7. β-D-Fructofuranose (left) and 2,5-dihydroxylmethyl 3,4-dihydroxypyrrolidine (DMDP) (right).

as they have also been located in the dendritic membranes of salt and water neurons in flies.[76,104] Conversely, sugar neurons may respond to salts, and there is evidence that the two stimuli operate via different stimulating mechanisms.[98] Thus insect chemosensory neurons may be equipped with more than one transduction mechanism, as has been conjectured for taste cells in vertebrates.

Some polyhydroxyalkaloids, which show antifeedant properties for a number of insect species,[16,96] are known to inhibit glycosidases.[36] They are also structurally similar to sugar molecules, and the configuration of the hydroxy substitutes on the most active polyhydroxyalkloid antifeedant, i.e., DMDP, is similar to those occurring in fructose (Figure 7). DMDP can reduce the neural response to fructose, and it is possible that it is binding to the pyranose sites on the "sugar" neurons of the maxillary sensilla styloconica of *S. littoralis*.[96] Alternatively, DMDP could act as an alkaloid, modulating neuron sensitivity by interaction with some other membrane structures.[16] Different behavioral responses to DMDP occur, even between related species of *Spodoptera,* and this may be due to differences in the importance of the neural input from sugar neurons in determining the feeding behavior of these species.

Sweet molecules, such as glucose, are believed to form H bonds with these receptor sites, and compounds that interfere with this binding could have potential as antifeedants. Thus, the antifeedant activity of hydrophilic compounds, such as DMDP, could reside in their ability to form hydrogen bonds. In order to develop antifeedants that disrupt the response to sugars, Frazier and Lam[40] have used fluorinated carbohydrates to probe the binding requirements for the "sugar" receptor sites on the "sugar" neurons in the maxillary sensilla of *Manduca sexta* . They showed that the C-1 and C-3 hydroxy sites on the glucose molecule are important hydrogen bond acceptors, interacting with the receptor sites of the "sugar" neuron in the medial sensillum. After characterizing the binding requirements of the receptor sites, Frazier and Lam[40] designed and tested a series of glucose derivatives to see if they would irreversibly inhibit the response to glucose. A C-1 fumarate derivative of glucose caused a 70% decrease in the response to glucose, and the inhibition lasted for 15 min.

Thus, the studies by Frazier and Lam[40] and by Simmonds et al.[96] have shown that an antifeedant can act by inhibiting "sugar" receptors. The chemicals used by these workers structurally resemble sugars, and it is postulated that the molecules might be binding with receptor sites for sugars and in some way inactivating them.

Other studies have shown that antifeedants with a diverse array of structural moieties can block or inhibit the response to sucrose.[10,44,46,93] These latter studies suggest that the "sugar" neuron is susceptible to interference by a wide range of compounds. For example, Dethier[29] showed that tannic acid, quinine, piperidine and caffeine inhibited the sensory response to sucrose in the larvae of *Lymantria dispar* and *Malacosoma americanum.* It is not known, however, whether these compounds have an unspecific disruptive effect on the receptor membrane, or whether they are interacting, in an as yet unknown way, with specific sugar receptor sites.

In the case of enedial antifeedants, such as muzigadial, there is evidence for interaction with receptor sites of the sugar-sensitive neurons via pyrrole formation with an NH_2 group,[55] rather than via a reaction with SH groups in a Michael-addition fashion, as proposed earlier.[60,61]

The chemical structures of substances that stimulate insect-deterrent neurons are often extremely diverse, and it is difficult to find common molecular features. At the same time, the efficacy of analogues with minor molecular changes is often markedly different.[17,18,95] Moreover, the stimulus spectrum of deterrent neurons in related insect species shows interspecific differences.[34,57,83] These facts are difficult to reconcile with the concept of one or a few types of specific receptor proteins, which bind deterrents. Therefore, several authors have suggested that deterrents, many of which are known to interfere with the normal functions of nerve cells, affect in a nonspecific way insect chemosensory neurons, for instance, by associating with nonpolar parts of the phospholipids in the cell membrane, disrupting the normal membrane structure, and altering ion conductances.[5,30,37,65] It has been shown that widely differing chemical structures can produce very similar changes in particular membrane ionic currents in insect nerve cells.[70] However, the assumption that many deterrents have one or more nonspecific modes of action does not solve the problem of how obvious differences between responses from different neurons arise within one insect, as well as those between different insect species. Quinine, for instance, inhibits the sugar neuron in blowflies,[67] but it stimulates in a dose-dependent way the deterrent neuron in larvae of *Bombyx mori,* without affecting other neurons in the same sensillum,[46] and it elicits the firing of all eight maxillary chemosensory neurons in *Danaus plexippus* larvae.[28] In adults of *Leptinotarsa decemlineata,* quinine inhibits the response to sucrose, while leaving the response of the same neuron to GABA wholly unaffected.[62]

Azadirachtin does not seem to affect the responsiveness of the sugar neuron in *S. exempta*, whereas it strongly inhibits the sugar response in a related species, *S. littoralis.*[93] Thus, there could be species-specific differences in membrane properties of what is termed the *deterrent neuron* as well as the *sugar neurons.*

The variety of known responses to deterrents does not seem to allow as yet any generalization about the type of transduction process at the subcellular level. A better understanding of receptor functioning at the molecular level will hopefully enhance our ability to interpret and predict structure-function interactions.

VII. STRUCTURE-ACTIVITY RELATIONSHIPS

Many behavioral studies have investigated aspects of the structure-activity relationships of the diverse range of molecules that deter insects from feeding, but there have been very few comparable electrophysiological investigations. Many of the behavioral studies have shown that a molecule, by simple one-step chemical changes, can be converted from a potent antifeedant to an inactive compound.[3,17,18,19,42] The modifications in the behavioral response to these compounds are presumably mediated by changes in neural responses, but the nature of these changes and the means by which they are effected remain unclear.

Electrophysiological and behavioral structure-function studies with clerodanes,[18] drimanes,[17,90] and flavonoids[95] have shown that small changes in the functional groups on these compounds can alter both the neural response and the behavior of larvae of species of *Spodoptera*, *Heliothis,* and *Pieris.* The studies with clerodanes showed that antifeedant activity lies in the configuration of both the furofuran unit and its decalin moiety. Small changes to the functional groups on the decalin ring caused important changes in the ability of the compound to stimulate neurons in the medial sensilla of species of Lepidoptera, including *S. exempta*. The medial sensillum of *S. exempta* is also stimulated by a group of flavonoids extracted from species of *Lonchocarpus* and *Tephrosia*. This study showed that the flavanones were generally more active antifeedants than the chalcones, and the activity was associated with the positions of the methoxy and hydroxy substitutions.[95] Overall these studies suggest that the chemicals are stimulating specific receptor sites on the "deterrent neuron". However, we do not know enough about the configuration of the compounds that stimulate sites on this neuron to decide whether the apparent diversity of structure shown by effective antifeedants precludes specific interactions with membrane receptors, or whether some unknown feature of the molecular conformation, common to all active molecules, allows such specificity. Alternatively, it could be that deterrent compounds act by disrupting the dendrite membrane.

Studying the molecular structure of compounds that elicit types of neural response associated with rejection behavior[84] should enable us to categorize the molecular parameters, such as chirality, lipophilicity, functional groups, and size, that elicit each particular type of response. However, although this approach can help to define the type of molecule that is active, it would appear from the available data that there are few if any structural similarities between the compounds that elicit a specific type of neural response.

VIII. EVOLUTION

Undoubtedly, deterrent receptors that serve to detect unacceptable and/or toxic plant food are under strong selection pressure. Their origin probably goes back to a very elementary type of chemoreceptor. Dethier[28] has suggested that deterrent receptors of herbivorous insects stem from the "common chemical sense" or salt receptors as the most common chemoreceptor type in animals.

Schoonhoven[87] has proposed a slightly different origin and has hypothesized that they have evolved from ancestral nerve cells retaining their sensitivity to noxious plant compounds, whereas most other chemoreceptor cell types have developed a relative insensitivity to such basically noxious chemicals. If some receptor neurons have retained their primeval sensitivity for different kinds of secondary plant compounds, they would be ideally suited to signal the presence of chemicals to be avoided.[87] Therefore, the primitive, unmodified taste cell may be considered as the primordial deterrent receptor, which still possesses a sensitivity to secondary plant compounds originally shown by all primitive neurons. That is not to say that the present-day deterrent receptors are unchanged and wholly identical to their ancestral neuron type. Furthermore, present-day deterrent receptors, although retaining a sensitivity to various secondary plant substances, have developed a physiological mechanism that protects them against the harmful effects of their adequate stimuli. Not only has the basic sensitivity to a number of secondary plant substances been preserved in these receptors, it has also been connected to the action-potential generating system, resulting in a change of impulse frequency upon stimulation.

Many deterrent receptors respond to a broad spectrum of chemically unrelated compounds.[57,85] The fact that related insect species often show marked differences in the reaction spectra of their deterrent cells is an indication that these receptors can be modified through time and most likely, to a certain extent, have been adapted to the present-day needs of each particular insect species.

Some authors[6,83] have noticed the paradoxical fact that deterrent cells often respond to compounds that the insect species concerned cannot have experienced in its recent evolution. It is then difficult to conjecture that sensitivity to these particular chemicals has resulted from direct natural selection. Alternatively, it seems quite likely that sensitivity to these compounds is part of a general sensitivity to particular classes of compounds to which they belong. Natural selection, then, may not reduce or increase the sensitivity to individual chemicals, but to classes of compounds, though exceptions of this rule may occur. The finding that the deterrent cells in two different strains of *Mamestra brassicae* show sensitivity differences for classes of stimuli rather than for individual chemicals[103] agrees with this view. An analysis of the deterrent receptors of insects resulting from crosses between different strains or species, as first attempted by van Drongelen and van Loon,[35] will provide indispensable information needed to answer the question of whether the genetic system codes for sensitivity to classes of compounds or for separate responsiveness to individual compounds.

IX. CONCLUSIONS

Phytophagous insects are confronted with thousands of compounds in their potential food. In a recent review, 550 chemicals have been listed that possess antifeedant properties,[102] probably representing only a small fraction of the whole gamut of such behaviorally active compounds. The intriguing question as to how phytophagous insects with a limited number of taste neurons are able to perceive and differentiate between the multitude of chemicals in plants is difficult to answer. It is clear, however, that several neural mechanisms are operative in the detection of deterrents, often in combination with each other. The different modes of action, which were recognized some time ago,[84] and which also form the basis of the present review, show a remarkable similarity with those found in hematophagous insects.[25] This similarity indicates the general occurrence of the various types of neural responses described here and argues for their basic role in insect chemoreception.

The complexity of the neurophysiological processes involved in the perception of deterrents, which we are only beginning to fathom, may be considered as another indication of the crucial role these plant substances play in relation to insects, allowing them to discriminate with unerring accuracy between host and nonhost plants. The mystery of the *botanical sense*, a term once used to denote an insect's inexplicable host selection behavior, still exists but now seems to be localized in the still largely obscure molecular interactions taking place at the insect's chemoreceptor membranes.[86] The elucidation of the nature of the sensory responses to feeding deterrents no doubt will reveal some of the basic principles that underlie insect–plant relationships.

ACKNOWLEDGMENTS

We thank J. J. A. van Loon (Wageningen, The Netherlands) and E. Städler (Wädenswil, Switzerland) for helpful suggestions and review of the manuscript.

REFERENCES

1. **Akabas, M. H., Dodd, J., and Al-Awquat, Q.,** A bitter substance induces a rise in intracellular calcium in a subpopulation of rat taste cells, *Science*, 242, 1047, 1988.
2. **Amakawa, T., Ozaki, M., and Kawata, K.,** Effects of cyclic GMP on the sugar taste receptor cell of the fly *Phormia regina, J. Insect Physiol.*, 36, 281, 1990.
3. **Belles, X., Camps, F., Coll, J., and Piulachs, M. D.,** Insect antifeedant activity of clerodane diterpenoids against larvae of *Spodoptera littoralis* (Boisd.) (Lepidoptera), *J. Chem. Ecol.*, 11, 1439, 1985.
4. **Bernays, E. A. and Chapman, R. F.,** Deterrent chemicals as a basis of oligophagy in *Locusta migratoria* (L.), *Ecol. Entomol.*, 2, 1, 1976.
5. **Bernays, E. A. and Chapman, R. F.,** Chemical deterrence of plants, in *Molecular Entomology,* Law, J. H., Ed., Alan R. Liss, New York, 1987, 107.
6. **Bernays, E. A. and Chapman, R. F.,** The evolution of deterrent responses in plant-feeding insects, in *Perspectives in Chemoreception and Behavior,* Chapman, R. F., Bernays, E. A. and Stoffolano, J. G., Eds., Springer-Verlag, New York, 1987, 159.
7. **Birch, N., Southgate, B. J., and Fellows, L. E.,** Wild and semi-cultivated legumes as potential sources of resistance to bruchid beetles for crop breeder: a study of *Vigna/Phaseolus*, in *Plants for Arid Lands,* Wickens, G. E., Goodin, J. R. and Fields, D. V., Eds., Allen and Unwin, London, 1985, 303.
8. **Blaney, W. M.,** Behavioural and electrophysiological studies of taste discrimination by the maxillary palps of larvae of *Locusta migratoria* (L.), *J. Exp. Biol.*, 62, 555, 1975.
9. **Blaney, W. M.,** Chemoreception and food selection by locusts, in *Olfaction and Taste* 7, van der Starre, H., Ed., IRL Press, London, 1980, 127.
10. **Blaney, W. M.,** Chemoreception and food selection in locusts, *Trends Neurosc.,* 4, 35, 1981.
11. **Blaney, W. M. and Duckett, A. M.,** The significance of palpation by the maxillary palps of *Locusta migratoria* (L.): an electrophysiological and behavioural study. *J. Exp. Biol.*, 63, 701, 1975.
12. **Blaney, W. M. and Simmonds, M. S. J.,** Food selection in adults and larvae of three species of Lepidoptera: a behavioural and electrophysiological study, *Entomol. Exp. Appl.*, 49, 111, 1988.
13. **Blaney, W. M. and Simmonds, M. S. J.,** A behavioural and electrophysiological study of the role of tarsal chemoreceptors in feeding by adults of *Spodoptera, Heliothis virescens* and *Helcicoverpa armigera, J. Insect Physiol.*, 36, 743, 1990.
14. **Blaney, W. M. and Simmonds, M. S. J.,** The chemoreceptors, in *Biology of Grasshoppers,* Chapman, R. F. and Joern, A., Eds., Wiley, New York, 1990, 1.
15. **Blaney, W. M., Chapman, R. F., and Cook, A. G.,** The structure of the terminal sensilla of the maxillary palps of *Locusta migratoria* (L.), and changes associated with moulting, *Z. Zellforsch. Mikrosk. Anat.*, 121, 48, 1971.
16. **Blaney, W. M., Simmonds, M. S. J., Evans, S. V., and Fellows, L.,** The role of the secondary plant compound 2,5-dihydroxydimethyl 3,4-dihydroxylpyrrolidine as a feeding inhibitor for insects, *Entomol. Exp. Appl.*, 36, 209, 1984.
17. **Blaney, W. M., Simmonds, M. S. J., Ley, S. V., and Katz, R. B.,** An electrophysiological and behavioural study of insect antifeedant properties of natural and synthetic drimane-related compounds, *Physiol. Entomol.*, 12, 281, 1987.
18. **Blaney, W. M., Simmonds, M. S. J., Ley, S. V., and Jones, P. S.,** Insect antifeedants: a behavioural and electrophysiological investigation of natural and synthetically derived clerodane diterpenoids, *Entomol. Exp. Appl.,* 46, 267, 1988.
19. **Blaney, W. M., Simmonds, M. S. J., Ley, S. V., Anderson, J. C., and Toogood, P. L.,** Antifeedant effects of azadirachtin and structurally related compounds on lepidopterous larvae, *Entomol. Exp. Appl.*, 55, 149, 1990.
20. **Blom, F.,** Sensory activity and food intake: a study of input-output relationships in two phytophagous insects, *Neth. J. Zool.,* 28, 277, 1978.
21. **Boeckh, J.,** Ways of nervous coding of chemosensory quality at the input level, in *Olfaction and Taste* 7, van der Starre, H., Ed., IRL Press, London, 1980, 113.
22. **Chapman, R. F.,** Chemoreception. The significance of sensillum numbers, *Adv. Insect Physiol.*, 16, 247, 1982.
23. **Chapman, R. F.,** Sensory aspects of host-plant recognition by Acridoidea: questions associated with the multiplicity of receptors and variability of response, *J. Insect Physiol.*, 34, 167, 1988.

24. **Clark, J. V.**, Feeding deterrent receptors in the last instar African armyworm *Spodoptera exempta*: a study using salicin and caffein, *Entomol. Exp. Appl.*, 29, 189, 1981.
25. **Davis, E. E.**, Insect repellents: concepts of their mode of action relative to potential sensory mechanisms in mosquitoes (Diptera: Culicidae), *J. Med. Entomol.*, 22, 237, 1985.
26. **Derby, C. D., Ache, B. W., and Kennel, E. W.**, Mixture suppression in olfaction: electrophysiological evaluation of the contribution of peripheral and central neural components, *Chem. Senses*, 10, 301, 1985.
27. **Dethier, V. G.**, Chemosensory input and taste discrimination in the blowfly, *Science*, 161, 389, 1968.
28. **Dethier, V. G.**, Evolution of receptor sensitivity to secondary plant substances with special reference to deterrents, *Am. Nat.*, 115, 45, 1980.
29. **Dethier, V. G.**, Mechanisms of host-plant recognition, *Entomol. Exp. Appl.*, 31, 49, 1982.
30. **Dethier, V. G.**, Discriminative taste inhibitors affecting insects, *Chem. Senses*, 12, 251, 1987.
31. **Dethier, V. G. and Crnjar, R. M.**, Candidate codes in the gustatory system of caterpillars, *J. Gen. Physiol.*, 79, 549, 1982.
32. **Dethier, V. G. and Kuch, J. H.**, Electrophysiological studies of gustation in lepidopterous larvae. I. Comparative sensitivity to sugars, amino acids, and glycosides, *Z. Vergl. Physiol.*, 72, 343, 1971.
33. **Dethier, V. G. and Schoonhoven, L. M.**, Olfactory coding by lepidopterous larvae, *Entomol. Exp. Appl.*, 12, 535, 1969.
34. **van Drongelen, W.**, Contact chemoreception of host plant specific chemicals in larvae of various *Yponomeuta* species (Lepidoptera), *J. Comp. Physiol.*, 134, 265, 1979.
35. **van Drongelen, W. and van Loon, J. J. A.**, Inheritance of gustatory sensitivity in F1 progeny of crosses between *Yponomeuta cagnagellus* and *Y. malinellus* (Lepidoptera), *Entomol. Exp. Appl.*, 28, 199, 1980.
36. **Fellows, L. E., Kite, G. C., Nash, R. J., Simmonds, M. S. J., and Schofield, A. M.**, Castanospermine, swainsonine and related polyhydroxy alkaloids: structure, distribution and biological activity, in *Plant Nitrogen Metabolism*, Poulton, J. E., Romeo, J. T., and Conn, E. E., Eds. Plenum, New York, 1989, 395.
37. **Frazier, J. L.**, The perception of plant allelochemicals that inhibit feeding, in *Molecular Aspects of Insect-Plant Associations*, Brattsten, L. B. and Ahmad, S., Eds., Plenum, New York, 1986, 1.
38. **Frazier, J. L. and Hanson, F. E.**, Electrophysiological recording and analysis of insect chemosensory responses, in *Insect-Plant Interactions*, Miller, J. R. and Miller, T. A., Eds., Springer, New York, 1986, 285.
39. **Frazier, J. L. and Lam, P. Y.-S.**, The effects of sesquiterpene dialdehydes on the styloconic taste cells of the tobacco hornworm larva, *Chem. Senses*, 11, 600, 1986.
40. **Frazier, J. L., and Lam, P. Y.-S.**, Three dimensional mapping of insect taste receptor sites as an aid to novel antifeedant development, in *Recent Advances in the Chemistry of Insect Control II*, Crombie, L., Ed., Roy. Soc. Chem. U.K., 1990, 247.
41. **Fredman, S. M.**, Peripheral and central interactions between sugar, water and salt receptors of the blowfly, *Phormia regina*, *J. Insect Physiol.*, 21, 265, 1975.
42. **Geuskens, R. B. M., Luteijn, J. M., and Schoonhoven, L. M.**, Antifeedant activity of some ajugarin derivatives in three lepidopterous species, *Experientia*, 39, 403, 1983.
43. **Hanson, F. E. and Peterson, S. C.**, Sensory coding in *Manduca sexta* for deterrence by a non-host plant, *Canna generalis*, (submitted) 1991.
44. **Harrison, G. D. and Mitchell, B. K.**, Host-plant acceptance by geographic populations of the Colorado potato beetle, *Leptinotarsa decemlineata*: the role of solanaceous alkaloids as sensory deterrents, *J. Chem. Ecol.*, 14, 777, 1988.
45. **Haskell, P. T. and Schoonhoven, L. M.**, The function of certain mouthpart receptors in relation to feeding in *Schistocerca gregaria* and *Locusta migratoria migratorioides*, *Entomol. Exp. Appl.*, 12, 423, 1969.
46. **Ishikawa, S.**, Electrical response and function of a bitter substance receptor associated with the maxillary sensilla of the silkworm, *Bombyx mori* L., *J. Cell. Physiol.*, 67, 1, 1966.
47. **Jermy, T.**, Untersuchungen über Auffinden und Wahl der Nahrung beim Kartoffelkäfer (*Leptinotarsa decemlineata* Say), *Entomol. Exp. Appl.*, 1, 197, 1958.
48. **Jermy, T.**, On the nature of oligophagy in *Leptinotarsa decemlineata* Say (Coleoptera: Chrysomelidae), *Acta Zool. Acad. Sci. Hung.*, 7, 119, 1961.

49. **Jermy, T.,** Feeding inhibitors and food preference in chewing phytophagous insects, *Entomol. Exp. Appl.*, 9, 1, 1966.
50. **Jermy, T.,** Multiplicity of insect antifeedants in plants, in *Natural Products for Innovative Pest Management,* Whitehead, D. L. and Bowers, W. S., Eds., Pergamon Press, Oxford, 1983, 223.
51. **Jermy, T.,** Prospects of the antifeedant approach to pest control, *J. Chem. Ecol.*, 16, 3151, 1990.
52. **de Jong, R. and Visser, J. H.,** Specificity-related suppression of response to binary mixtures in olfactory receptors of the Colorado potato beetle, *Brain Res.*, 447, 18, 1988.
53. **Kinnamon, S. C.,** Taste transduction: a diversity of mechanisms, *Trends Neurosci.*, 11, 491, 1988.
54. **Kurihara, K., Yoshii, K., and Kashiwayanagi, M.,** Transduction mechanisms in chemoreception, *Comp. Biochem. Physiol.*, 85A, 1, 1986.
55. **Lam, P. Y.-S. and Frazier, J. L.,** Model study on the mode of action of muzigadial antifeedant, *Tetrahedron Lett.*, 28, 4577, 1987.
56. **Luo, L.-E., Liao, C.-Y., and Zhou, P.-A.,** Electrophysiological study of the antifeedant action of toosendanin to the armyworm larvae, *Acta Entomol. Sinica,* 32, 257, 1989.
57. **van Loon, J. J. A.,** Chemoreception of phenolic acids and flavonoids in larvae of two species of *Pieris, J. Comp. Physiol.*, A166, 889, 1990.
58. **Ma, W.-C.,** Some properties of gustation in the larva of *Pieris brassicae, Entomol. Exp. Appl.*, 12, 584, 1969.
59. **Ma, W.-C.**, Dynamics of feeding responses in *Pieris brassicae* L. as a function of chemosensory input: a behavioural, ultrastructural and electrophysiological study, *Med. Landbouwhogeschool Wageningen,* 72-11, 1, 1972.
60. **Ma, W.-C.**, Alterations of chemoreceptor function in armyworm larvae (*Spodoptera exempta*) by a plant-derived sesquiterpenoid and by sulfhydryl reagents, *Physiol. Entomol.*, 2, 199, 1977.
61. **Ma, W.-C.,** Receptor membrane function in olfaction and gustation: implications from modification by reagents and drugs, in *Perception of Behavioral Chemicals,* Norris, D. M., Ed., Elsevier, Amsterdam, 1981, 267.
62. **Mitchell, B. K.,** Interactions of alkaloids with galeal chemosensory cells in the adult Colorado potato beetle, *Leptinotarsa decemlineata, J. Chem. Ecol.*, 13, 2009, 1987.
63. **Mitchell, B. K. and Harrison, G. D.,** Characterization of galeal chemosensilla in the adult Colorado potato beetle, *Leptinotarsa decemlineata, Physiol. Entomol.*, 9, 46, 1984.
64. **Mitchell, B. K. and Harrison, G. D.,** Effects of *Solanum* glycoalkaloids on chemosensilla in the Colorado potato beetle. A mechanism of feeding deterrence?, *J. Chem. Ecol.*, 11, 77, 1985.
65. **Mitchell, B. K. and Sutcliffe, J. F.,** Sensory inhibition as a mechanism of feeding deterrence: effects of three alkaloids on leaf beetle feeding, *Physiol. Entomol.*, 9, 57, 1984.
66. **Mitchell, B. K., Rolseth, B. M., and McCashin, B. G.,** Differential responses of galeal sensilla of the adult colorado potato beetle, *Leptinotarsa decemlineata* (Say), to leaf saps from host and non-host plants, *Physiol. Entomol.*, 15, 61, 1990.
67. **Morita, H.,** Initiation of spike potentials in contact chemosensory hairs of insects. III. D. C. stimulation and generator potentials of labellar chemoreceptor of *Calliphora, J. Cell. Comp. Physiol.*, 54, 189, 1959.
68. **Morita, H. and Shiraishi, A.,** Chemoreception physiology, in *Comprehensive Insect Physiology, Biochemistry and Pharmacology,* Vol. 6, Kerkut, G. A. and Gilbert, L. I., Eds., Pergamon Press, Oxford, 1985, 133.
69. **O'Connell, R. J.,** Responses to pheromone blends in insect olfactory neurons, *J. Comp. Physiol.*, 156, 747, 1985.
70. **Pelhate, M. and Sattelle, D. B.,** Pharmacological properties of insect axons: a review, *J. Insect Physiol.*, 28, 889, 1982.
71. **van der Pers, J. N. C., Thomas, G., and den Otter, C. J.,** Interactions between plant odours and pheromone reception in small ermine moths (Lepidoptera: Yponomeutidae), *Chem. Senses,* 5, 367, 1980.
72. **Peterson, S. C. and Hanson, F. E.,** Sensory coding for feeding deterrence in *Manduca:* a gustatory cell mediating rejection of a non-host plant (*Canna generalis* L.), *Physiol. Entomol.*, (submitted) 1991.
73. **Rees, C. J. C.,** Chemoreceptor specificity associated with choice of feeding site by the beetle, *Chrysolina brunsvicensis* on its foodplant, *Hypericum hirsutum, Entomol. Exp. Appl.*, 12, 565, 1969.

74. **Roessingh, P., Schöni, R., Städler, E., Feeny, P., and Sachdev, K.,** Tarsal contact chemoreceptors of the black swallowtail butterfly *Papilio polyxenes:* responses to phytochemicals from host- and non-host plants, *Physiol. Entomol.*, 16, 485, 1991.
75. **Rothschild, M., Alborn, H., Stenhagen, G., and Schoonhoven, L. M.,** A strophanthidine glycoside in the Siberian wallflower: a contact deterrent for the large white butterfly, *Phytochemistry,* 27, 101, 1988.
76. **Schnuck, M. and Hansen, K.,** Sugar sensitivity of a labellar salt receptor of the blowfly *Protophormia terranovae, J. Insect Physiol.*, 36, 409, 1990.
77. **Schoonhoven, L. M.,** Chemoreception of mustard oil glucosides in larvae of *Pieris brassicae, Proc. Kon. Ned. Akad. Wet. Amsterdam,* C70, 556, 1967.
78. **Schoonhoven, L. M.**, Chemosensory bases of host plant selection, *Annu. Rev. Entomol.*, 13, 115, 1968.
79. **Schoonhoven, L. M.,** Gustation and foodplant selection in some lepidopterous larvae, *Entomol. Exp. Appl.*, 12, 555, 1969.
80. **Schoonhoven, L. M.,** Plant recognition by lepidopterous larvae, *Symp. R. Entomol. Soc. London,* 6, 87, 1973.
81. **Schoonhoven, L. M.,** On the individuality of insect feeding behavior, *Proc. Kon. Ned. Akad. Wet. Amsterdam.* 80C, 341, 1977.
82. **Schoonhoven, L. M.,** Perception of azadirachtin by some lepidopterous larvae, in *Proc. 1st Int. Neem Conf., Rottach-Egern,* Schmutterer, H., Ascher, K. R. S., and Rembold, H., Eds., GTZ, Eschborn, 1981, 105.
83. **Schoonhoven, L. M.,** Chemical mediators between plants and phytophagous insects, in *Semiochemicals: Their Role in Pest Control,* Nordlund, D. A., Jones, R. L., and Lewis, W. J., Eds., Wiley, New York, 1981, 31.
84. **Schoonhoven, L. M.**, Biological aspects of antifeedants, *Entomol. Exp. Appl.*, 31, 57, 1982.
85. **Schoonhoven, L. M.,** What makes a caterpillar eat? The sensory code underlying feeding behavior, in *Perspectives in Chemoreception and Behavior,* Chapman, R. F., Bernays, E. A., and Stoffolano, J. G., Eds., Springer-Verlag, New York, 1987, 69.
86. **Schoonhoven, L. M.,** Insects and host plants: 100 years of "botanical instinct", *Symp. Biol. Hung.*, 39, 3, 1990.
87. **Schoonhoven, L. M.,** The sense of distaste in plant-feeding insects — A reflection on its evolution, *Phytoparasitica,* 19, 3, 1991.
88. **Schoonhoven, L. M. and Blom, F.,** Chemoreception and feeding behaviour in a caterpillar: towards a model of brain functioning in insects, *Entomol. Exp. Appl.*, 49, 123, 1988.
89. **Schoonhoven, L. M. and Jermy, T.,** A behavioural and electrophysiological analysis of insect feeding deterrents, in *Crop Protection Agents — Their Biological Evaluation,* McFarlane, N. R., Ed., Academic Press, London, 1977, 133.
90. **Schoonhoven, L. M. and Yan F.-S.**, Interference with normal chemoreceptor activity by some sesquiterpenoid antifeedants in an herbivorous insect *Pieris brassicae, J. Insect Physiol.*, 35, 725, 1989.
91. **Schreiber, K.**, Natürliche pflanzliche Resistenzstoffe gegen den Kartoffelkäfer und ihr möglicher Wirkungsmechanismus, *Züchter,* 27, 289, 1957.
92. **Shi, Y.-L., Wang, W.-P., Liao, C.-Y., and Chiu, S.-F.,** Effect of toosendanin on the sensory inputs of chemoreceptors of the armyworm larvae *Mythimna separata, Acta Entomol. Sinica,* 29, 233, 1986.
93. **Simmonds, M. S. J. and Blaney, W. M.,** Some neurophysiological effects of azadirachtine on lepidopterous larvae and their feeding response, in *Natural Pesticides from the Neem Tree and Other Tropical Plants*, Schmutterer, H. and Ascher, K. R. S., Eds., GTZ, Eschborn, 1984, 163.
94. **Simmonds, M. S. J. and Blaney, W. M.,** Gustatory codes in lepidopterous larvae, *Symp. Biol. Hung.*, 39, 17, 1990.
95. **Simmonds, M. S. J., Blaney, W. M., Delle Monache, F., and Marini Bettolo, G. B.,** Insect antifeedant activity associated with compounds isolated from species of *Lonchocarpus* and *Tephrosia, J. Chem. Ecol.*, 16, 365, 1990.
96. **Simmonds, M. S. J., Blaney, W. M., and Fellows, L. E.,** Behavioral and electrophysiological study of antifeedant mechanisms associated with polyhydroxy alkaloids, *J. Chem. Ecol.*, 16, 3167, 1990.
97. **Smith, J. J. B., Mitchell, B. K., Rolseth, B. M., Whitehead, A. T., and Albert, P. J.,** SAPID tools: microcomputer programs for analysis of multi-unit nerve recordings, *Chem. Senses,* 15, 253, 1990.
98. **Städler, E. and Schöni, R.**, High sensitivity to sodium in the sugar chemoreceptor of the cherry fruit fly after emergence, *Physiol. Entomol.*, 16, 117, 1991.

99. **Stürckow, B.,** Über den Geschmackssinn und den Tastsinn von *Leptinotarsa decemlineata* Say (Chrysomelidae), *Z. Vergl. Physiol.,* 42, 255, 1959.
100. **Teeter, J. H. and Brand, J. G.,** Peripheral mechanisms of gustation: physiology and biochemistry, in *Neuro-biology of Taste and Smell,* Finger, T. E. and Silver, W. L., Eds., Wiley, New York, 1987, 299.
101. **Waladde, S. M., Hassanali, A., and Ochieng, S. A.,** Taste sensilla responses to limonoids, natural insect antifeedants, *Insect Sci. Appl.,* 10, 301, 1989.
102. **Warthen, J. D. and Morgan, E. D.,** Insect feeding deterrents, in *CRC Handbook of Natural Pesticides, Vol. IV, Insect Attractants and Repellents,* Morgan, E. D. and Mandava, N. B., Eds., CRC Press, Boca Raton, FL, 1990, 23.
103. **Wieczorek, H.,** The glycoside receptor of the larvae of *Mamestra brassicae* L. (Lepidoptera: Noctuidae), *J. Comp. Physiol.,* A106, 153, 1976.
104. **Wieczorek, H., Shimada, I., and Hopperdietzel, C.,** Treatment with pronase uncouples water and sugar reception in the labellar water receptor of the blowfly, *J. Comp. Physiol.,* A163, 413, 1988.
105. **White, P. R. and Chapman, R. F.,** Tarsal chemoreception in the polyphagous grasshopper *Schistocerca americana:* behavioural assays, sensilla distributions and electrophysiology, *Physiol. Entomol.,* 15, 105, 1990.
106. **White, P. R., Chapman, R. F. and Ascoli-Christensen, A.,** Interactions between two neurons in contact chemosensilla of the grasshopper, *Schistocerca americana, J. Comp. Physiol.,* A167, 431, 1990.
107. **Yan, F.-S., Evans, K. A., Stevens, L. H., van Beek, T. A., and Schoonhoven, L. M.,** Deterrents extracted from the leaves of *Ginkgo biloba:* effects on feeding and contact chemoreceptors, *Entomol. Exp. Appl.,* 54, 57, 1990.
108. **Zacharuk, R. Y.,** Antennae and sensilla, in *Comprehensive Insect Physiology, Biochemistry and Pharmacology,* Vol. 6, Kerkut, G. A. and Gilbert, L. I., Eds., Pergamon Press, Oxford, 1985, 1.
109. **Zhou, P.-A., Luo, L.-E., Bai, D.-L., and Liao, C.-Y.,** Toosendanin, a natural product, as a neurotoxin and an antifeedant, in *Abstracts Sixth IUPAC Congress of Pesticide Chemistry, Ottawa,* 2D/E-16, 1986.

4 Extrafloral Nectary-Mediated Interactions Between Insects and Plants

Suzanne Koptur
Department of Biological Sciences
Florida International University
Miami, Florida
and
Fairchild Tropical Garden
Miami, Florida

TABLE OF CONTENTS

I. NECTAR, NECTARIES, AND EXTRAFLORAL NECTARIES

A. DEFINITIONS

Nectaries are plant secretory structures of diverse morphology and anatomy. They can be located on virtually any part of the plant body (nectary locations are usually characteristic of the species or genus), but the most familiar nectaries are those located in flowers.

The nectar produced in floral nectaries serves as a reward for floral visitors and is the primary physiological cost paid by the plant in nectar-based pollination systems. The animals that imbibe and collect nectar inadvertently transfer pollen (thus benefitting the plant), but visit flowers for the purpose of meeting their energetic and nutritional needs.

The remainder of nectaries are located outside the flowers and are, therefore, termed *extrafloral* nectaries. The majority of extrafloral nectaries do not involve pollination, and their function is not as uniform as that of floral nectaries, as this review will examine. They are visited by a wide variety of animals for the energy and nutrition considerations mentioned above; however, the associated effects on the plants range from beneficial (patrolling of plant surfaces and disturbance of herbivores; enhanced predation and parasitism of plant feeders) to harmful (attraction of herbivorous insect adults who, in turn, oviposit on the plant; distraction of nectar-collecting pollinators from flowers), depending on the ecological context.

This paper will examine the importance of extrafloral nectaries in mediating interactions between insects and plants. First, I will consider the function and distribution of extrafloral nectaries in the plant kingdom. Second, I will examine the reciprocal effects of nectar on insects, and insects on nectar. Third, I will consider the myriad interactions between plants and insects based on extrafloral nectaries. I will then proceed to examine the complex interactions between insects visiting nectaries on plants with extrafloral nectaries and the resulting effects on the plants.

B. MORPHOLOGY AND ANATOMY

1. Locations on the Plant Body

Extrafloral nectaries may be found on virtually every vegetative and reproductive structure.[106] On leaves, nectaries occur on the petiole, the rachis (of compound leaves), the upper and lower surfaces of the blade, the leaf margin, and on stipules (outgrowths of the leaf base, common in Leguminosae and Rosaceae). Nectaries may occur on young stems, especially in the vicinity of nodes (e.g., *Polygonum* spp.[305]). In a few species, nectaries are located on the cotyledons (e.g., *Ricinus communis*[359]).

Extrafloral nectaries on inflorescences and on the external part of flowers, but not directly involved with pollination, are common. Nectaries may be on the outer surface of bracts or involucres (e.g., *Helianthella quinquenervis*[163]), on sepals (e.g., *Bixa orellana*[32]), or petals (e.g., *Paeonia* spp.[277,359]). Some plants have extrafloral nectaries that are produced outside the flowers but still function in attracting pollinators (see Section IIIA).

In some plants the floral nectaries continue to function after the corolla abscises. This phenomenon has given rise to the term *postfloral* nectaries (e.g., *Mentzelia nuda*;[185] *Thevetia ovata*, personal observation). Some plants have glands on the surface of developing fruit (e.g., *Crescentia* and other Bignoniaceae;[110] *Annona glabra*, personal. observation).

Other structures on the plant body may look like nectaries but produce other materials (e.g., hydathodes, oil glands, and resin glands). It is important to analyze secretions and to determine that nectar is being secreted rather than resins,[162] oil, or water.[209] The presence of sooty molds can indicate whether or not sugar secretions are present (Pemberton, personal communication), but even here, sugar may be from Homoptera rather than nectaries.

2. Structure, Complexity, and Types

The levels of complexity vary among extrafloral nectaries. The simplest nectaries are nonvascularized, without a well-defined internal structure. These may be externally formless, pit nectaries, or most commonly, scalelike nectaries.

All nectaries that are known to be vascularized have a well-defined external structure, ranging from sessile to raised to stalked glands (trough, saucerlike, to cupular in shape). Most nectaries are vascularized by both phloem[124] and xylem, the vascular strands terminating beneath the secretory cells.[101] Some nectaries are only fed by phloem; nectaries are distinguished from hydathodes by vascularization, at least in part, by phloem. The secretory cells are located below the epidermis (one to three layers thick), and have conspicuous nuclei and dense cytoplasm, numerous mitochondria, and abundant endoplasmic reticula, indicative of high metabolic activity.[100-102]

Extrafloral nectaries range in size and complexity from a simple pore (or single cell delivering its contents) to large, stalked, cup-shaped structures centimeters in diameter. Many botanists have sought to classify extrafloral nectaries in various groups of plants using a variety of criteria: developmental patterns, organographic relationships, and

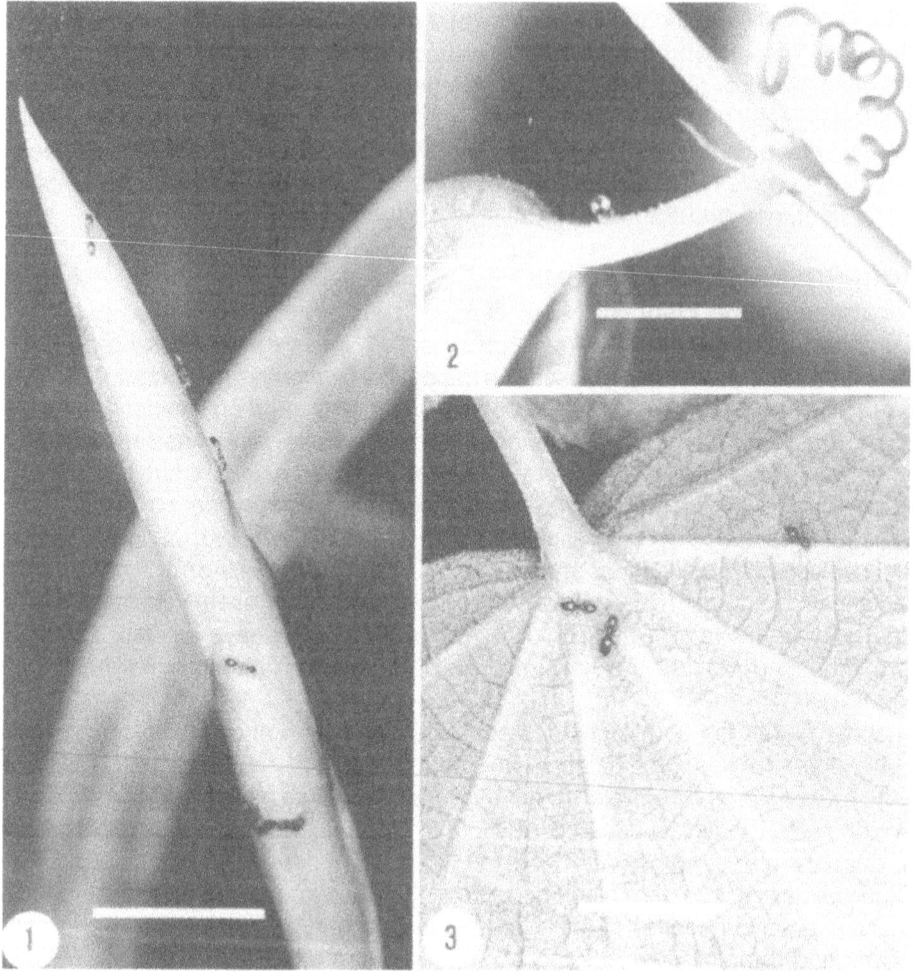

PLATE 1. Bar scales = 1 cm.
FIGURE 1. Ants visiting formless nectaries on young inflorescence of *Tillandsia balbisiana* (Bromeliaceae).
FIGURE 2. Nectar droplet on elevated nectary of *Passiflora suberosa* (Passifloraceae).
FIGURE 3. Ants visiting pit nectaries on abaxial leaf surface of *Urena lobata* (Malvaceae).

phylogenetic distribution.[106] A variety of types of extrafloral nectaries are illustrated in Figures 1 to 7.

Zimmermann[359] categorized extrafloral nectaries according to their structure and position into six basic groups (a to f below); Elias[106] added a seventh group (g below).

1. Formless nectaries — with no structural specialization, but may be colored in contrast to background (*Costus; Paeonia; Tillandsia,* Figure 1; *Myrsine,* Figure 4)
2. Flattened nectaries — glandular tissue closely pressed against underlying tissue, common on leaf surfaces (*Chrysobalanus,* Figure 5; *Dipterocarpus*)
3. Pit nectaries — glandular tissue sunken in tissue of other organs, in depressions as large as or larger than the nectary (*Urena*, Figure 3; *Inga punctata)*
4. Hollow nectaries — cavities in other organs with a narrow channel extending to the surface; the cavities are lined with secretory trichomes (*Conocarpus*)
5. Scalelike nectaries — glandular trichomes modified for nectar production and secretion (*Mucuna, Bauhinia* stipules, *Plumeria*)

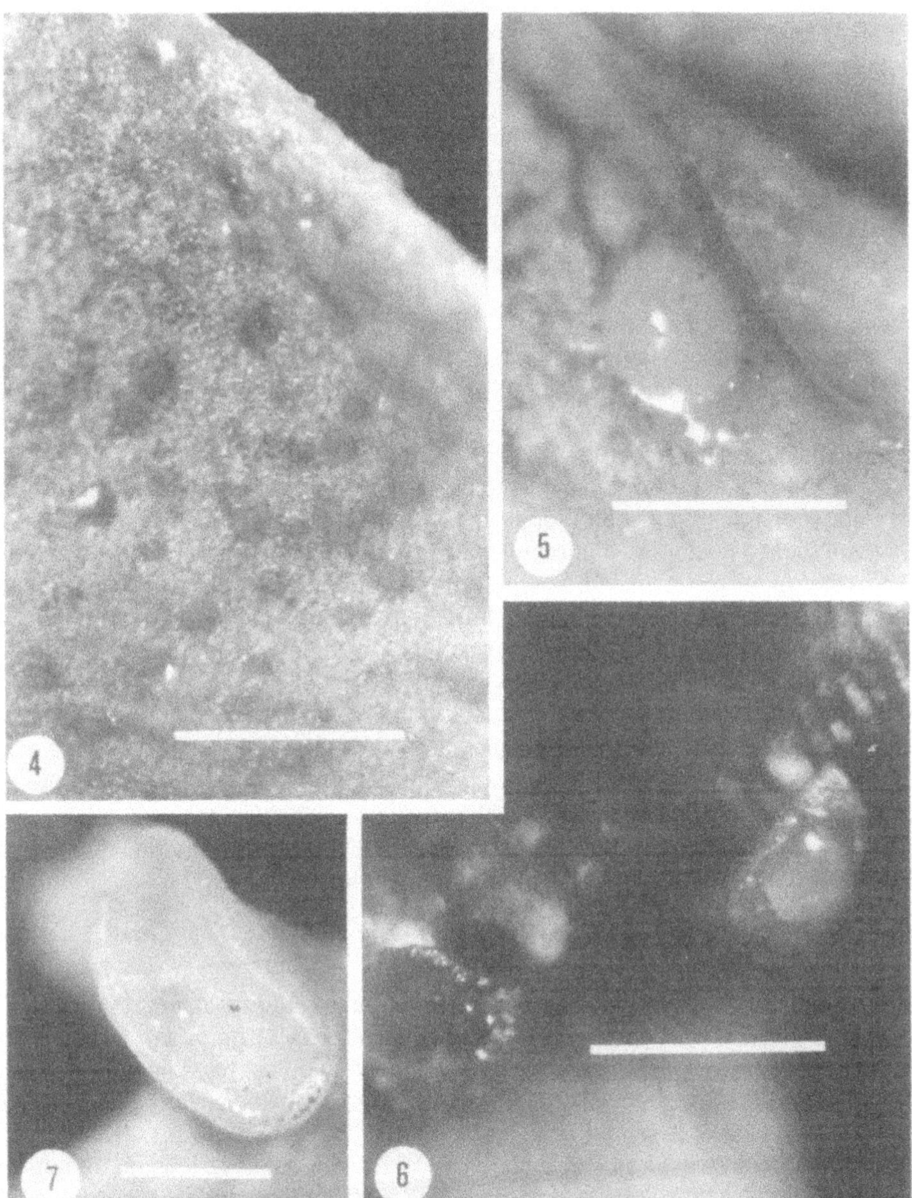

PLATE 2. Bar scales = 1 mm.

FIGURE 4. Formless nectaries of *Myrsine floridana* (Myrsinaceae) located on abaxial leaf surface at base of leaf on either side of midrib.

FIGURE 5. Flattened nectary on new red leaf of *Chrysobalanus icaco* (Chrysobalanaceae) located on abaxial leaf surface near leaf base.

FIGURE 6. Paired, elevated nectaries on stipels of young leaves of *Erythrina herbacea* (Fabaceae).

FIGURE 7. One of a pair of elevated, cupular nectaries on petiole of *Passiflora alata* (Passifloraceae).

6. Elevated nectaries — nectaries that have glandular tissue raised above ground tissue (except the scalelike, which are in their own category) (*Erythrina,* Figure 6; *Passiflora,* Figures 2 and 7)
7. Embedded nectaries — secretory tissue totally embedded in tissues of other organs (*Leonardoxa, Vigna* inflorescence nectaries)

C. EXTRAFLORAL NECTAR

1. Secretion

Nectar secretion has been described as a passive process by some researchers,[33] and some have purported to show nectaries to function as "sap valves," eliminating excess carbohydrates from growing points of the plant when they shift from sink to source status. An experimental test of the "sap-valve" hypothesis provided no evidence that nectaries function in this way,[17] while evidence for ecological functions is mounting (see Sections II and III), and active secretory mechanisms certainly exist.

2. Composition and Constitution

Extrafloral nectars, like floral nectars, are aqueous solutions of sugars, amino acids, lipids, and other organic compounds.[12,15] The other compounds include antioxidants and substances (alkaloids, phenols, and nonprotein amino acids) that may be deterrents or are potentially toxic to nectar drinkers.[15,115,140,141,169]

Sugar is the largest proportion of the solutes in nectars, and extrafloral nectars show a wide range of sugar concentrations, largely influenced by the microenvironment around the exposed nectaries. In a rainy, misty climate, the extrafloral nectar may have only 5 to 10% sugar (on a wt/wt basis), but in a sunny, dry climate extrafloral nectars are frequently in excess of 50% sugar. The sugar concentrations of extrafloral nectars show a much wider range than floral nectars of the same species, even when collected from nectaries protected by bags.[205]

Amino acid complements of floral and extrafloral nectar produced on the same plant usually differ,[11] but there are species in which both have the same essential amino acids (e.g., *Inga brenesii*[205]). Certain amino acids that are only sometimes represented in floral nectars are more common in extrafloral nectars, especially cysteine/cystine and lysine, asparagine, and tyrosine,[16] presumably relating to the different nutritional requirements of ant or wasp "guards", and of pollinators.

D. DISTRIBUTION OF EXTRAFLORAL NECTARIES

1. Taxonomic

While not common, nectaries occur in ferns and are equivalent to extrafloral nectaries of angiosperms. Nectaries have been described in 11 genera of ferns.[209] The term *extrafloral* cannot really be used for ferns, since they do not have flowers; the term *extrasoral* has been suggested,[209] but perhaps the best moniker is simply *nectaries* for ferns.

Nectaries are unknown from gymnosperms.[188] Bentley's review[33] listed 39 families of flowering plants in which there are species with extrafloral nectaries. Elias' review[106] shows extrafloral nectaries to occur in 68 angiosperm familes in 35 orders (approximately one fourth of all flowering plant families). New reports include the presence of extrafloral nectaries in the Myrsinaceae,[266] Bromeliaceae, and a variety of subtropical and tropical species that have only recently been discovered to have extrafloral rewards (by observing live plants in the lab or greenhouse, rather than plants in the field with formless, nonobvious nectaries with all reward removed [87,88,311] (personal observation). A list of 93 families and 332 genera in which extrafloral nectaries have been reported to occur is presented in Table 1.

Floral nectaries are not a prerequisite for extrafloral nectaries in a plant species; extrafloral nectaries occur both on plants with floral nectaries (*Passiflora, Inga, Vicia*) and on plants without them (*Acacia, Mimosa, Anadenanthera, Cassia*).

In addition to nectaries that are morphologically part of the plant, arthropod-induced galls can also produce nectar.[348] Pathogens can also induce secretions similar to extrafloral nectar: Percival[278] includes the ergot fungus (*Claviceps purpurea*) and the pycnidia of the rust *Puccinia poarum* in a summary of nectar sources.

2. Life Form

Extrafloral nectaries are found on representatives of all plant life forms: herbs, shrubs, trees, vines, lianas, and epiphytes. They are commonly active on developing parts (e.g., new leaves, flowers, and fruits).

Annual plants (with only one growing season prior to reproduction and death) tend to have extrafloral nectaries in positions that can directly enhance the protection of flowers and fruit (see Section III.F.3); an annual plant may thereby maximize seed production during the single reproductive event.[206] Examples of this are *Vicia sativa*[199] and *Cassia fasciculata*.[19,189] Perennial herbs also often have inflorescence-associated extrafloral nectaries: *Paeonia*,[277] *Costus* spp.,[307,308] and *Calathea ovandensis*.[157]

Vines use other plants for support and, therefore, contact vegetation at many points. Bentley[34] concluded that this life form was more likely than others to capitalize on protection by crawling ants visiting extrafloral nectaries (see Section III.F). Studies of *Passiflora*[320,321] and *Byttneria*[153] vines have revealed that a variety of potentially beneficial insects visit the extrafloral nectaries, many of them flying to the nectaries (parasitoid wasps and flies, see Section III.G), in addition to crawling ants.

A variety of epiphytic flowering plants[87] have extrafloral nectaries; some of these are inhabitants of ant gardens,[196] providing ants with nectar (see Sections III.C. and III.D). Some epiphytic ferns[209] have nectaries, though their ecological role has not yet, to my knowledge, been studied.

3. Geographic

Extrafloral nectaries can be found on plants in both the tropics and the temperate zone, though they are more common in tropical areas (Table 2). Within a geographic area, the cover of plants with extrafloral nectaries usually varies among[182,204,250,276] and within[277] habitats. Some biologists have sought to correlate extrafloral nectary-bearing plant abundance with the abundance or activity of ants at different sites,[31,179,204,207] and in general, have found fewer nectary-bearing individuals in areas with few or no ants.

II. RECIPROCAL EFFECTS

A. EFFECTS OF NECTAR ON INSECTS

1. Nectar as an Attractant and Reward

Since sugars provide a quick and easily utilized source of energy for many kinds of animals (invertebrate and vertebrate), nectar is an appealing and attractive commodity when presented by a plant. The term *attractant* has been used by some biologists to describe the function nectar serves initially for pollinators to flowers,[112] but since the nectar is usually discovered after the animals are attracted by some other means (visual displays, odors, or movement), the term *reward* is more appropriate.

The discovery of the presence of nectar is a compelling force for animals to seek out and use more of the nectar. Agronomists have found that crop cultivars without nectaries will receive better pollination if interplanted with cultivars with nectaries.[44] Ecologists have found that crop plants without nectaries can benefit from beneficial insects attracted and supported by nectar-producing weeds growing around the fields.[5] In addition, ants will recruit to new nectar sources, e.g., artifical nectaries on beans.[31]

Many extrafloral nectaries are colored, providing a visual cue for animals to locate the nectar (*Vicia*, *Pteridium*, *Costus*, *Cassia*, *Lysiloma*). In many, the presence of

Table 1
TAXONOMIC DISTRIBUTION OF PLANTS WITH EXTRAFLORAL NECTARIES

Family (order): Genus (position of nectaries)	Ref.
PTERIDOPHYTA all families combined	
Angiopteris (1)	45
Cyathea (1)	45, 310
Drynaria (1)	45, 83, 131
Hemitelia (Cnemidaria) (1)	45, 310
Holostachyum (1)	358
Merinthosorus (1)	358
Platycerium (1)	233
Photinopteris (1)	233
Polybotrya (1)	209
Polypodium (1)	209
Pteridium (1)	149, 150, 151, 213, 230, 215, 217, 271, 288, 336
ACANTHACEAE (Asteridae: Scrophulariales)	
Aphelandra (7)	92, 102
AGAVACEAE (Liliidae: Liliales)	
Sansevieria (5)	16, 335
Yucca (8)	178
ANACARDIACEAE (Rosaceae: Sapindales)	
Family	106
Holigavna (1)	359
ANCISTROCLADACEAE (Dilleniidae: Theales)	
Ancistrocladus (1)	245
ANNONACEAE (Magnoliidae: Magnoliales)	
Annona (8)	personal observation
APOCYNACEAE (Asteridae: Gentianales)	
Plumeria (2)	248
Thevetia (9)	personal observation
ARACEAE (Arecidae: Arales)	
Philodendron (2)	87, 88
Spathodea	293
ASCLEPIADACEAE (Asteridae: Gentianales)	
Calotropis (1,2)	9, 136
Ceropegia (1)	310
Dregea (= *Wattakaka*)(1,2)	9, 136
ASTERACEAE (Asteridae: Asterales)	
Centaurea (1)	178, 359
Helichrysum (1,7)	261
Helianthella (7)	163
Helianthus (1,7)	178, 277
BALSAMINACEAE (Rosidae: Geraniales)	
Impatiens (1,2,4)	16, 109, 277, 323

Table 1 (continued)
TAXONOMIC DISTRIBUTION OF PLANTS WITH EXTRAFLORAL NECTARIES

Family (order): Genus (position of nectaries)	Ref.
BIGNONIACEAE (Asteridae: Scrophulariales)	
Family	97
Adenocalymma (8)	110
Amphilophium (8)	110
Amphitecna (8)	110
Anemopaegma (6,8)	16, 110
Arrabidaea (1,8)	110, 266
Bignonia (1,6)	160, 310, 331
Campsis (1,5,6,8)	107, 108, 328
Catalpa (1,5,8)	277, 310, 326
Chilopsis (1,6,7)	276
Crescentia (8)	110
Cybistax (1)	266
Distictella (8)	110
Haplophragma (8)	110
Incarvillea (8)	110
Jacaranda (1,8)	250
Kigelia (3)	82, 331
Melloa (8)	110
Memora (8)	110
Newbouldia (1)	310
Pachyptera (7,8)	110
Spathodea (1,8)	110, 310
Tabebuia (1)	268
Tanaecium (8)	110
BIXACEAE (Dilleniidae: Violales)	
Bixa (4,6)	32
BOMBACACEAE (Dilleniidae: Malvales)	
Family	337
Eriotheca (2)	266
Ochroma (1)	259, 260
BROMELIACEAE	
Tillandsia (5)	Keeler, pers. comm.
BURSERACEAE (Rosidae: Sapindales)	
Protium	250
CACTACEAE (Caryophyllidae: Caryophyllales)	
Ancistrocactus (4)	239
Epiphyllum (8)	87, 88
Ferocactus (4, areolar)	42, 277, 304
Opuntia (4, areolar)	276, 279, 359
CAPPARIDACEAE (Dilleniidae: Capparales)	
Family	106
Capparis (1)	359
Cleome (6)	16
CAPRIFOLIACEAE (Asteridae: Dipsacales)	
Sambucus (1,4)	81, 113, 128
Viburnum (1,2,3)	277

Table 1 (continued)
TAXONOMIC DISTRIBUTION OF PLANTS WITH EXTRAFLORAL NECTARIES

Family (order): Genus (position of nectaries)	Ref.
CARYOCARACEAE (Dilleniidae: Theales)	
Caryocar (6)	263, 266
CARYOPHYLLACEAE (Caryophyllidae: Caryophyllales)	
Silene	80 (suspect — Pemberton)
CEPHALOTACEAE (Rosidae: Rosales)	
Cephalotus (1)	172, 173
CHRYSOBALANACEAE (Rosidae: Rosales)	
Chrysobalanus (1)	340
Hirtella (1)	268, 310
Licania (1)	266
Parinari (1)	310
COMBRETACEAE (Rosidae: Myrtales)	
Conocarpus (2)	190
Laguncularia (2)	340, pers. obs.
Terminalia (2)	10, 266, 268, 359
CONVOLVULACEAE (Asteridae: Polemoniales)	
Cuscuta (1,2,4)	306
Ipomoea (2,6)	23, 177, 181, 188, 329
Pharbitis (*Ipomoea*) (2)	277
COSTACEAE (Commelinidae: Zingiberales)	
Costus (7)	307, 308
CRASSULACEAE (Rosidae: Rosales)	
Adromischus (7)	16
CUCURBITACEAE (Dilleniidae: Violales)	
Family	262
Cayaponia (2)	310
Cucurbita (1)	158, 159, 277
Lagenaria (2)	277, 359
Luffa (1,6)	277, 359
Momordica (1,5)	106, 277, 359
Siolmatra (2)	310
Trichosanthes (1)	277
DICHAPETALACEAE (Rosidae: Celastrales)	
Family	106
Dichapetalum (1)	106, 359
DIONCOPHYLLACEAE (Dilleniidae: Violales)	
Family	245
Dioncophyllum	324
Triophyphyllum (1)	245, 310
DIOSCOREACEAE (Liliidae: Liliales)	
Dioscorea (2)	36, 73, 138, 269, 270, 277, 346, 310

Table 1 (continued)
TAXONOMIC DISTRIBUTION OF PLANTS WITH EXTRAFLORAL NECTARIES

Family (order): Genus (position of nectaries)	Ref.
DIPTEROCARPACEAE (Dilleniidae: Theales)	
Anisoptera (1)	310
Dipterocarpus (1)	129
Dryobalanops (1)	129
Hopea (1)	129, 310
Monotes (1)	310
Parashorea (1)	310
Shorea (1)	129, 310
Upuna (1)	310
Vateria (1)	310
Vatica (1)	129
EBENACEAE (Dilleniidae: Ebenales)	
Diospyros (1)	10, 70, 268, 277, 359
Euclea (1)	70
Maba (1)	70
Royena (1)	70
ERICACEAE (Dilleniidae: Ericales)	
Vaccinium (1)	277
EUPHORBIACEAE (Rosidae: Euphorbiales)	
Acalypha	310
Alchornea (2)	25, 250
Aleurites (1,2)	28, 29, 310
Aparisthmium	250
Bertya	310
Caperonia	310
Cephalocroton	310
Claoxylon	310
Conceveiba	250
Cnidoscolus (1,2)	Pemberton, pers. comm.
Croton (1,2)	38, 93, 96, 250, 310
Crotonogyne (1)	310
Excoecaria	310
Hevea	79, 123, 272, 310
Hieronyma (1)	268
Hippomane (1)	pers. obs.
Homolanthus (1,2)	352
Hura (1)	106, 359
Jatropha (1,3)	310
Macaranga (1,2)	310, 352
Mallotus (1,2)	26, 277
Manihot (1)	Pemberton, pers. comm.
Mareya	310
Micrandra	310
Pedilanthus (1)	86
Poinsettia (5)	345
Richeria (2)	268
Ricinodendron	310
Ricinus (2,5)	17, 139, 176, 277, 310, 359
Sapium (1,2)	277, 310
Sebastiana (3)	310
Tragia	310

Table 1 (continued)
TAXONOMIC DISTRIBUTION OF PLANTS WITH EXTRAFLORAL NECTARIES

Family (order): Genus (position of nectaries)	Ref.
FAGACEAE (Hamamelidae: Fagales)	
Quercus alba (1)	Robert J. Marquis, pers. comm.
FLACOURTIACEAE (Dilleniidae: Violales)	
Idesia (2)	277, 359
FOUQUIERIACEAE	
Fouqueria splendens (6)	276
GESNERIACEAE (Asteridae: Scrophulariaceae)	
Codonanthe (1,4,6,8)	87, 88, 195, 357
GOODENIACEAE	
Scaevola (1)	Pemberton, pers. comm.
HUMIRIACEAE (Rosidae: Linales)	
Family	106
Vantanea (1)	359
HYDRANGEACEAE (Rosidae: Rosales)	
Family (rare)	106, doubtful—Pemberton
ICACINACEAE (Rosidae: Celastrales)	
Emmotum	250
IRIDACEAE (Liliidae: Liliales)	
Iris (6)	277
LEGUMINOSAE	
Acacia (1)	48, 49, 52, 57, 167, 236, 237, 238
Affonsea (1)	359
Albizzia (1)	213, 277, 310
Anadenanthera (1)	266, 268
Bauhinia (1,3)	16, 266
Brownea (7)	62, 241
Caesalpinia (1)	10, 277, 310
Calliandra (1)	359
Calpocalyx	241, 310
Canavalia (2,5)	10, 359
Cassia (1,2,5)	19, 43, 189, 213, 241, 266
Copaifera	310
Crotalaria (3,10)	10, 16, 241, 359
Cylicodiscus	310
Delonix (2)	241, 359
Desmanthus (1)	178
Dichrostachys (2)	310
Dipteryx	250
Dolichos (3,5)	277, 359
Entada (3,4)	10, 220, 310
Enterolobium (1)	266, 268, 310
Erythrina (3,6)	220, 316, H. Hernandez, pers. comm.

Table 1 (continued)
TAXONOMIC DISTRIBUTION OF PLANTS WITH EXTRAFLORAL NECTARIES

Family (order): Genus (position of nectaries)	Ref.
LEGUMINOSAE (continued)	
Gagnebina (1)	359
Gilbertiodendron (1)	310
Glottidium (1)	241
Glycyrrhiza (1)	178
Glycine (1)	241, 359
Hardenbergia (3)	213
Hippocrepis (3)	241
Humboldtia (1,3)	241
Hymenaea (1)	310
Inga (1)	201, 202, 250, 310
Lablab (3)	241, 359
Leucaena (2)	241
Leonardoxa (1)	105, 240
Lotus (3)	241, 285
Lysiloma (1)	359
Macrolobium (1)	310
Mimosa (1)	250, 310, 359
Mucuna (3)	220
Newtonia	310
Parkia (1)	310
Pellegriniodendron (1)	310
Pentaclethra (1,4)	16, 30, 241
Phaseolus (3,5)	277, 359
Physostigma (5)	241, 359
Piptadenia (1)	310, 359
Pithecellobium (1)	85, 104, 250, 310
Plathymenia (1,4)	266, 268, 359
Prosopis (1,2)	276
Robinia (3)	277
Senna (1,2)	241, 359
Serianthes (1)	359
Sesbania (1)	241
Stryphnodendron (1)	266, 268, 359
Vicia (3,6)	117, 199, 208, 211
Vigna (3,5)	210, 273
Xylia (1)	359
LILIACEAE (Liliidae: Liliales)	
Hemerocallis (6)	277
LINACEAE (Rosidae: Linales)	
Linum	106 — doubtful
LOASACEAE (Dilleniidae: Violales)	
Mentzelia (9)	184
LOBELIACEAE	
Centropogon (6)	B. Stein, pers. comm.
LOGANIACEAE (Asteridae: Gentianales)	
Family (rare)	106
LYTHRACEAE (Rosidae: Myrtales)	
Lafoensia (1)	266

Table 1 (continued)
TAXONOMIC DISTRIBUTION OF PLANTS WITH EXTRAFLORAL NECTARIES

Family (order): Genus (position of nectaries)	Ref.
MALPIGHIACEAE (Rosidae: Polygalales)	
Acridocarpus (1)	310
Aspicarpa (1)	310
Gaudichaudia (1)	310
Heteropteris (1)	266, 359
Stigmaphyllon (3)	16, 359
MALVACEAE (Dilleniidae: Malvales)	
Family	166
Alyogyne (1)	313
Cienfuegosia (1)	310
Decaschistia (1)	310
Dicellostyles (1)	310
Gossypium (1,7)	1, 2, 161, 251, 310, 344
Hibiscus (1,6)	10, 277, 310
Julostylis (1)	310
Kydia (1)	310
Thespesia (1)	310
Urena (1)	310
MARANTACEAE (Commelinidae: Zingiberales)	
Calathea (7)	157, 309
Ischnosiphon (7)	16
MARCGRAVIACEAE (Dilleniidae: Theales)	
Marcgravia (5)	290
Norantea (1,5)	268
MELASTOMATACEAE (Rosidae: Myrtales)	
Family (rare)	106
MELIACEAE (Rosidae: Sapindales)	
Carapa (1)	222, 287
Cipadessa (1)	221
Swietenia (1)	222
MORACEAE (Hamamelidae: Urticales)	
Ficus (8)	69, 250
MUSACEAE (Commelinidae: Zingiberales)	
Family (occasional)	106
MYRSINACEAE (Dilleniidae: Primulales)	
Myrsine (1)	Pemberton, pers. comm.
Rapanea (1)	266
NEPENTHACEAE (Dilleniidae: Sarraceniales)	
Nepenthes (1)	172, 173, 359
OCHNACEAE (Dilleniidae: Theales)	
Ouratea (1,3)	266
OLEACEAE (Asteridae: Scrophulariaceae)	
Ligustrum (1)	277
Osmanthus (1)	248, 277
Syringa (2)	277

Table 1 (continued)
TAXONOMIC DISTRIBUTION OF PLANTS WITH EXTRAFLORAL NECTARIES

Family (order): Genus (position of nectaries)	Ref.
ONAGRACEAE (Rosidae: Myrtales)	
Oenothera (1,5,8)	178
ORCHIDACEAE (Liliidae: Orchidales)	
Family	170
Ansellia (4,5)	20
Arundina (6,8)	187
Aspasia (5,6)	119
Brassavola (4,5,6)	119
Catasetum (4,5,6)	119
Caularthron (4,5,6,8)	119
Dimerandra (4,5)	119
Encyclia (4,5)	119
Epidendrum (4,5)	20, 119
Eulophiella (8)	16
Gongora (4,5)	119
Notylia (5,6)	119
Oncidium (4,5)	119
Scaphyglottis (5)	119
Schomburgkia (5,6,8)	298, 299, 300
Spathoglotis (6,8)	165
Trigonidium (4,5)	119
Vanilla (5)	119
PAEONIACEAE (Dilleniidae: Dilleniales)	
Paeonia (6)	277, 359
PASSIFLORACEAE (Dilleniidae: Violales)	
Adenia (1)	310, 359
Crossostema (1)	310, 359
Deidamia (1)	310, 359
Hollrungia (1)	310, 359
Passiflora (1,2,7)	100, 242, 310, 319, 320, 321, 359
Smeathmannia (1)	310
PEDALIACEAE (Asteridae: Scrophulariales)	
Ceratotheca (10)	249
Sesamum (10)	249, 277
Rogeria (10)	249
Sesamothamnus (10)	249
PLUMBAGINACEAE (Caryophyllidae: Plumbaginales)	
Family	106
POACEAE (Commelinidae: Cyperales)	
Andropogon (1)	41, 50, 51
Eragrostis (1,4)	277, 359
POLYGALACEAE (Rosidae: Polygalales)	
Family	106
POLYGONACEAE (Caryophyllidae: Polygonales)	
Polygonum (2)	178, 277, 305
Reynoutria (2)	277

Table 1 (continued)
TAXONOMIC DISTRIBUTION OF PLANTS WITH EXTRAFLORAL NECTARIES

Family (order): Genus (position of nectaries)	Ref.
PROTEACEAE (Rosidae: Proteales)	
Adenanthos (1)	213
Leucadendron (1)	213
Leucospermum (1)	213
PUNICACEAE (Rosidae: Myrtales)	
Punica (1)	277
RHAMNACEAE (Rosidae: Rhamnaceae)	
Family	106
Colubrina (1)	340
RHIZOPHORACEAE	
Rhizophora (1)	Pemberton, pers. comm.
ROSACEAE (Rosidae: Rosales)	
Crataegus (1)	310
Cydonia (1)	310
Prunus (1,6,7)	98, 198, 274, 277, 291, 310, 339
Pyrus (1)	310
Rosa (2,3,5)	178, 277
Sorbus (1)	310
RUBIACEAE (Asteridae: Rubiales)	
Alibertia	250
Amaioua	250
Faramea	250
Hamelia (9)	33
Morinda (9)	Pemberton & Koptur, pers. obs.
Myrmecodia (9)	88
Sickingia (8)	J. Stout, pers. comm.
Tocoyena (6)	266, 268
RUTACEAE (Rosidae: Sapindales)	
Zanthoxylum (1)	277
SALICACEAE (Dilleniidae: Salicales)	
Populus (1,2)	76, 77, 277, 343
Salix (1,2)	178
SAPINDACEAE (Rosidae: Sapindales)	
Family (rare)	106
SARRACENIACEAE (Dilleniidae: Sarraceniales)	
Darlingtonia (1)	172, 173
Drosera (1)	137, 342
Heliamphora (1)	172, 173
Sarracenia (1)	172, 173
SCROPHULARICEAE (Asteridae: Scrophulariales)	
Melampyrum (1)	310
Paulownia (1,6)	277, 359

Table 1 (continued)
TAXONOMIC DISTRIBUTION OF PLANTS WITH EXTRAFLORAL NECTARIES

Family (order): Genus (position of nectaries)	Ref.
SIMAROUBACEAE (Rosidae: Sapindales)	
Ailanthus (1)	46, 67, 178, 277
Simarouba	250
SMILACACEAE (Liliidae: Liliales)	
Smilax (1)	10
SOLANACEAE (Asteridae: Solanales)	
Markea (6,8)	87,88
Solanum (1,2,4)	6, 178
STERCULIACEAE (Dilleniidae: Malvales)	
Ayenia (1)	74
Byttneria (1)	8, 153
STYLIDIACEAE (Asteridae: Campanulales)	
Stylidium (1)	212
TILIACEAE (Dilleniidae: Malvales)	
Grewia (1)	277
Triumfetta (1)	pers. obs.
TURNERACEAE (Dilleniidae: Violales)	
Turnera (2)	111
URTICACEAE (Hamamelidae: Urticales)	
Pilea (1)	27
VERBENACEAE (Asteridae: Lamiales)	
Aegiphila (1)	266
Amasonia (1)	235, 246
Avicennia (2)	114, 297
Baillonia (1)	235, 246
Callicarpa (1)	235, 277
Casselia (1)	235, 246
Citharexylum (1)	235, 246
Clerodendron (1)	10, 235, 246, 277
Dipyrena (1)	235
Duranta (1)	235
Faradaya (1)	10, 235
Lampaya (1)	235
Monochilus (1)	235
Raphithamnus (1)	235
Stachytarpheta (1)	235
Vitex (1)	250
VITACEAE (Rosidae: Rhamnales)	
Cissus (3)	16
VOCHYSIACEAE (Rosidae: Polygalales)	
Callisthene (4)	268
Qualea (4,5)	265

Table 1 (continued)
TAXONOMIC DISTRIBUTION OF PLANTS WITH EXTRAFLORAL NECTARIES

Note: Families are listed alphabetically, with order designated afterward. Genera in which nectaries have been reported in at least one species are listed; however, not all species in these genera necessarily have extrafloral nectaries. References for each species are not exhaustive, but only examples. The position of nectaries, if known, is indicated by numbers: 1 = on leaf; 2 = on petiole; 3 = on stipules and/or stipels; 4 = on stems; 5 = on pedicels, peduncles, or stems of inflorescence; 6 = on petals or sepals; 7 = on bracts; 8 = on fruit, capsule, or pod; 9 = on ovary (postfloral); 10 = aborted flowers or bud scars.

glistening droplets of nectar may provide a visual stimulus. While many flowers have odors that serve to attract pollinators, floral and extrafloral nectars themselves do not have obvious odors. Olfaction is probably not involved in the detection of locations of extrafloral nectaries.

2. Insect Nutrition

The same chemicals required for growth of a protozoan or cells of a vertebrate are also required by the cells of an insect.[144] Phytophagous insects require the following components in their diets: water, carbohydrates, proteins and amino acids, lipids, water-soluble vitamins, minerals, water, nucleic acids, purines, and pyrimidines.[125] In addition, sterols and certain fatty acids are required by most species.[144] Many insect predators (members of the Neuroptera, Heteroptera, Coleoptera, Lepidoptera, and Hymenoptera) eat both insects and noninsect foods.[143] Predators and parasitoids have the same qualitative nutritional requirements for larval growth and development as phytophagous insects,[143] but quantitatively predators require a much higher ratio of protein to carbohydrate in their diet (1:1 in species studied) than plant feeders.

Insects from 10 orders have been documented taking nectar or visiting nectaries (Table 3). The most commonly noted insect visitors to extrafloral nectaries are Hymenoptera, followed by Diptera and Lepidoptera. Nectars and honeydew are foods for many entomophagous insects[142] and are important supplements to encourage the control of plant pests by beneficial insects. Some insects may eat only plant or animal tissue as larvae, but nectar as adults; and others get nutrition from other sources, in addition to nectar (pollen, fruit, animal tissue, secretions, blood) throughout their lives. The extent to which an insect depends upon nectar in its diet is related in part to the nutritional value of the nectar. For example, floral nectars used by long-lived adult Sphingidae (Lepidoptera), which feed on only nectar, have lipids that may replenish the fat body and prolong the life of the moth.[205] Certain Lepidoptera (*Heliconius*) enrich their diet of nectar by soaking pollen grains in nectar, increasing the amino acid concentration of the solutions they imbibe.[132]

Nectar use increases the longevity of many beneficial insects. The longevity of the parasitoid *Campoletis sonorensis* was reduced, and its host was parasitized less on nectarless than nectar-producing cotton.[229] *Geocoris* spp. (big-eyed bugs) survived longer when caged on nectar-bearing *Polygonum* as opposed to nectarless plants.[64] When prey was scarce, the longevity of *Geocoris pallens* fluctuated according to the availability of extrafloral nectar on cotton.[90] Predatory mites may take plant exudates[94] but rely primarily on pollen and phytophagous mites for their nutrition.

Nectar provides an important source of nutrients for adult Lepidoptera. The size of the fat body in adult Lepidoptera is related to larval size and larval nutrition, and larval nutrition may generally determine the reserves available for oogenesis.[66] Nonetheless, the longevity of many adult Lepidoptera is positively influenced by nectar consumption, thereby prolonging reproductive life; water and nutrients assimilated by adult females

Table 2
COVER OF PLANTS WITH EXTRAFLORAL NECTARIES IN DIFFERENT LOCATIONS – TEMPERATE TO TROPICAL

% Mean cover	Location	Vegetation types	Ref.
0%	N. California, USA	Native grassland 0% Riparian forest 0% Deciduous forest 0% Chapparal 0%	180
0 – 14.2%	Nebraska, USA	Deciduous forest 1.8% Riparian forest 1.3% Tallgrass prairie 0% Sandhills prairie 8.3–14.2%	182
0 – 27.8%	S. California, USA	Desert bush scrub 0.1–6.6% Desert wash 23.9–27.8% Sand dunes 0–1.4% Yucca-agave 3.7%	276
7.5 – 55%	Korea	Deciduous forest 7.5%, 23%, 55%	277
2 – 34%	Everglades Florida, USA	Sawgrass prairie 2% Rockledge pinelands 34% Hardwood hammock 23%	204
0 – 28%	Jamaica	Forested second growth (0 m) 28% Same, at 1310 m 0%	179
7.6 – 20.3%	SE Brazil	Cerrado (woody spp. only)	266
21.6–31.2%	SW Brazil	Cerrado (woody spp.)	266, 268
17.6–53.3%	Amazonian Brazil	Terra firme forest 19.1% Successional forest 42.6% Buritirana 29.7% Shrub canga 50%	250
10–80%	Costa Rica	Tropical dryforest hillside 40–80% Tropical riparian forest 10–40%	31
0–22%	Costa Rica	Lowland rain forest (0 m) 1–8% Lower montane cloud forest (1500 m) 3–22% high montane oak forest (3000 m) 0–3%	207

support metabolism for egg maturation.[66] For example, sugars in the adult diet enhance female longevity and the number of egg clusters in *Euphydras editha*.[103] DeVries and Baker[93] found that the ant-tended caterpillars of the myrmecophilous riodinid butterfly *Thisbe* that drank extrafloral nectar grew substantially larger than their counterparts that did not drink nectar. Presumably these larvae gave rise to larger adult females that could lay more eggs.

Table 3
ARTHROPOD NECTAR DRINKERS

	Family (or superfamily)	Ref.
ARANEAE		
	Oxyopidae	1
	Salticidae	153
DIPTERA		
	Agromyzidae	153
	Aulacigastridae	153
	Blepharoceridae	99
	Cecidomyiidae	92, 153
	Ceratopogonidae	99, 153
	Chironomidae	153
	Chloropidae	153
	Clusiidae	153
	Culicidae	99
	Dolichopodidae	153, 177
	Drosophilidae	153, 254
	Lauxaniidae	153
	Milichiidae	153
	Muscidae	177
	Otitidae	177
	Platystomatidae	177
	Psychodidae	99
	Richardiidae	153
	Sarcophagidae	177, 286
	Sciaridae	153
	Sciomyzidae	177
	Sepsidae	153, 177
	Simuliidae	99
	Somatiidae	153
	Stratiomyidae	177
	Syrphidae	63
	Tabanidae	99, 177, 349
	Tachinidae	177, 218, 315, 341
ORTHOPTERA		
	Blattidae	153
HEMIPTERA/HOMOPTERA		
	Cicadellidae	92, 153
	Fulgoroidea	153
	Membracidae	92
HEMIPTERA/HETEROPTERA		
	Anthocoridae	1, 63, 356
	Lygaeidae	1, 64, 177, 356
	Miridae	1,153
	Nabidae	1
	Pentatomidae	1, 177
	Reduviidae	153
NEUROPTERA		
	Chrysopidae	1, 63, 177, 333
	Hemerobiidae	63
	Mantispidae	177

Table 3 (continued)
ARTHROPOD NECTAR DRINKERS

	Family (or superfamily)	Ref.
COLEOPTERA		
	Brentidae	177
	Buprestidae	153
	Cantharidae	153, 177
	Carabidae	153
	Cerambycidae	177
	Chrysomelidae	153, 177
	Coccinellidae	1, 63, 153, 276
	Curculionidae	153, 208
	Dermestidae	254
	Elateridae	177
	Lampyridae	177
	Malachiidae	254
	Melyridae	1
	Nitidulidae	153
	Ostomatidae	177
	Scarabidae	338
	Tenebrionidae	177
DERMAPTERA		
	Earwigs	63
HYMENOPTERA		
	Apidae	89, 92, 121, 148, 177, 338
	Braconidae	1, 153, 202, 254
	Chalcididae	63, 153, 177, 254
	Cynipidae	153
	Diapriidae	63
	Encyrtidae	153
	Eulophidae	153
	Eumenidae	96, 254
	Eurytomidae	153
	Formicidae	
	Dolichoderinae	177, 199
	Formicinae	163, 177, 201
	Myrmecinae	163, 177, 201, 208
	Ponerinae	30, 177
	Pseudomyrmecinae	167, 177
	Halictidae	338
	Ichneumonidae	1, 65, 153, 208, 229
	Megachilidae	96
	Meliponidae	177
	Proctotrupoidea	153
	Pteromalidae	153
	Scelionidae	153
	Sphecidae	153, 177
	Thynninae	338
	Vespidae sensu lato	96, 177
LEPIDOPTERA		
	Lycaenidae	93, 96
	Morphidae	G. E. Martinez, pers. comm.
	Noctuidae	1, 2
	Pyralidae	1
	Riodinidae	93, 177
	Tortricidae	1

Most of the compounds dissolved or suspended in nectar are of nutritive value to insects. The most important of these components are sugars, amino acids, and lipids. Lipids in nectars may be metabolized by nectar-drinking insects, providing energy,[133] and in some cases they may be incorporated into the fat body, providing reserves that may prolong adult life. However, most extrafloral nectars do not contain much, if any, lipids. Some of the other organic constituents of nectar have been found to be of nutritional significance to certain herbivores when the compounds are contained in foliage (e.g., phenols[37]), but these compounds are present in relatively low concentrations in nectar. The water of the nectar can be of great value to insect visitors, especially in xeric habitats.[304]

a. Sugars

Carbohydrates provide energy and organic carbon for all nectar imbibers. When insects drink both nectar and blood, the blood is directed to the midgut, and the nectar sugars are directed to the crop.[47] Sugar solutions stimulate the chemoreceptors of many insects[91] and stimulate further feeding behavior.[55,247] In one study using stable isotope analysis, Fisher et al.[118] found that the carbon contribution of extrafloral nectar to ant diets ranged from 11 to 48%, differing with the ant species and season.

The three most common sugars in extrafloral nectar are the monosaccharides glucose and fructose, and the disaccharide sucrose;[15] other sugars found in nectars are arabinose, galactose, maltose, mannose, melezitose, melibiose, raffinose, and ribose (also cellobiose and stachyose in orchids).[15] Sucrose is the primary translocate in the phloem sap of plants, and the occurrence of other sugars in nectars indicates that nectar secretion is an active selective process, not just passive diffusion. Freshly secreted floral nectar of a plant species has a characteristic sugar composition and correlates with the primary pollinator type.[12,15] Freshly secreted extrafloral nectar may also be characteristic of a species, but since the nectar is immediately exposed to the drying and contaminating effects of the environment (microbes may hydrolyze sucrose to the component hexoses), the ratio of sucrose to hexose sugars is much more variable within a species. In plants with sucrose-dominated floral nectars, extrafloral nectars generally have much lower sucrose/hexose ratios. The sucrose/hexose ratio is, in general, lower for nectars more exposed to the environment, perhaps due to microbial degradation of sucrose: extrafloral nectars are generally not sucrose dominated, but have equal or greater amounts of the hexose sugars.

The preferences of extrafloral nectar drinkers for various sugars has been investigated in several species: fire ants,[295] argentine ants,[199] various ants,[140] and *Paraponera* ants.[30] Ants prefer higher concentrations of sugars over lower concentrations, and they perfer sucrose and glucose over fructose. *Heliothis* moths are most responsive to sucrose followed by fructose and glucose, but do not respond to ribose, rhamnose, and raffinose.[292] Honeybees prefer sucrose to fructose, and fructose to glucose.[347] Bees can utilize most sugars, but certain ones are not nutritionally beneficial (e.g., rhamnose, lactose, melibiose), and mannose is poisonous to honeybees.[148]

b. Amino acids

Amino acids in nectar provide the building blocks for proteins that may be valuable to insects that subsist primarily or exclusively on a nectar diet. All insects that have been studied require the same 10 amino acids (found also to be necessary for the rat): arginine, histidine, isoleucine, leucine, lysine, methionine, phenylalanine, threonine, tryptophan, and valine; and the same amino acids are required by adult insects for egg production.[144] However, disproportionately abnormal amounts of a single amino acid can inhibit growth.[275] Studies on honeybees have revealed 10 amino acids to be required by adult honey bees; early in life, these amino acids are obtained from pollen, and later they may be supplied by nectar.[148]

The significance of amino acids in the diet of nectar drinkers has been experimentally investigated[350] with *Colias* butterflies; they preferentially visit flowers that produce nectars rich in amino acids. The availability of amino acids extends butterfly production of maximum-sized eggs in later clutches,[253] while in lepidopteran species whose adults do not eat nitrogen-rich foods, age-dependent decreases in egg size reflect the depletion of nitrogen stores.[66]

The concentration of amino acids in nectar undoubtedly affects the taste of nectar.[16] Amino acids are detectable by the chemosensors of insects such as flies.[317] Flesh flies prefer sucrose solutions containing amino acids at concentrations comparable to those in floral nectar over plain sucrose;[286] they avoid sucrose solutions with histidine and lysine, but prefer glycine, at abnormally high concentrations. Similarly, fire ants prefer certain amino acids over others.[295] When given a choice between nectar solutions containing only sugar and nectar solutions containing sugars and amino acids, some insects have shown a preference for those containing both sugars and amino acids[164,214] (Koptur unpublished data).

Herbert and Irene Baker have observed that there seem to be two types of extrafloral nectars: those with high concentrations of amino acids and those with low concentrations. They suggested that this may correspond with the dietary habits of the ants that have coevolved with these plants: ants that are exclusively nectar drinkers would benefit from nectar rich in amino acids, whereas ants that eat insects and other animals would get their amino acids elsewhere and use nectar only as a source of liquid and ready energy. Hagen[142] postulated that the presence of some, but not all, the essential amino acids in some nectars may have evolved to ensure the presence of certain kinds of insects, but to prevent them from being totally dependent on nectar. When ants are not supplied all the amino acids they require for their broods, they become more aggressive in their patrolling and prey seeking.

3. Oviposition

A variety of Lepidoptera have been found to increase oviposition in proximity to nectar. Butterflies that oviposit on flowers may be stimulated to lay eggs when nectar is present[66] and may therefore lay more eggs in the vicinity of nectar sources.[252] Females of lycaenids and riodinids preferentially oviposit in the vicinity of ants,[10] whose presence may be regulated by nectar secretion. *Heliothis* moths that feed at cotton extrafloral nectaries are likely to oviposit on the plants.[232] Caterpillars of this species drink extrafloral nectar and may chew into cotton squares after nectar is consumed.[24] Cotton genotypes lacking nectaries are resistant to major lepidopterous pests.[231,232,354]

Insect predators often find their prey by first finding the suitable prey habitat and secondly the prey.[143] If adult parasitoids with entomophagous larvae are attracted to feed on extrafloral nectar, they may be more likely to discover suitable hosts for their offspring in the herbivore eggs and larvae feeding on the host plants[202,206] (see Section III.G).

4. Location

Ant nests are not located at random: while competition may be the overriding force in determining spatial dispersion of ant nests,[227] the location of food resources (e.g., nectar) may also be important determinants.[30,228,280,281,318] It is not unusual to find ant nests at the base of plants bearing extrafloral nectaries (e.g., *Chrysobalanus icaco* in south Florida) or plants that support Homoptera tended by ants (e.g., *Carex intumescens* in northern Michigan).

Bentley[31] demonstrated that by adding nectar droplets to bean plants, a protective ant force was recruited. A number of experiments have demonstrated that blocking or removing plant nectaries effectively decreases ant numbers on those plants.[43,149,199,336]

The presence of ants on plants with nectaries also influences the establishment of myrmecophytic epiphytes in which ants nest (e.g., *Solenopteris* ferns[134]). Davidson and Epstein[88] found that the principal plant inhabitants of ant gardens are over-represented on trees with extrafloral nectaries or dependable homopteran nectary analogs.

5. Population Biology

In order for beneficial insects that regulate problem pest insects to be established or recruited to an area, it may be necessary to have nectar resources to attract them and to support them through low pest times. Crops bordered by weeds or in weedy fields can fare better than well-manicured plantings because predators and parasitoids are more abundant.[4] Polycultures have more beneficial insects than monocultures,[224-226] due, in part, to the greater diversity of nectar rewards for these insects. Populations of 15 entomophagous arthropods were substantially larger in cotton with extrafloral nectaries compared with nectarless cotton cultivars.[1] However, removing nectaries of cotton has greatly reduced populations of certain homopteran[312] and lepidopteran[1,146,232] pests that are also attracted by extrafloral nectar.

6. Insect Behavior

The discovery of nectar resources by ants prompts a variety of responses: feeding, collecting, recruitment, and territoriality. Some ants are solitary foragers, and when they encounter nectaries will drink until replete; encounters with other insects during that time may result in some territorial aggressiveness (e.g., *Paraponera* on *Pentaclethra*[30]) but not necessarily. Worker ants treat extrafloral nectaries in the same way that they treat honeydew-producing insects; the more aggressive species defend the active nectaries, in many cases extending their territory over the entire plant.[156] In group foraging species, when one worker discovers nectar, it will drink until replete and then return to the nest and/or mark a scent trail to the resource, recruiting more individuals to take advantage of the nectar. Discovery of and recruitment to baits has been used to indicate ant activity and abundance at different sites.[31,179,202,204,208]

Resources of adequate quantities (determined by ant species and temperament) justify defending against interlopers, and this is the basis for ant defense of plants. While other insects alighting on plants may have no interest at all in nectar, the ants will remove them from the plant surface to protect their nectar resources. Many ant species may also utilize other insects on plants as food; the nectar serves to attract the ants to the foliage, where they wander around and discover insect prey. Ant-tended insects may not always benefit from ant care;[78] when times are tough, ants may eat their aphids rather than simply tend them for their honeydew (B. Edinger, personal communication).

Certain folivorous caterpillars exhibit the unusual behavior of positioning themselves on a plant not only to feed on leaves, but with their head in the extrafloral nectary.[93] Tiny *Apion* weevils feeding on foliage of *Vicia sativa* frequently drink nectar[208] and are camouflaged against the nectary, which is dark colored in contrast to the leaf.

B. EFFECTS OF INSECTS ON NECTAR

1. Nectar Quantity

There are two strategies in plants using floral nectar as a reward for pollinators: nectar is produced one time, and when removed, it is gone; or nectar is replenished as it is used, for the duration of the functional life of the flower. Animal visitors to flowers of the second strategy, therefore, have an effect on nectar quantity: the more the flowers are visited, the more nectar they produce.[75,200] Extrafloral nectars usually function for a much longer time than floral nectaries, e.g., over the period of expansion and maturation of new foliage (3 to 4 weeks) vs. 1 to 3 d of average flower duration. Most plants with extrafloral nectaries produce nectar in the absence of insects (e.g., in insect-free greenhouse conditions), and some accumulate large viscous quantities at the nectaries, obviously not dependent on insect removal or stimulation for continued secretion (e.g., *Turnera*). Other plants secrete small quantities of nectar to fill the nectaries, and replenish these small amounts when the nectar is removed (e.g., *Inga*).

Homoptera may decrease the amounts of extrafloral nectar produced, since a substantial amount of translocate is being diverted through these insects.[95] Ant-tended Homoptera are insect analogs of extrafloral nectaries, over which the host plant has no control, unlike real extrafloral nectaries.[206]

Herbivorous insects may increase the volume of extrafloral nectar secretion by their foliage-feeding activity. Sphinx caterpillars feeding on *Catalpa* resulted in greater nectar secretion on damaged leaves than on leaves without caterpillars.[326] Moderate levels of simulated damage to leaves of *Vicia* resulted in increased nectar production in stipular nectaries the day following the damage.[203] Experimental herbivory on leaves of *Campsis radicans* caused an increase in extrafloral nectar production on adjacent flower buds.[328] This propensity may constitute an inducible defense for certain plants that can respond to damage from herbivores by attracting more biotic protection (see Sections III.F and III.G).

2. Nectar Quality

Certain pollinators can introduce agents into floral nectar that alters the composition of the nectar, making it more suitable for a different guild of pollinator. Insect visitors to flowers may add amino acids by direct contact, salivation, damaging tissue causing cell leakage, and dislodging pollen into the nectar.[353] Hawkmoths that take the sucrose-rich nectar of *Inga vera* in the early evening may introduce fungi and bacteria into the flowers, which hydrolyze the sucrose into its hexose components, making the flowers more suitable for the bat visitors that come later.[15] Insects carry spores of yeasts and bacteria; *Drosophila* are primary vectors of yeasts, while muscoids, dolichopodids, and other insects frequently carry molds.[325] Bumblebees disperse spores of *Ustilago* fungus, the causal agent of anther smut disease, while gathering nectar and pollen from *Viscaria*.[171]

Extrafloral nectaries are, in general, more open and exposed than floral nectar, and they are more subject to the effects of the surrounding microclimate; in low humidity, nectars can be very highly concentrated. Perhaps because they are open to the introduction of spores and microbes from the air, which cause the hydrolysis of sucrose, extrafloral nectars tend to be hexose dominated when sampled. When extrafloral nectar is allowed to accumulate for days unsampled by visitors and open to the evaporative effects of the environment, nectar can become quite highly concentrated and viscous, and may even become so sticky that certain insects scrape it from the nectaries (personal observation).

Careful nectar analyses have revealed that certain plants may alter the composition of their extrafloral nectar in response to herbivore damage. The amino acids of *Impatiens sultani* nectar change in their relative concentrations subsequent to artificial defoliation,[323] perhaps to make the nectar more attractive to insect protectors of the plant.

3. Insect-Induced Secretory Structures

Certain internally feeding, gall-forming insects occupy plant parts that subsequently develop nectaries. Cynipid gall wasps attacking oaks occupy round galls covered with glands,[348] and the ants visiting these glands protect the wasp larvae from their parasitoids.

Feeding lesions due to insects or other animals may weep plant sap, which, if well supplied by phloem, is a nectarlike substance. Sapsucker holes in tree trunks are visited by hummingbirds (which normally visit flowers) and a variety of insects. Wasps visit wounds on the stems of *Baccharis* to collect the nectarlike sap.[254]

III. INTERACTIONS BETWEEN INSECTS AND PLANTS

A. POLLINATION

1. Extrafloral Nectar as Pollinator Reward

Although extrafloral nectaries are defined as usually not being involved with pollination, there are some exceptions. In some plant genera, the nectaries are outside of the individual flowers, but are a part of the inflorescence, and visitors to these nectaries contact floral organs, carry pollen, and pollinate flowers. The best-known example of this is *Poinsettia* (Euphorbiaceae), the showy blossoms of which are made up of brightly colored bracts subtending a small inflorescence of unisexual flowers, including some large nectaries.[345] *Norantea* (Marcgraviaceae) is an epiphytic shrub with a pendant inflorescence, the individual flowers of which have long pedicels; at the base of each pedicel is a large, cup-shaped nectary.[268] The birds, bats, or other mammals that visit these nectaries must touch the flowers as they poke through to the nectaries.

Foliar nectaries serve to attract visitors that inadvertently contact the floral parts of some plant species. Floral nectaries have not been found in Australian *Acacia* species.[39] Bernhardt[40] concludes that *Acacia* are generally entomophilous, with an emphasis on bees associated with pollen reward and secondarily on extrafloral nectar. The Australian *Acacia terminalis*[191,192,197] has large, flattened foliar nectaries that are reddish in color and are visited by passerine birds that pick up pollen in their feathers and inadvertently pollinate flowers in the process. Bees and flies serve as active transporters of pollen in *A. terminalis* when they move to flowers after feeding at extrafloral nectaries.[191] Thorp and Sugden[338] observed passive vectoring of pollen by honeybees on *Acacia longifolia* while insects were foraging for extrafloral nectar on the phyllodes; the proximity of inflorescences to nectaries results in insects of that size (including native, presumably coevolved, Hymenoptera) brushing the blossoms as they feed at the nectaries. Birds were suggested as possible pollinators of Australian *Acacia pycnantha*,[122] because the nectaries function during flowering, and only on phyllodes among the inflorescences.

2. Pollinator Distraction

Extrafloral nectaries may serve to distract pollinators from flowers of a plant, a drawback as far as pollination is concerned. Darwin[84] noted that "humblebees" were more interested in the extrafloral nectaries of *Vicia sativa* than the flowers, and collected the nectar assiduously during the sunny morning hours, when it flowed most freely. Honeybees collect extrafloral nectar from the outer petal bases of *Iris pallasii* in Korea (Pemberton, personal commmunication) rather than enter the flowers (though they may not be good pollinators of this species).

3. Diversion of Flower Plunderers

The role of extrafloral nectaries as objects of diversion for animals that might visit flowers but not serve as pollinators was suggested by Kerner.[193] Niewenhuis[258] considered their role as deterrents to flower robbing, but found that the frequency of robbing in Buitenzorg Garden depended more on weather and the proximity of other plants in flower than on the protection of flowers by ants. The antibiotic compounds on the body surface of ants renders them ineffective pollen transporters for most plant species.[21] That ants like to visit flowers has been debated[14,115,140,169] and well documented,[141] and nectaries outside of flowers might indeed interrupt the progress of these usually nonpollinating insects to the flowers. But their presence at the extrafloral nectaries and vicinity can have a variety of benefits to the plants, as will be discussed below.

B. PLANT FEEDING BY ANTS — MYRMECOPHYTES WITH EXTRAFLORAL NECTARIES

Myrmecophytes can be defined as plants that live regularly and often exclusively in association with ants;[88] these are plants that house ants and exist symbiotically with their resident ant fauna. Many authors have restricted the use of the term ant plant to only true myrmecophytes, vs. any plant interacting with ants.[240] Most myrmecophytes offer not only shelter for their ants, but also some sort of nutritional supplement. Most common is the presentation of food bodies for the resident ants to harvest (Mullerian bodies on *Cecropia;*[296] food bodies on *Macaranga;*[116] and food bodies produced inside hollow nodes of *Piper coenocladum.*[223,301] Some myrmecophytes have extrafloral nectaries in addition to food bodies (*Acacia cornigera*[167]); others have only extrafloral nectar as a nutritional bonus for their resident ants (*Leonardoxa africana;*[240] *Platycerium* spp;[88,209] *Schomburgkia tibicinis*[300]).

C. ANT GARDENS — EXTRAFLORAL NECTARIES AND MYRMECOCHORY

The taxonomically diverse angiosperm flora characteristic of nest sites of arboreal ants has been dubbed "ant garden", due to the fact that the seeds of these plants are ant dispersed and carried to these nests, and the seeds are discarded (thereby planted) on the outside of the nest. All flowering plants that are ant garden epiphytes are ant dispersed, and the majority offer a visible food reward for the ants (aril, sticky pulp, or gelatinous matrix), though not in all ant-garden species is the attractiveness of seeds based wholly on nutritional concerns.[87] Epiphyte seeds with no obvious food substances are preferred by ants over some seeds with nutritional rewards;[88] these seeds, however, contain chemical attractants that may serve fungicidal purposes in controlling pathogens of ant nests.[314] Roughly 60% of ant-garden epiphyte species summarized by Davidson[87] have extrafloral nectaries (e.g., *Codonanthe*, *Epiphyllum*, *Markea*, *Myrmecodia*, and *Philodendron*), making them especially nice plants for the ants to have around as a food source close to home.[88,195,196]

D. HERBIVORE ATTRACTION — A DRAWBACK OF NECTARIES

Extrafloral nectar may serve as an attractant for insects that will feed on foliage after drinking nectar, including adult Lepidoptera and Coleoptera, which in turn may oviposit on the plant providing the nectar. This problem has been reported in cultivated cotton, where, under conventional cultivation conditions, nectariless varieties have less pest damage and greater yield than varieties with nectaries.[1,2] Adult *Apion* weevils, which feed on the foliage of *Vicia sativa* in England,[208] can be seen drinking nectar from the stipular nectaries. Riodinid and lycaenid butterflies usually use plant species with extrafloral nectaries as hosts, perhaps because they provide an enemy-free space[10] for their myrmecophilous larvae (tending ants ward off parasitoids[284]); the larvae can add "insult to herbivory" by drinking extrafloral nectar as well as consuming foliage.[93]

E. ANT GUARDS AND PLANT PROTECTION

Evidence for ants visiting extrafloral nectaries of plants and providing protection for the plants against their herbivores is ever increasing (many references contained in Bentley,[33] Buckley,[56] and Jolivet[174]), but it is not universally reported.[43,149,261] Ants can most easily be effective against herbivores that are on the surface of plants, and can provide a generalized defense against most insect herbivores that an introduced plant may encounter in a new habitat. The historical development of the "protectionist" and "exploitationist" views of the significance of extrafloral nectar production are reviewed by Bentley.[33] In her review, substantial evidence is cited for the "protectionist" hypothesis; as research continues in this field, there are additional demonstrations of ant protection, some negative evidence, and some more complex situations.

1. Obligate vs. Facultative Mutualisms

Janzen's pioneering experimental work[167] demonstrated that *Acacia cornigera* could not survive herbivore predation when their resident ant protectors, *Pseudomyrmex ferruginea*, were removed. These plants are myrmecophytes, housing the ants in the swollen stipular thorns, and providing nourishment for the ants via Beltian bodies and extrafloral nectar. A number of myrmecophytic ant-plant associations do not involve nectar and appear to be more facultative than the ant acacias: *Cecropia*,[296] *Piper coenocladum*,[301] and *Macaranga*.[116]

Most plants with extrafloral nectaries do not have resident ants, and their mutualistic relationship with ants is facultative. In these facultative associations, a plant's fitness increases when the plant is associated with ants, but plants can survive without ants. The nonspecialized nature of ants visiting extrafloral nectaries is demonstrated by Schemske's quantification of ant species assemblages on *Costus*: high ant species richness and similarity in the nectar composition of *Costus* spp. indicate generalized mutualisms with limited coevolution between the plants and ants.[308]

2. Protection of Vegetative Structures

Ants attracted to nectaries encounter and remove herbivores from the plant surface, providing protection for foliage in a variety of systems. Bentley[31] created artificial extrafloral nectaries on bean plants, attracting significantly more ants and resulting in significantly less damage than on control bean plants without nectar. The exclusion of ants from entire trees of *Acacia saligna* in Australia resulted in significantly greater numbers of herbivorous insects than on control trees; trees with ants excluded apparently gained less height and lost crown canopy area compared with controls.[236] Similar results, plus increased seed set, were found in *Acacia decurrens*[53] in the absence of membracids (see Section IV.A.). Foliar nectaries of *Prunus serotina* in Michigan attract *Formica obscuripes* ants, which control early instars of tent caterpillars.[339] Areolar nectaries of *Opuntia acanthocarpa* attract *Crematogaster opuntiae* ants, which defend the plants against cactus-feeding insects.[279] Argentine ants, *Iridiomyrmex humilis*, visiting stipular nectaries of weedy *Vicia sativa* in California reduced leaf damage, resulting in an increase in fruit and seed set.[199] Foliar and sepal nectaries of *Ipomoea leptophylla* are visited by ants, which decrease damage to flowers and seeds from grasshoppers and beetles.[181] Five species of ants visit the foliar nectaries of *Catalpa speciosa*, limiting herbivory by sphingid caterpillars and increasing fruit set.[326] A temperate passion vine, *Passiflora incarnata*, experienced greater herbivory and set less fruit when extrafloral nectaries were experimentally removed.[242] The survival and mortality of *Heliconius* caterpillars on neotropical *Passiflora* spp. are significantly influenced by ants visiting the foliar nectaries.[320,321] Saplings of neotropical *Inga* species are protected from foliage-feeding herbivores by a variety of ants in the lowlands of Costa Rica.[201] Nectaries on bracken fern, *Pteridium aquilinum*, attract ants that repel nonadapted herbivores from the plant surface in Yorkshire, England.[217] The foliar nectaries of *Cassia fasiculata* support protective ants that reduce damage[189] and increase plant fecundity.[19] Ant foraging is promoted on Brazilian *Qualea grandiflora* by nectaries on the stems, resulting in greater attack rates of live insect baits on foliage.[265] *Macaranga aleuritoides* saplings have foliar nectaries supporting 13 species of ants that limit damage from insect herbivores in Papua New Guinea.[352]

3. Protection of Reproductive Structures

Nectaries on or near plant reproductive structures can provide ant protection of developing flowers, ovules, and seeds. Nectaries on inflorescences of *Schomburgkia tibicinis* (Orchidaceae) attract ants that disturb the activities of curculionid beetles that bore holes at the tip of the growing inflorescence spike, which can kill the spike or decrease flower number (depending on the timing of the attack); ants can thus decrease the number of dead spikes and increase the number of mature fruit.[299] A similar interaction has been hypothesized for the orchid *Spathoglotis plicata*, but has not yet tested.[165] The sepal nectaries of *Bixa orellana* support ants that protect the developing seeds.[32] Ants visiting the nectaries on inflorescence bracts of *Aphelandra deppeana* enhance seed capsule formation.[92] *Campsis radicans* has five extrafloral nectary systems,[107] potentially providing not only protection for foliage but developing flowers and fruit as well.[328] Suppression of the boll weevil by the imported fire ant was observed in cotton (a crop with extrafloral nectaries) in Texas.[327] The phyllary nectaries of *Helianthella quinquenervis* attract five species of ants in the Rocky Mountains of Colorado that protect the inflorescences from predispersal seed predators.[163] Neotropical *Costus woodsonii* have nectaries on the bracts of the inflorescences that attract *Ectatomma* and other ants, which vary in their ability to protect the inflorescences from Diptera that oviposit in developing ovules.[307,308] Floral nectaries of *Mentzelia nuda* function after the corolla abscises ("postfloral nectaries"), attracting ants that protect the developing fruits from herbivores.[185] Neotropical *Calathea ovandensis* have nectaries on bracts of the inflorescence, a variety of ants that visit the nectaries, and many inflorescence herbivores that are controlled by the ants and specialized (ant-tended) herbivores that are not.[157]

4. Negative Findings — No Ant Protection Evident

Some studies have demonstrated no apparent benefit to plants from insects visiting their extrafloral nectaries. *Helichrysum* spp. (Asteraceae) in Australia have extrafloral nectaries on the bracts of the inflorescence, but ants visiting the nectaries had no significant effect on herbivore damage to developing seeds or parasitism of seed predators by parasitoids.[261] These authors found seasonal differences in seed damage and herbivore parasitization, not correlated with ant abundance, and therefore cautioned against generalization of the protection hypothesis to all plants with extrafloral nectaries.

Ant-exclusion experiments performed on bracken, *Pteridium aquilinum*, at three sites in New Jersey showed no significant differences in damage between fronds with ants and those without.[336] Substantial damage was inflicted on both groups by a variety of herbivores equally. Similar results were obtained in Australia (D. O'Dowd, personal communication) and England,[149,217] though the interpretation was that the ants kept any new species of herbivores from colonizing bracken (a taxon-level defense system); the contemporary herbivores were all specialized and adapted to survive with ants.[150] Only the large, aggressive wood ant, *Formica lugubris*, was found to have an effect on contemporary herbivores of bracken in the North York Moors, England.[151]

Three species of euphorbiaceous saplings in Papua New Guinea with foliar nectaries were subjected to ant-exclusion experiments; two showed no difference in the damage sustained between ant-access and ant-excluded branches.[352] These species had as many ants as did *Macaranga aleuritoides*, for which ant defense was demonstrated; but unlike *M. aleuritoides*, did not have lepidopteran larvae, the most damaging of the associated insects, as their dominant herbivores.

Overall, the proportion of studies that have examined the effects and found benefit for plants from ants visiting extrafloral nectaries are in the majority, representing a significant effect in general, since specialized herbivores are not necessarily important at all times. It is important, however, to not assume a wholly beneficial function until the particulars of a given system are examined.

5. Complex Situations — Temporal and Spatial Variation

a. Temporal Variation

In cool temperate habitats, nectaries are visited by ants during daylight hours only (e.g., British *Vicia sativa*)[208] whereas in more clement temperate and many tropical localities, nectaries secrete nectar and ants are actively visiting them around the clock[60,201] (Pemberton, personal communication). In Britain, this ant-less window permits oviposition by *Cydia* moths, whose larvae feed inside pods of *V. sativa* protected from ants and from their enemies by ants.

Many plant species have nectaries active only on young, rapidly developing tissues, allowing ant protection at a time when the tissues are especially vulnerable to herbivores.[241] For example, extrafloral nectaries of black cherry are most active during the first few weeks after budbreak, attracting large numbers of *Formica obscuripes,* which attack young instars of the tent caterpillar that are present during that same peak of nectar secretion.[339]

b. Spatial Variation

A plant species encounters a variety of interactive situations throughout its distribution. Various studies have shown relatively small-scale spatial variation in the effects of ants on plants, depending on factors such as the proximity to the ant nest,[163] the intensity of herbivory,[19] ant activity,[189,320,321] and the behavior of different ant species.[157,175,201,308] The particular combination of ants, other protective agents, and herbivores will determine how effective a plant's defenses are and how much damage the plants will sustain.[208]

The distance of plants from ant nests results in significant variation in the magnitude of the beneficial effects of ants.[163,177,339] In a study of the neotropical herb, *Calathea ovandensis*, ants were not equally distributed over sites, and spatial heterogeneity in seed production reflected differences in ant communities among sites.[157] Barton[19] found that where the abundance of both ants and potential herbivores was high, ants visiting extrafloral nectaries strongly increased the reproductive success of *Cassia fasciculata* in Florida; the strength of the protective effect varied widely among local populations, influenced by the density of both ants and herbivores. In Yorkshire, England, ant abundance varies widely between local sites where *V. sativa* occurs, as does the density of herbivores. Where surface-feeding herbivores predominate, ants benefit the plants; but where internal feeders are abundant, the ants deter parasitoids of these herbivores and the vetches suffer, rather than benefit from ant protection.[208]

c. Specialized Herbivores

In native environments, specialist herbivores evolve that can short circuit plant defenses, including ants at extrafloral nectaries. There are many cases of specialized herbivores using the plants despite the ants;[10,150,157,201,307] these specialized herbivores are most likely to occur in a plant's native habitat.

On ant acacias, caterpillars have been observed to drop down on a thread when disturbed by ants, and to climb back up later (K. Keeler and W. Haber, personal communication). Rolling leaves into protective shelters is common among caterpillars, and two of the three most common caterpillars eating the nectary-bearing *Inga* in lowland Costa Rica are "leaf sealers" (Hesperiidae and Gelechiidae), whose shelters probably provide them with some degree of protection from ants.[201]

Many lycaenid butterflies, whose larvae have nectar glands and are tended by ants, have host plants that have extrafloral nectaries. Atsatt[10] hypothesizes that this is because they find "enemy free space" in the midst of ants and nectaries (see Section IV.A).

Although the extrafloral nectaries on bracts of *Calathea ovandensis* attract ants that deter dipteran seed predators, a specialized herbivore, *Eurybia elvina*, is tended by ants and can feed unmolested on the inflorescences.[157] Path analysis was employed[309] to estimate the direct and indirect effects of various insect-plant interactions on flower and fruit production in *Calathea ovandensis.* Ant guards had a direct positive effect on flower number, and a resulting positive indirect effect on fruit set. Ant guards have a positive direct effect on the specialized, ant-tended herbivore, *Eurybia*, although damage from *Eurybia* is greater in the absence of ants than when ants are present.[157] *Eurybia* had a negative direct effect on flower and fruit numbers.

On bracken, externally feeding sawfly larvae regurgitate toxic juices from the host plant when disturbed by ants, and the ants are repelled by the emissions. Other herbivores feed internally and are protected from the ants.[150,217]

In a facultative mutualism between alien *V. sativa* and non-native ants, ant visitation to extrafloral nectaries reduced damage from folivores and increased seed set for plants attended by ants.[199] In those California plant populations, no herbivores attacked flowers and fruits, and overall herbivore pressure was low; the only herbivores seen fed externally on leaves and flowers. In the native habitat, more guilds of herbivores feed on *V. sativa*, including insects whose larvae feed and develop inside developing pods. *Cydia* moths apparently exploit ant protection: when in pods on plants with ants, they are protected from their natural enemies. Where these herbivores are abundant, ant protection can be a problem: vetches in areas of high ant activity and high *Cydia* abundance get damaged heavily, and set much less viable seed than their counterparts in areas either with low ant activity and low *Cydia* abundance, or with low ant and high *Cydia* abundance.[208]

6. Ant Behavior — Aggressive vs. Passive Ants

Bentley[33] observed that many ants species common on extrafloral nectaries are notoriously aggressive and suggested that this is very important in their protective function on plants, involving ownership behavior around nest sites and food resources, predatory behavior, and the pheromone-mediated attack response. Indeed, in many experimental demonstrations of ant protection, ants have been observed attacking insect herbivores.[31,150,201,208,217]

In Nigeria, *Pachysima* ants occupy *Barteria* plants, resulting in more vegetative growth and less damage to shoot tips than in unoccupied plants.[168] The ants were observed to attack plants adjacent to their *Barteria* and to clean the large leaves of debris and epiphyllae, the first suggestion of the importance of "housekeeping" by ants benefitting the plants. Letourneau[223] was the first to discover the importance of nonaggressive ants in plant protection: *Pheidole bicornis*, which are small and sluggish ants, were previously thought to function only in nutrient procurement of the myrmecophytic *Piper* spp. Her experiments demonstrated that the ants clean the surface of leaves from insect eggs and early instar larvae, thereby significantly reducing damage from foliage-feeding insects.

When studies have been conducted in sufficient detail, it is frequently revealed that some ant species are better than others in protecting plants. The eight ant species associated with *Calathea* differed greatly in the magnitude of their beneficial effects on seed production;[157] the smallest ants (*Wasmannia auropunctata*) were the best defenders, and plants with these ants set the greatest amount of seed (more than twice as much as plants defended by *Pheidole gouldi*). While the four species of ants commonly associated with *Inga densiflora* and *I. punctata* were all capable of removing some herbivores, *Pheidole biconstricta* was the most effective overall against all the major guilds of herbivores on the plants.[201] Of four ant species visiting extrafloral nectaries on reproductive structures of the orchid *Schomburgkia tibicinis*, the larger species (*Ectatomma tuberculatum* and *Camponotus rectangularis*) were best at reducing damage to inflorescenses by beetles and thus increasing fruit set.[299]

Preliminary observations suggest that a loose association exists between the aggressive bullet ant *Paraponera clavata* and *Pentaclethra macroloba*:[30] the trees provide suitable nest sites for the ants and a food source in extrafloral nectar; the ants provide some amount of protection by collecting herbivorous insects. In some systems, ants protect plants against other ants. *Azteca* ants deter *Atta* defoliators from citrus trees in Trinidad.[175]

7. Trade-Offs with Other Defenses

Once ant defense has been demonstrated, a related hypothesis is engendered: are there evolutionary trade-offs between ant defense and other types of antiherbivore defense? The diversion of resources from one type of defense to another may be termed a *trade-off.* Trade-offs could occur between types of defenses because plants have finite resources, and natural selection should act to optimize the pattern of resource allocation.[7]

A comparison of ant acacias and nonmyrmecophytic acacias revealed the species not protected by ants to have greater chemical protection,[294] though not for all types of chemical defense (D. Seigler personal communication). Bentley[32] observed that when *Bixa orellana* occur at higher elevations where ants are not abundant, the sepal nectaries are not evident and the sepals are much thicker, suggesting compensation for lack of ant defense with mechanical protection of ovules and developing seeds.

The bracken fern *Pteridium aquilinum* has nectaries that function primarily on developing fronds,[271,336] providing potential ant protection of tender leaf tissue. Mature, fully hardened leaves do not have functional nectaries; perhaps mechanical defense makes ant protection less important. There is an obvious trade-off in chemical defenses as fronds age: young fronds have cyanogenic glycosides, whereas mature fronds have high levels of tannins.[71]

The same species of *Inga* that are protected by ants in the lowland wet forest occur at higher elevations without ant defense; higher elevation *Inga* have more phenolic compounds in their foliage than their lower elevation counterparts. Alternative defenses do not fully compensate for the lack of ant defense, however, as upland trees sustain more damage than lowland trees.[202]

A comparison of the defenses of *Ipomoea* species (potential chemical defense from indole alkaloids, potential biotic defense from nectaries, and potential mechanical defense from pubescence and woodiness) gave no evidence of trade-offs.[329] Despite this negative result, it is certain that the generalized nature of ant defense is in some systems superior to any combination of alternative defenses.[206]

F. OTHER PREDATORS AND PARASITOIDS FEEDING AT NECTARIES

The importance of extrafloral nectar in supporting predators (in addition to ants) is well documented for cultivated cotton.[3,302,356] Many arthropod predator species are much more abundant on cotton with extrafloral nectaries than on nectarless cotton.[1] Lacewings and several coccinellid species were observed feeding at peach leaf nectaries.[291] Floral nectaries in wind-pollinated *Croton suberosus* function to attract *Polistes* wasp predators of the herbivores.[96] Overall, however, less work has been directed to nonant predators than to parasitoids supported by extrafloral nectaries.

Adult parasitoids require carbohydrates and water to achieve maximum longevity.[120] The short mouthparts of parasitic Hymenoptera limit their use of nectar to sources that are easily accessible, such as unspecialized flowers or extrafloral nectar.[64] Sixty species of Ichneumonidae fed at the extrafloral nectaries of faba bean, *Vicia faba*.[65]

The foliar nectaries of *Byttneria aculeata* (Sterculiaceae) attract a variety of chalcid wasps, in addition to *Ectatomma* ants, thereby serving as an "insectary plant", affecting the insect community beyond serving as a host for its own herbivores.[152] Censuses of visitors to *Byttneria* nectaries reveal ants and parasitic Hymenoptera to be about equally frequent at the nectaries; ants spend more time at nectaries than in patrolling the plants; and seven species of parasitoid Hymenoptera, including some collected from nectaries, were reared from leaf-feeding herbivores. Hespenheide[153] suggests that these plants may benefit more from parasitoids visiting its nectaries than ants. A variety of parasitic wasps in the families Braconidae and Chalcidae were observed feeding at extrafloral nectary analogs, insect-induced wounds, on stems of *Baccharis sarothroides* in Arizona.[254]

At sites where ants are not present in great numbers, the nectaries can serve as attractants to parasitoids. Legume trees in the genus *Inga* have extrafloral nectaries and ant protection in Costa Rican lowlands; the same species of trees, occurring at upland sites where ants are scarce, have nectaries visited by parasitoids, and a significantly greater proportion of caterpillars from upland trees are parasitized than in the lowlands.[202]

In a study of vetches in England,[208] the evidence of parasitoid abundance from baiting did not demonstrate clear differences between the subsites where overall ant abundance was high. Data on parasitoids at baits may not provide an accurate estimate of parasitoid activity, because visits by flying insects are often shorter than those of ants. It is likely that once ants are at a bait, parasitoids will not visit that bait or will be chased away if they try. More solid evidence of the presence of parasitoids comes from their actual capture at nectaries and of parasitization of herbivore larvae. In the same study, there was evidence that ants kept parasitoids away and caused the reduced levels of parasitization of herbivore larvae in the pods at sites where ants were abundant.

IV. INTERACTIONS BETWEEN NECTARY VISITORS AND EFFECTS ON PLANTS

A. ANT-TENDED INSECTS — INSECT ANALOGS OF EXTRAFLORAL NECTARIES

A variety of insects that feed on plants exude secretions and are tended by ants. Many Homoptera excrete honeydew as they feed on phloem sap,[58,59,351] including aphids,[194] psyllids, scale insects,[18] and treehoppers.[130,355] Lepidopteran larvae of the families Lycaenidae[10] and Riodinidae[93] have specialized nectar glands with associated eversible tentacles. A hypothesis for the evolution of extrafloral nectaries[22] proposes that nectaries can divert the ants from tending Homoptera, thereby reducing the recruitment of ant-tended Homoptera and damage to plants from these herbivores, and selecting for extrafloral nectar production by the plant. They reason that nectar collection does not require as much energy from ants as Homoptera husbandry does; therefore, if extrafloral nectar provides an equivalent resource with less work, the Homoptera may be abandoned. However, a great many species with extrafloral nectaries also have Homoptera tended by ants (e.g., *Populus* and *Vicia*). A survey of Homoptera and host plants, divided into hosts with and without extrafloral nectaries, might reveal interesting patterns. If species with nectaries have markedly fewer ant-tended Homoptera, the Becerra and Venable[22] theory would have some support. It may be, however, that species with extrafloral nectaries have more than their share of Homoptera, as might be predicted if the Homoptera are exploiting "enemy free space".[10]

The presence of ants stimulates the production of honeydew from tended insects.[154] Ant attendance increased honeydew excretion by *Aphis craccivora* more than fivefold and altered the posture of the aphid larvae as well.[334] Exclusion of *Formica rufa* from *Periphyllus testudinaceus* aphids on branches of host trees in Britain resulted in a marked drop in the numbers of aphids.[318]

In some cases, ants tending herbivores that are insect analogs of extrafloral nectaries can provide protection for host plants from other herbivores and increase plant fitness over conspecifics without tended herbivores. In order for the outcome of this interaction to be positive with respect to plant fitness, the ant-tended herbivore should not be the primary herbivore of the plant, and the ants should be effective against nonhomopteran herbivores. *Crematogaster africana* tending Homoptera on *Tapinanthus*, a mistletoe that grows on cocoa in Ghana, protect the mistletoe from other herbivores and allow greater shoot growth on plants with Homoptera and ants.[303] Stout[330] discovered Homoptera living in association with *Myrmelachista* ants inside hollowed-out stems of lauraceous understory trees in Costa Rica, and after observing the swift removal of insect eggs placed on young stems and leaves by the ants, hypothesized that the benefits to the plant in protection from herbivory by the ants outweigh the losses incurred by mealy-bug feeding. *Formica* ants tending *Publilia* treehoppers protect stems of goldenrod, *Solidago altissima*, from defoliation by chrysomelid beetles (*Trirhabda* spp.), resulting in greater seed production from plants with ants.[243] Fritz[127] found no apparent harm or benefit to black locust in having *Vanduzea arquata* treehoppers and attendant *Formica subsericea* ants, because additional predators were important in the ant–herbivore interaction: while ants reduced adult density and oviposition of a leaf-mining beetle, *Odontota dorsalis*, they also excluded an important hemipteran predator of these beetles, *Nabicula subcoleoptrata,* indirectly providing protection for the beetle larvae from this predator. Resident ants in myrmecophytic *Macaranga* in Southeast Asia tend scale insects inside the stems, and aid their host plants by removing young herbivores. In addition, they cleared them of other pioneer plant species by biting off foreign plant parts that come in contact with host trees.[116]

More typically, ants tending herbivores that secrete nectar decrease the fitness of the host plant. On Aldabra Atoll, the introduced coccid, *Icerya seychellarum,* utilizes a wide variety of plant taxa[155] and is tended by small numbers of various ant species during the day, and by large numbers of *Camponotus maculatus* during the night; ants reduce the numbers of a dispid scale predator, *Chilocorus nigritus,* providing protection for the *Icerya* scale;[154] larger trees had heavier scale infestations and decreased shoot vigor.[255-257] The interaction between ants and membracid treehoppers *Sextius virescens* has a negative effect on growth and seed set of Australian *Acacia decurrens,*[57] a host plant with extrafloral nectaries. Ants tending mealybugs on *Schomburgkia tibicinis* increase peduncle damage and inflorescence death, and reduce flowering and fruit set of the orchid.[299]

In a complex interaction, ants tending *Hilda patruelis* (Homoptera: Tettigometridae) on fig-producing branches of *Ficus sur* in south Africa reduce seed predation and parasitism of pollinating wasps by repelling nonpollinating wasps from the figs.[69] The net effect of the homopteran and its associated ants is beneficial for fig trees, as parasitism of the pollinator is significantly reduced wherever ant densities are high.

B. ANTS VS. PARASITOIDS AND OTHER PREDATORS

Negative interactions between ants and parasitoids or predators of herbivores have been demonstrated in various nectar reward situations. In systems with extrafloral nectaries, and even more in extrafloral nectary analog situations, ants have been shown to protect certain herbivores (specialized either for ant resistance or for ant associations) from their enemies.

Herbivores that feed internally on plants with nectaries bear some resemblance to herbivores that have honeydew glands themselves, or form structures that exude honeydew, and may benefit in a similar way from the presence of ants at the secretions, since the aggressive behavior of ants toward other plant visitors will serve to drive away parasitoids. Cynipid gall wasps in Californian oaks are protected from their parasitoids by Argentine ants feeding on honeydew produced by the galls.[348] Caterpillars eating leaves of lowland *Inga* trees (which are protected by ants visiting their foliar nectaries against many herbivores) were parasitized much less frequently than their higher elevation counterparts on *Inga* without ants.[202] Concealed feeders on *Vicia sativa* were parasitized less frequently in areas of high ant activity.[208]

Lycaenid caterpillars have glands on their bodies that exude honeydew, and tending ants that take this exudate protect the larvae from braconid wasp parasitoids.[284] The larvae of the lycaenid butterfly, *Jalmenus evagoras*, are sometimes heavily parasitized by a braconid wasp, despite the presence of swarms of small attending *Iridomyrmex* ants.[68] Field experiments[282] demonstrated, however, that without the attendant ants, parasitism and predation of the immature stages are so intense that populations would not survive. Females use ants as an ovipositional cue,[283] and the resulting local distribution of the butterfly is strongly influenced by the distribution of ants on their host plants.[322]

Many Homoptera derive benefits from their association with ants.[351] Nymphs of *Vanduzea arquata*, a treehopper that feeds on the sap of black locust in Maryland, are protected from their predators by aggressive *Formica subsericia.*[126] The saltbush scale, *Pulvinaria maskelli*, is tended by *Iridomyrmex,* which protect the scale from predators and climatic extremes, and enhance the damage suffered by the Australian host plant, *Atriplex vesicaria.*[53] Both aphid and membracid herbivores of New York ironweed (*Vernonia noveboracensis*) are protected by ants from their coccinellid and chrysopid larvae predators, but the aphids benefit more from association with *Tapinoma* ants, and membracids benefit more from association with *Myrmica* species.[54] In Arizona, *Formica altipetens* tend the membracid *Publilia modesta*, and during some seasons they protect the nymphs from a predatory salticid spider.[78] Ants aggressively protect green scale from their coccinellid predators in coffee plantations.[145] More aggressive ant species provide better protection for soft scales and mealybugs against predators and parasitoids.[61] The continuous day-and-night tending of Homoptera in the tropics may be especially important in protection from nocturnally feeding spider predators.[60]

Of obvious importance to the interaction between ants and the other predators and parasites of herbivores is whether the herbivores are surface feeders or are concealed in various ways.[72] For plants with extrafloral nectaries, ants are potentially effective at removing all external herbivores, except for those that are specialized to avoid ant defense or to benefit from it themselves (the various honeydew producers discussed above). Insects without the specializations to produce honeydew may also gain protection from ants tending extrafloral nectaries by living inside the plant tissues (in developing fruits, tied leaves, boring in stems, or occupying galls). The architectural complexity of plants is correlated with the diversity of herbivore species that utilize the plants;[216] at the next two trophic levels, parasitoid species richness is determined largely by host-plant architecture and herbivore feeding niche.[147] It would be of great interest to contrast herbivores from host plants with and without extrafloral nectaries to see if the richness of herbivores and parasitoids is influenced by the presence or absence of nectaries. I predict that plant species with extrafloral nectaries will have greater numbers of concealed or internal feeders, and may have lower parasitoid species richness as a consequence.

V. CONCLUSIONS

The myriad relationships between insects and plants mediated by extrafloral nectaries range from simple opportunism to mutual benefit and on to complex multiple interactions (both positive and negative). It is time now for biologists specializing in various groups of plants and arthropods involved in such interactions to test some hypotheses and to make generalizations where applicable.

The discoveries that herbivory can affect both the volume and composition of extrafloral nectar indicates a potential avenue of investigation for inducible defenses in plants with extrafloral nectaries. Does increased nectar volume result in greater visitation by insects? Does this result in more biotic protection against herbivores? Are increases in certain amino acids influencing visitation by various guilds of insects? Does this increase plant protection?

Extrafloral nectaries involved in pollination should be under selective pressures, as are floral nectaries, but some considerations differ. Do extrafloral nectaries respond to visitors by secreting more nectar? Or, like some floral nectaries, do they only produce a finite amount of nectar, not changing if visited?

We need to examine the distribution of honeydew-producing Homoptera on plants with and without extrafloral nectaries. Do plants with extrafloral nectaries have fewer Homoptera than congeners without? Is it possible that extrafloral nectaries were selected as ways of eliminating ant-tended Homoptera from plants?[22] Or do Homoptera exploit plants with extrafloral nectaries as enemy-free space, like ant-tended Lepidoptera do?[10]

It would be of great interest to look at food webs based on extrafloral nectar. Nectar drinking is a special type of herbivory, yet the role of nectar in amplifying insect abundance in agricultural systems has been noted many times. Which of these insects are directly supported by nectar, or by insects that drink the nectar?

Whether or not a plant species has extrafloral nectaries may influence the structure of the herbivore community that feeds upon it. Do plant species with extrafloral nectaries have fewer herbivores than congeners without nectaries? Or perhaps a greater proportion of herbivores of certain guilds (e.g., concealed feeders)? And what about diversity of higher trophic levels (e.g., parasitoids) on congeners with and without nectaries?

ACKNOWLEDGMENTS

I thank Robert Pemberton for his constructive comments on an earlier draft of this review, Roger Gunther for his input, and Elizabeth Bernays for her skillful editing. L. Scott Quackenbush furnished bibliographic advice, and Nancy Koptur, Esq., provided invaluable clerical assistance.

REFERENCES

1. **Adjei-Maafo, I. K. and Wilson, L. T.,** Factors affecting the relative abundance of arthropods on nectaried and nectariless cotton, *Environ. Entomol.*, 12(2), 349, 1983.
2. **Adjei-Maafo, I. K. and Wilson, L. T.,** Association of cotton nectar production with *Heliothis punctigera* (Lepidoptera: Noctuidae) oviposition, *Environ. Entomol.*, 12(4), 1166, 1983.
3. **Agnew, C. W., Sterling, W. L., and Dean D. A.,** Influence of cotton nectar on red imported fire ants and other predators, *Environ. Entomol.*, 11(3), 629, 1982.
4. **Altieri, M. A.,** Crop-weed-insect interactions and the development of pest-stable cropping systems, in Thresh, J. M., Ed., *Pests, Pathogen, and Vegetation,* Pitman, Boston, 1981,459.
5. **Altieri, M. A., von Schoonhoven, A. and Doll, J.,** The ecological role of weeds in insect pest management systems: a review illustrated by bean cropping systems, *PANS*, 23(2), 195, 1977.
6. **Anderson, G. J. and Symon, D. E.,** Extrafloral nectaries in *Solanum, Biotropica,* 17(1), 40, 1985.
7. **Antonovics, J.,** Concepts of resource allocation and partitioning in plants, Stadden, J. E. R., Ed., *Limits to action. The allocation of individual behavior,* Academic Press, New York, 1980, 1.
8. **Arbo, M. M.,** Estructura y ontogenia de los nectarios foliares del genero *Byttneria* (Sterculiaceae), *Darwiniana,*17, 104, 1972.
9. **Arekal, G. D. and Ramakrishna, T. M.,** Extrafloral nectaries of *Calotropis gigantea* and *Wattakaka volubilis, Phytomorphology,* 30(2-3), 303, 1980.
10. **Atsatt, P. R.,** Lycaenid butterflies and ants: selection for enemy-free space, *Am. Nat.,* 118, 638, 1981.
11. **Baker, H. G.,** Non-sugar chemical constituents of nectar, *Apidologie*, 8(4), 349, 1977.
12. **Baker, H. G. and Baker, I.,** Nectar Constitution, in *Coevolution of Animals and Plants,* Gilbert, L. and Raven, P. H., Eds., University of Texas Press, Austin, 1975, 100.
13. **Baker, H. G. and Baker, I.,** Intraspecific constancy of floral nectar amino acid components, *Bot. Gaz.*, 138, 183, 1977.
14. **Baker, H.G. and Baker I.,** Ants and flowers, *Biotropica,* 10, 80, 1978.
15. **Baker, H.G. and Baker, I.,** A brief historical review of the chemistry of floral nectar, in *The Biology of Nectaries,* Elias T. S. and Bentley, B. L., Eds., Columbia University Press, New York, 1983, 153.
16. **Baker, H. G., Opler, P.A. and Baker, I.,** A comparison of the amino acid complements of floral and extrafloral nectars, *Bot. Gaz.,* 139, 322, 1978.
17. **Baker, D. A., Hall, J. L., and Thorpe, J. R.,** A study of the extrafloral nectaries of *Ricinus communis, New Phytol.*, 81, 129, 1978.

18. **Bartlett, B. R.,** The influence of ants upon parasites, predators, and scale insects, *Ann. Entomol. Soc. Am.,* 54, 543, 1961.
19. **Barton, A. M.,** Spatial variation in the effect of ants on an extrafloral nectary plant, *Ecology,* 67(2), 495, 1986.
20. **Baskin, S. I. and Bliss, C. A.,** Sugar content in extrafloral exudates from orchids, *Phytochemistry,* 8, 1139, 1969.
21. **Beattie, A. J.,** *The Evolutionary Ecology of Ant-Plant Mutualisms,* Cambridge University Press, London, 1984.
22. **Becerra, J. X. I. and Venable D. L.,** Extrafloral nectaries: a defense against ant-Homoptera mutualisms?, *Oikos,* 55(2), 276, 1989.
23. **Beckmann, R. L. and Stucky J. M.,** Extrafloral nectaries and plant guarding in *Ipomoea pandurata* (L.) G. R. W. May (Convolv.), *Am. J. Bot.,* 68(1), 72, 1981.
24. **Belcher, D. W., Schneider, J. C., and Hedin, P. A.,** Impact of extrafloral cotton nectaries on feeding behavior of young *Heliothis virescens* Lepidoptera Noctuidae larvae, *Environ. Entomol.,* 13(6), 1588, 1984.
25. **Belin-Depoux, M.,** Introduction à l'étude des glandes foliaires de *l'Alchornea cordata* (Juss.) Muell. Arg. (Euphorbiaceae), *Rev. Gén. Bot.,* 84, 127, 1977.
26. **Belin-Depoux, M.,** Introduction à l'étude des glandes foliaires et des hydathodes du *Mallotus glaberrimus* Muell. Arg. (Euphorbiaceae), *Rev. Gén. Bot.,* 88, 313, 1980.
27. **Belin-Depoux, M.,** Aspects morphologies et histologiques des hydathodes du *Pilea semidentata* (Juss.) Wedd. et du *P. microphylla* (L.) Liebm. (Urticaceae), *Rev. Gén. Bot.,* 90, 117, 1983.
28. **Belin-Depoux, M., and Clair-Maczulajtys, D.,** Introduction à l'étude des glandes foliares de l'*Aleurites moluccana* Willd. (Euphorbiacée) I. La glande et son ontogenèse, *Rev. Gèn. Bot.,* 81, 335 1974.
29. **Belin-Depoux, M., and Clair-Maczulajtys, D.,** Introduction a l'étude des glandes foliaires de l'Aleurites moluccana Willd. (Euphorb.) II. Aspects histologique et cytologique de la glande petiolaire fonctionnelle, *Rev. Gén. Bot.,* 82, 119 1975.
30. **Bennett, B. and Breed M. D.,** On the association between *Pentaclethra macroloba* (Mimosaceae) and *Paraponera clavata* (Hymenoptera: Formicidae) colonies, *Biotropica,* 17(3), 253, 1985.
31. **Bentley, B. L.,** Plants bearing extrafloral nectaries and the associated ant community: interhabitat differences in the reduction of herbivore damage, *Ecology,* 57, 815, 1976.
32. **Bentley, B. L.,** The protective function of ants visiting the extrafloral nectaries of *Bixa orellana* (Bixaceae), *J. Ecol.,* 65, 27, 1977.
33. **Bentley, B. L.,** Extrafloral nectaries and protection by pugnacious bodyguards, *Ann. Rev. Ecol. Syst.,* 8, 407, 1977.
34. **Bentley, B. L.,** Ants, extrafloral nectaries and the vine life-form: an interaction, *Trop. Ecol.,* 22(1), 127, 1981.
35. **Bentley, B. L.,** Nectaries in agriculture, with an emphasis on the tropics, in *The Biology of Nectaries,* Bentley, B. L. and Elias, T. L. Eds., Columbia University Press, New York, 1983, 204.
36. **Bergmann, E.,** Die Entwicklungsgeschichte der extranuptialen Nektarien in *Dioscorea discolor,* Munster, 1913.
37. **Bernays, E. A. and Woodhead, S.,** Incorporation of dietary phenols into the cuticle in the tree locust *Anacridium melanorhodon, J. Insect Physiol.,* 28(7), 601, 1982.
38. **Bernhard, F.,** Contribution à l'étude des glandes foliaires chez les Crotonoidees (Euphorbiacee) Mémoires de l'I, F. A. N., 75, 70, 1966.
39. **Bernhardt, P.,** Insect pollination of Australian *Acacia,* in *Pollination '82,* E. G. Williams, R. B. Knox, J. H. Gilbert, and P. Bernhardt, Eds., School of Botany, University of Melbourne, 1983, 84.
40. **Bernhardt, P.,** A comparison of the diversity, density, and foraging behavior of bees and wasps on Australian *Acacia, Ann. Mo. Bot. Gard.,* 74, 42, 1987.
41. **Bessey, C. E.,** Glands on a grass, *Am. Nat.* 18, 420, 1884.
42. **Blom, P. E. and Clark, W. H.,** Observations of ants (Hymenoptera: Formicidae) visiting extrafloral nectaries of the barrel cactus, *Ferocactus gracilis* Gates (Cactaceae) in Baja California, Mexico, *Southwest. Nat.,* 25(2), 181, 1980.
43. **Boecklen, W. J.,** The role of extrafloral nectaries in the herbivore defence of *Cassia fasiculata, Ecol. Ent.,* 9(3), 243, 1984.
44. **Bohn, G. W. and Davis, G. N.,** Insect pollination is necessary for the production of muskmelons (*Cucumis melo vs. reticulatus*), *J. Apic. Res.,* 3, 61, 1964.
45. **Bonnier, G.,** *Les nectaires.* Etude critique, anatomique, et physiologique, *Ann. Sci. Natur. Bot. Ser. VI,* 8, 5, 1879.
46. **Bory, G. and Clair-Maczulajtys, D.,** Composition of nectar and role of extrafloral nectars in *Ailanthus glandulosa, Can. J. Bot.,* 64(1), 247, 1986.
47. **Bosler, E. M. and Elton, J. H.,** Natural feeding behavior of adult saltmarsh greenheads and its relation to oogenesis, *Ann. Entomol. Soc. Am.,* 67(3), 321, 1974.

48. **Boughton, V. H.,** Extrafloral nectaries of some Australian phyllodineous Acacias, *Aust. J. Bot.*, 29(6), 653, 1981.
49. **Boughton, V. H.,** Extrafloral nectaries of some Australian bipinnate Acacias, *Aust. J. Bot.*, 33(2), 175, 1985.
50. **Bowden, B. N.,** The sugars in the extrafloral nectar of *Andropogon gayanus* var. *bisquamulatus*, *Phytochemistry,* 9, 2315, 1970.
51. **Bowden, B. N.,** Studies on *Andropogon gayanus* VI. The leaf nectaries of *Andropogon gayanus* (Hochst.) Hack. (Gramineae), *Bot. J. Linn. Soc.*, 64, 77, 1970.
52. **Brahmachary, R. L.,** The interrelationship between *Acacia* thorns and insects in East Africa, *J. Arid Environ.*, 5, 319, 1982.
53. **Briese, D. T.,** Damage to saltbush by the coccid *Pulvinaria maskelli* Olliff, and the role played by an attendant ant, *J. Aust. Entomol. Soc.*, 21(4), 293, 1982.
54. **Bristow, C. M.,** Differential benefits from ant attendance to two species of Homoptera on New York ironweed, *J. Anim. Ecol.*, 53, 715, 1984.
55. **Browne, L. B., Moorhouse, J. E., and Van Gerwen, A. C. M.,** A relationship between weight loss during food deprivation and subsequent meal size in the locust, *Chortoicetes terminifera*, *J. Insect Physiol.*, 22, 89, 1976.
56. **Buckley, R.,** *Ant-Plant Interactions in Australia*, Junk, The Hague, 1982.
57. **Buckley, R.,** Interaction between ants and membracid bugs decreases growth and seed set of host plant bearing extrafloral nectaries, *Oecologia (Berl.),* 58, 132, 1983.
58. **Buckley, R. C.,** Interactions involving plants, Homoptera and ants, *Ann. Rev. Ecol. Syst.*, 18, 111, 1987.
59. **Buckley, R. C.,** Ant-homopteran interactions, *Adv. Ecol. Res.*, 16, 53, 1987.
60. **Buckley, R.,** Ants protect tropical Homoptera against nocturnal spider predation, *Biotropica*, 22(2), 207, 1990.
61. **Buckley, R. and Gullan, P.,** More aggressive ant species (Hymenoptera: Formicidae) provide better protection for soft scales and mealybugs (Homoptera: Coccidae, Pseudococcidae), *Biotropica,* 23(3), 282, 1991.
62. **Buesgen, M.,** Einige Wachstumsbeobachtungen aus den Tropen, *Ber. Durtsch. Bot. Ges.*, 21, 435, 1903.
63. **Bugg, R. L.,** Observations on insects associated with a nectar-bearing Chilean tree, *Quillaja saponaria* Molina (Rosaceae), Pan-Pac., *Entomol.*, 63(1), 60, 1987.
64. **Bugg, R. L., Ehler, L. E., and Wilson, L. T.,** Effect of common knotweed (*Polygonum aviculare*) on abundance and efficiency of insect predators of crop pests, *Hilgardia*, 55(7), 1, 1987.
65. **Bugg, R. L., Ellis, R. T., and Carlson, R. W.,** Ichneumonidae (Hymenoptera) using extrafloral nectar of faba bean (*Vicia faba* L., Fabaceae) in Massachusetts, *Biol. Agric. Hortic.*, 6, 107, 1989.
66. **Chew, F. and Robbins, R. K.,** Egg-laying in butterflies, in *The Biology of Butterflies*, Vane-Wright, R. I. and Ackery, P. R. Eds., Princeton University Press, Princeton, New Jersey, 1984, 65.
67. **Clair-Maczulajtys, D. and Bory, G.,** Pediceled extrafloral nectaries in *Ailanthus glandulosa*, *Can. J. Bot.*, 61(3), 683, 1983.
68. **Common, I. F. B. and Waterhouse, D. F.,** *Butterflies of Australia*, Angus and Robertson, Sydney, Australia, 1972.
69. **Compton, S. G. and Robertson, H. G.,** Complex interactions between mutualisms: ants tending homopterans protect fig seeds and pollinators, *Ecology*, 69(4), 1302, 1988.
70. **Contreras, L. S. and Lersten, N. R.,** Extrafloral nectaries in Ebenaceae: Anatomy, morphology, and distribution, *Am. J. Bot.*, 71(6), 865, 1984.
71. **Cooper-Driver, G., Finch, S., Swain, T., and Bernays, E.,** Seasonal variation in secondary plant compounds in relation to the palatability of *Pteridium aquilinum*, *Biochem. Syst. Ecol.*, 5, 177, 1977.
72. **Cornell, H. V.,** Endophage-ectophage ratios and plant defense, *Evol. Ecol.*, 3, 64, 1989.
73. **Correns,** Zur Anatomie und Entwicklungsgeschichte der extranuptialien Nektarien von *Dioscorea*, Sitzungsber. Math-Naturw. Kl. Kais. Akad. Wiss. Wien. XCVII, 651, 1888.
74. **Cristobal, C. L. and Arbo, M. M.,** On the species of *Ayenia* (Sterculiaceae) with leaf nectaries, *Darwiniana (B. Aires),* 16(3/4), 603, 1971.
75. **Cruden, R. W. and Hermann, S. M.,** Studying nectar? Some observations on the art, in *The Biology of Nectaries*, Bentley, B. L. and T. S. Elias Eds., Columbia University Press, New York, 1983, 223.
76. **Curtis, J. D. and Lersten, N. R.,** Morphology, seasonal variation, and function of resin glands on buds and leaves of *Populus deltoides* (Salicaceae), *Am. J. Bot.*, 61(8), 835, 1974.
77. **Curtis, J. D. and Lersten, N. R.,** Heterophylly in *Populus grandidentata* (Salicaceae) with emphasis on resin glands and extrafloral nectaries, *Am. J. Bot.*, 65(9), 1003, 1978.

78. **Cushman, J. H. and Whitham, T. G.,** Conditional mutualism in a membracid-ant association: temporal, age-specific, and density dependent effects, *Ecology*, 70(4), 1040, 1989.
79. **Daguillon, A. P.,** Observation sur la structure des glandes petiolaires d' *Hevea brasiliensis*, *Rev. Gén. Bot.*, 16, 81, 1904.
80. **Damanti, P.,** Rapporti tra i nettarii estranuziali della *Silene fustea* L. e le formiche, Giorn. Soc. accl. et agr. Sicilia, 5, XXV, 101, 1885.
81. **Dammer, U.,** Die extrafloralen Nektarien an *Sambucus nigra, Oest. Bot. Zeitschr,* 1890, 264.
82. **Damstra, K. S. J.,** Ant activity on extrafloral nectaries of *Kigelia africana*, abstract for Ant-Plant Symposium, Oxford University, July 6 to 8, 1989.
83. **Daniels, G.,** The life history of *Hypochrysops theon medocus* (Fruhstorfer) (Lepidoptera: Lycaenidae), *J. Aust. Entomol. Soc.*, 15, 197, 1976.
84. **Darwin, C. R.,** *The effects of cross and self-fertilization in the vegetable kingdom*, 2nd ed., J. Murray, London, 1899.
85. **Dave, Y. S. and Menon, A. R. S.,** Structure origin and development of the extra floral nectaries of *Pithecellobium dulce* Benth, (Mimosaceae). *Acta Bot. Hung.*, 33(1-2), 117, 1987.
86. **Dave, Y. S. and Patel, N. D.,** A developmental study of extrafloral nectaries in slipper spurge (*Pedilanthus tithymaloides*, Euphorbiaceae), *Am. J. Bot.*, 62(8), 808, 1975.
87. **Davidson, D. W.,** Ecological studies of neotropical ant gardens, *Ecology*, 69(4), 1138, 1988.
88. **Davidson, D. W. and Epstein, W. W.,** Epiphytic associations with ants, in *Vascular Plants as Epiphytes*, U. Lüttge, Ed., Springer-Verlag, Berlin, 1989.
89. **Degroot, A. P.,** Protein and amino acid requirements of the honeybee (*Apis mellifera* L.), *Comp. Physiol. Ecol.*, 3(2/3), 1, 1953.
90. **Delima, J. O. G. and Leigh, T. F.,** Effect of cotton genotypes on the western bigeyed bug (Heteroptera: Miridae), *J. Econ. Entomol.*, 77, 898, 1984.
91. **Dethier, V. G.,** *The physiology of insect senses*, Methuen & Co., London, 1963.
92. **Deuth, D.,** The function of extrafloral nectaries in *Aphelandra deppeana* Schl. and Cham. (Acanthaceae), *Brenesia*, 10/11, 135, 1977.
93. **Devries, P. J. and Baker, I.,** Butterfly exploitation of an ant-plant mutualism: adding insult to herbivory, *J. N. Y. Ent. Soc.*, 97(3), 332, 1989.
94. **Dicke, M. and Sabelis, M. W.,** How plants obtain predatory mites as bodyguards, *Neth. J. Zool.*, 38(2-4), 148, 1988.
95. **Dixon, A. F. G.,** *Aphid Ecology*, Blackie & Son, London, 1985.
96. **Dominguez, C. A., Dirzo, R. and Bullock, S. H.,** On the function of floral nectar in *Croton suberosus* (Euphorbiaceae), *Oikos*, 56, 109, 1989.
97. **Dop, P.,** Les glandes florales externes de Bignoniacées, *Bull. Soc. Hist. Nat. Toulouse*, LVI, 189, 1927.
98. **Dorsey, M. J. and Weiss, F.,** Petiolar glands in the plum, *Bot. Gaz.*, 69, 391, 1920.
99. **Downes, J. A.,** The feeding habits of biting flies and their significance in classification, *Ann. Rev. Ent.*, 3, 249, 1958.
100. **Durkee, L. T.,** The floral and extrafloral nectaries of *Passiflora* II. the extra-floral nectary, *Am. J. Bot.*, 69, 1420, 1982.
101. **Durkee, L. T.,** The ultrastructure of floral and extrafloral nectaries,in *The Biology of Nectaries*, Bentley, B. and Elias, T. Eds., Columbia University Press, New York, 1983, 1.
102. **Durkee, L. T.,** Ultrastructure of extrafloral nectaries in *Aphelandra* spp. (Acanthaceae), *Proc. Iowa Acad. Sci.*, 94(572), 78, 1987.
103. **Ehrlich, P. R.,** The structure and dynamics of butterfly populations, in *The Biology of Butterflies*, Vane-Wright, R. I. and Ackery, P. R. Eds., Princeton University Press, Princeton, N.J., 1984.
104. **Elias, T. S.,** Morphology and anatomy of foliar nectaries of *Pithecellobium macradenium* (Leguminosae), *Bot. Gaz.*, 133, 38, 1972.
105. **Elias, T. S.,** Foliar nectaries of unusual structure in *Leonardoxa africana*, an African obligate myrmecophyte, *Am. J. Bot.*, 67, 423, 1980.
106. **Elias, T. S.,** Extrafloral nectaries: their structure and distribution, in *The Biology of Nectaries*, Bentley, B. L. and Elias, T. S. Eds., Columbia University Press, New York, 1983, 174.
107. **Elias, T. and Gelband, H.,** Nectar: its production and functions in the trumpet creeper, *Science*, 189, 289, 1975.
108. **Elias, T. S. and Gelband, H.,** Morphology and anatomy of floral and extrafloral nectaries in *Campsis* (Bignoniaceae), *Am. J. Bot.*, 63(10), 1349, 1976.
109. **Elias, T. S. and Gelband, H.,** Morphology, anatomy and relationship of extrafloral nectaries and hydathodes in two species of *Impatiens* (Balsaminaceae), *Bot. Gaz.*, 138(2), 206, 1977.

110. **Elias, T. S. and Prance, G. T.,** Nectaries on the fruit of *Crescentia* and other Bignoniaceae, *Brittonia*, 30(2), 175, 1978.
111. **Elias, T., Rozich, W. and Newcombe, L.,** The foliar and floral nectaries of *Turnera ulmifolia, L. Am. J. Bot.*, 62, 570, 1975.
112. **Faegri, K. and L. Van Der Pijl, L.,** *The Principles of Pollination Ecology*, Pergamon Press, New York, 1979.
113. **Fahn, A.,** The extrafloral nectaries of Sambucus nigra, *Ann. Bot. (London)*, 60(3), 299, 1987.
114. **Fahn, A. and Shimony, C.,** Develpment of the glandular and non-glandular leaf hairs of *Avicennia marina*, *Bot. J. Linn. Soc.*, 74(1), 37, 1977.
115. **Feinsinger, P. and Swarm, L. A.,** How common are ant-repellent nectars?, *Biotropica*, 10, 238, 1978.
116. **Fiala, B., Maschwitz, U., Pong, T. Y., and Helbig, A. J.,** Studies of a South East Asian ant-plant association: protection of *Macaranga* trees by *Crematogaster borneensis*, *Oecologia*, 79, 463, 1989.
117. **Figier, J.,** Etude infrastucturale de la stipule de *Vicia faba* L. au niveau du nectaire, *Planta*, 98(1), 31, 1971.
118. **Fisher, B. L., Sternberg, L. S. L., and Price, D.,** Variation in the use of orchid extrafloral nectar by ants, *Oecologia (Berl.)*, 83(2), 263, 1990.
119. **Fisher, B. L. and Zimmerman, J. K.,** Ant/orchid associations in the Barro Colorado National Monument, Panama, *Lindleyana*, 3(1), 12, 1988.
120. **Flanders, R. V. and Oatman, E. R.,** Laboratory studies on the biology of *Orgilus jenniae* (Hymenoptera: Braconidae), a parasitoid of the potato tuberworm, *Phthorimaea operculella* (Lepidoptera: Gelechiidae), *Hilgardia*, 50(8), 1, 1982.
121. **Fonseca, V. L. I.,** Miscellaneous observations on the behavior of *Schwarziana quadripunctata* (Hymenoptera: Apidae; Meliponinae), *Bol. Zool. Biol. Mar. (Nova Ser.)*, 30, 633, 1974.
122. **Ford, H. A. and Forde, N.,** Birds as possible pollinators of *Acacia pycnantha, Austr. J. Bot.*, 24, 793, 1976.
123. **Frey-Wyssling, A.,** Uber die physiologische Bedentung der extrafloralen Nektarien von *Hevea brasiliensis*, *Ber. Schwerz. Bot. Ges.*, 42, 109, 1933.
124. **Frey-Wyssling, A.,** The phloem supply to the nectaries, *Acta Bot. Neerl.*, 4, 358, 1955.
125. **Friend, W. G.,** Nutritional requirements of phytophagous insects, *Ann. Rev. Ent.*, 3, 57, 1958.
126. **Fritz, R. S.,** An ant-treehopper mutualism: effects of *Formica subsericea* on the survival of *Vanduzea arquata*, *Ecol. Ent.*, 7, 267, 1982.
127. **Fritz, R. S.,** Ant protection of a host plant's defoliator: consequence of an ant-membracid mutualism, *Ecology*, 64(4), 789, 1983.
128. **Frost, S. W.,** Insects associated with the extrafloral nectaries of elderberry, *Fla. Entomol.*, 60(3), 186, 1977.
129. **Gadrinab, L. U. and Belin, M.,** Biology of the green spots in the leaves of some dipterocarps, *Malay. For.*, 44(2/3), 253, 1981.
130. **Gersani M. and Degen, A. A.,** Daily energy intake and expenditure of the weaver ant *Polyrhachis simplex* (Hymenoptera: Formicidae) collecting honeydew from the cicada *Oxyrrhachis versicolor* (Homoptera: Membracidae), *J. Arid Environ.*, 15, 75, 1988.
131. **Gerson, U.,** The associations between pteridophytes and arthropods, *Fern Gaz.*, 12, 29, 1979.
132. **Gilbert, L. E.,** Ecological consequences of a coevolved mutualism between butterflies and plants, in Gilbert, *Coevolution of Animals and Plants*, L. E. and Raven, P. H. Eds., University of Texas Press, Austin, 1975, 210.
133. **Gilby, A. R.,** Lipids and their metabolism in insects, *Ann. Rev. Entomol.*, 10, 141, 1965.
134. **Gomez, L. D.,** Biology of the potato-fern *Solanopteris brunei*, Brenesia, 4, 37, 1974.
135. **Gomez, L. D.,** The Azteca Ants of *Solanopteris brunei*, *Am. Fern J.*, 67, 31, 1977.
136. **Govindappa, D. A. and Ramakrishna, T. M.,** Extrafloral nectaries of *Calatropis gigantea* and *Wattakaka volubilis*, *Phytomorphology*, 30, 303, 1980.
137. **Groenland, J.,** Note sur les organes glanduleux des *Drosera*, *Bull. Soc. Bot. Fr.*, 11, 395, 1855.
138. **Grout, B. W. W. and Williams, A.,** Extrafloral nectaries of *Dioscorea rotundata*: their structure and secretions, *Ann. Bot.*, 46(3), 255, 1980.
139. **Guedes, M.,** Leaf morphology in *Ricinus* — meaning of extrafloral nectaries, *Phytomorphology*, 34, 147, 1984.
140. **Guerrant, E. O. and Fiedler, P. L.,** Flower defenses against nectar-pilferage by ants, *Biotropica*, 13(Suppl.), 238, 1981.
141. **Haber, W. A, Frankie, G. W., Baker, H. G., Baker, I., and Koptur, S.,** Ants like flower nectar, *Biotropica*, 13(3), 211, 1981.

142. **Hagen, K. S.,** Ecosystem analysis: plant cultivars (HPR), entomophagous species and food supplements, in *Interactions of Plant Resistance and Parasitoids and Predators of Insects*, Boethel, D. J. and Eikenbary, R. D. Eds., Halstead Press, New York, 1986, 151.
143. **Hagen, K. S.,** Nutritional Ecology of Terrestrial Insect Predators, in *Nutritional Ecology of Insects, Mites, and Spiders*, Slansky, F. and Rodriguez, J. G. Eds., John Wiley & Sons, New York, 1987, 533.
144. **Hagen, K. S., Dadd, R. H., and Reese, J.,** The food of insects, in *Ecological Entomology*, Huffaker, C. B. and Rabb, R. L. Eds., John Wiley & Sons, New York, 1984, 80.
145. **Hanks, L. M. and Sador C. S.,** The effect of ants on nymphal survivorship of *Coccus viridis* (Homoptera: Coccidae), *Biotropica*, 22(2), 210, 1990.
146. **Hanny, B. W. and Elmore, C. D.,** Amino acid composition of cotton nectar, *J. Agric. Food Chem.*, 22, 99, 1974.
147. **Hawkins, B. A.,** Species diversity in the third and fourth trophic levels: patterns and mechanisms, *J. Anim. Ecol.*, 57, 137, 1988.
148. **Haydak, M. H.,** Honey bee nutrition, *Ann. Rev. Entomol.*, 15, 143, 1970.
149. **Heads, P. A. and Lawton, J. H.,** Bracken, ants, and extrafloral nectaries. II. The effect of ants on the insect herbivores of bracken, *J. Anim. Ecol.*, 53(3), 1015, 1984.
150. **Heads, P. A. and Lawton, J. H.,** Bracken, ants, and extrafloral nectaries. III. How insect herbivores avoid ant predation, *Ecol. Ent.*, 10, 29, 1985.
151. **Heads, P. A.,** Bracken, ants and extrafloral nectaries. IV. Do wood ants (*Formica lugubris*) protect the plant against insect herbivores?, *J. Anim. Ecol.*, 55, 795, 1986.
152. **Hespenheide, H. A.,** *Agrilus xanthonotus*, in *Costa Rican Natural History*, Janzen, D. H. Ed., University of Chicago Press, Chicago, 1983, 682.
153. **Hespenheide, H. A.,** Insect visitors to extrafloral nectaries of *Byttneria aculeata* (Sterculiaceae): relative importance of roles, *Ecol. Entomol.*, 10, 191, 1985.
154. **Hill, M. G. and Blackmore, P. J. M.,** Interactions between ants and the coccid *Icerya seychellarum* on Aldabra Atoll, *Oecologia*, 45, 360, 1980.
155. **Hill, M. G. and Newberry, D. M. C.,** The distribution and abundance of the coccid *Icerya seychellarum* Westw. on Aldabra atoll, Ecol. Entomol., 5, 115, 1980.
156. **Holldobler, B. and Wilson, E. O.,** *The Ants*, Belknap Press, Harvard University, Cambridge, MA., 1990.
157. **Horvitz, C. C. and Schemske, D. W.,** Effects of ants and an ant-tended herbivore on seed production of a neotropical herb, *Ecology*, 65(5), 1369, 1984.
158. **Hunziker, A. T. and Subils, R.,** On the taxonomic importance of foliar nectaries in wild and cultivated species of *Cucurbita*, *Kurtziana*, 8, 43, 1976.
159. **Hunziker, A. T. and Subils, R.,** New data on leaf nectaries of *Cucurbita* and their taxonomic importance, *Kurtziana*, 14, 137, 1981.
160. **Inamdar, J. A.,** Stucture and ontogeny of foliar nectaries and stomata in *Bignonia chamberlagnii* Sims, *Proc. Indian Acad. Sci. Sec. B*, 70, 232, 1969.
161. **Inamdar, J. A. and Rao, V. S.,** Structure, ontogeny, classification, taxonomic significance of trichomes and extrafloral nectaries in cultivars of cotton *Feddes Repert.*, 92(7/8), 551, 1981.
162. **Inamdar, J. A., Subramanian, R. B., and Mohan, J. S. S.,** Studies on the resin glands of *Azadirachta indica* Meliaceae, *Ann. Bot. (Lond.)*, 58(3), 425, 1986.
163. **Inouye, D. W. and Taylor, O. R.,** A temperate region plant-ant-seed predator system: consequences of extrafloral nectary secretion by *Helianthella quinquenervis*, *Ecology*, 60, 1, 1979.
164. **Inouye, D. W. and Waller, G. D.,** Responses of honey bees (*Apis mellifera*) to amino acid solutions mimicking floral nectars, *Ecology*, 65, 618, 1984.
165. **Jaffe, K., Pavis, C., Van Suyt, G., and Kermarrec, A.,** Ants visit extrafloral nectaries of the orchid *Spathoglotis plicata* Blume, *Biotropica*, 21(3), 278, 1989.
166. **Janda, C.,** Die extranuptialen Nektarien der Malvaceen, *Osl. Bot. Zeitschrift*, 86(2), 81, 1937.
167. **Janzen, D. H.,** Coevolution of mutualism between ants and acacias in Central America, *Evolution*, 20, 249, 1966.
168. **Janzen, D. H.,** Protection of *Barteria* (Passifloraceae) by *Pachysima* ants (Pseudomyrmecinae) in a Nigerian rain forest, *Ecology*, 53, 885, 1972.
169. **Janzen, D. H.,** Why don't ants visit flowers?, *Biotropica*, 9, 252, 1977.
170. **Jeffrey, D. C., Arditti, J., and Koopowitz, H.,** Sugar content in floral and extrafloral exudates of orchids: pollination, myrmecology, and chemotaxonomy implications, *New Phytol.*, 69, 187, 1970.
171. **Jennersten, O.,** Insect dispersal of fungal disease: effects of *Ustilago* infection on pollinator attraction in *Viscaria vulgaris*, *Oikos*, 51, 163, 1988.

172. **Joel, D. M.,** Glandular structures in carnivorous plants: their role in mutual and unilateral exploitation of insects, in *Insects and the Plant Surface*, Juniper, B. E. and Southwood, T. R. E. Eds., Edward Arnold, London, 1986, 219.
173. **Joel, D. M.,** Mimicry and mutualism in carnivorous pitcher plants, *Biol. J. Linn. Soc.*, 35, 185, 1988.
174. **Jolivet, P.,** *Les fourmis et les plantes*, Societe Nouvelle des Editions Boubee, Paris, 1986.
175. **Jutsum, A. R., Cherrett, J. M., and Fisher, M.,** Interactions between the fauna of citrus trees in Trinidad and the ants *Atta cephalotes* and *Azteca* sp. *J. Appl. Ecol.*, 18, 187, 1981.
176. **Kalman, F. and Gulyas, S.,** Ultrastructure and mechanism of secretion in extrafloral nectaries of *Ricinus communis* L., *Acta Biol., Nova Ser. (Szeged)*, 20(1/4), 57, 1974.
177. **Keeler, K. H.,** Insects feeding at extrafloral nectaries of *Ipomoea carnea* (Convolvulaceae), *Entomol. News*, 89(7-8), 163, 1978.
178. **Keeler, K. H.,** Species with extrafloral nectaries in a temperate flora (Nebraska), *Prairie Nat.*, 11, 33, 1979.
179. **Keeler, K. H.,** Distribution of plants with extrafloral nectaries and ants at two elevations in Jamaica, *Biotropica*, 11, 152, 1979.
180. **Keeler, K. H.,** Cover of plants with extrafloral nectaries at four northern California sites, *Madroño*, 28(1), 26, 1980.
181. **Keeler, K. H.,** The extrafloral nectaries of *Ipomoea leptophylla* (Convolvulaceae), *Am. J. Bot.*, 67(2), 216, 1980.
182. **Keeler, K. H.,** Distribution of plants with extrafloral nectaries in temperate communities, *Am. Midl. Nat.*, 104, 274, 1980.
183. **Keeler, K. H.,** Infidelity by acacia ants, *Biotropica*, 13, 79, 1981.
184. **Keeler, K. H.,** Function of *Mentzelia nuda* (Loasaceae) postfloral nectaries in seed defense, *Am. J. Bot.*, 68(2), 295, 1981.
185. **Keeler, K. H.,** A model of selection for facultative nonsymbiotic mutualism, *Am. Nat.*, 118, 488, 1981.
186. **Keeler, K. H.,** Extrafloral nectaries on plants in communities without ants: Hawaii, [USA], *Oikos*, 44(3), 407, 1985.
187. **Keeler, K. H.,** Ant-Plant Interactions, in *Plant-Animal Interactions*, Abrahamson, W. G. Ed., Mc-Graw Hill, New York, 1989, 207.
188. **Keeler, K. and Kaul, R.,** Distribution of defense nectaries in *Ipomoea* (Convolvulaceae), *Am. J. Bot.*, 71(10), 1364, 1984.
189. **Kelly, C. A.,** Extrafloral nectaries: ants, herbivores and fecundity in *Cassia fasciculata*, *Oecologia (Berlin)*, 69, 600, 1986.
190. **Kemis, J. R.,** Petiolar glands in Combretaceae: new observations and an anatomical description of the extrafloral nectary of buttonwood *Conocarpus erectus*, AIBS-BSA (Abstr.), *Am. J. Bot.*, 71(5:2), 34, 1984.
191. **Kenrick, J., Bernhardt, P., Marginson, R., Beresford, G., Knox, R. B., Baker, I., and Baker, H. G.,** Pollination-related characteristics in the mimosoid legume *Acacia terminalis* (Leguminosae), *Plant Syst. Evol.*, 157, 49, 1987.
192. **Kenrick, J., Bernhardt, P., Marginson, R., Beresford, G., and Knox, R. B.,** Birds and pollination in *Acacia terminalis*, in *Pollination '82*, Williams, E. G., Knox, R. B., Gilbert, J. H. and Bernhardt, P. Eds., School of Botany, University of Melbourne, Australia, 1983, 102.
193. **Kerner, A.,** *Flowers and their unbidden guests*, C. K. Paul & Co., London, 1878, 164.
194. **Kiss, A.,** Melezitose, aphids and ants, *Oikos*, 37, 382, 1981.
195. **Kleinfeldt, S. E.,** Ant-gardens: the interaction of *Codonanthe crassifolia* (Gesneriaceae) and *Crematogaster longispina* (Formicidae), *Ecology*, 59, 449, 1978.
196. **Kleinfeldt, S. E.,** Ant-gardens: mutual exploitation, in *Insects and the Plant Surface*, Juniper, B. and Southwood, R., Eds., Arnold, London, 1986.
197. **Knox, R. B., Kenrick, J, Bernhardt, P., Marginson, R., Beresford, G., Baker, I., and Baker, H. G.,** Extrafloral nectaries as adaptations for bird pollination in *Acacia terminalis*, *Am. J. Bot.*, 72, 1185, 1985.
198. **Kolasa, Z.,** Morphologischer und anatomischer bau der extrafloralen nektarien bei der susskirsche, *Apidologie*, 8(1), 25, 1977.
199. **Koptur, S.,** Facultative mutualism between weedy vetches bearing extrafloral nectaries and weedy ants in California, *Am. J. Bot.*, 66(9), 1016, 1979.
200. **Koptur, S.,** Flowering phenology and floral biology of *Inga* (Fabaceae: Mimosoideae), *Syst. Bot.*, 8(4), 354, 1983.
201. **Koptur, S.,** Experimental evidence for defense of *Inga* (Mimosoideae) saplings by ants, *Ecology*, 65, 1787, 1984.
202. **Koptur, S.,** Alternative defenses against herbivores in *Inga* (Fabaceae: Mimosoideae) over an elevational gradient, *Ecology*, 66(5), 1639, 1985.

203. **Koptur, S.,** Is extrafloral nectar production an inducible defense?, in *Evolutionary Ecology of Plants*, Bock, J. and Linhart, Y. Eds., Westview Press, Boulder, CO, 1989, 323.
204. **Koptur, S.,** Plants with extrafloral nectaries and ants in Everglades habitats, *Fla. Entomol.*, 75(1), in press, 1992.
205. **Koptur, S.,** Floral and extrafloral nectars of neotropical *Inga* trees: a comparison of their constituents and composition, Abstract from 4th Int. Congr. Systematic and Evolutionary Biology, College Park, Maryland, 1990.
206. **Koptur, S.,** Extrafloral nectaries of herbs and trees: modelling the interaction with ants and parasitoids, in *Ant/Plant Interactions*, Cutler, D. and Huxley, C. Eds., Oxford University Press, New York, 1991, 213.
207. **Koptur, S., Dillon, P., Foster, C., Chaverri, A., and Chu, K.,** A comparison of ant activity and extrafloral nectaries at various sites in Costa Rica, O.T.S. Tropical Biology coursebook, 77-3, 299, 1977.
208. **Koptur, S. and Lawton, J. H.,** Interactions among vetches bearing extrafloral nectaries, their biotic protective agents, and herbivores, *Ecology*, 69, 278, 1988.
209. **Koptur, S., Smith, A. R., and Baker, I.,** Nectaries in some neotropical species of *Polypodium* (Polypodiaceae): preliminary observations and analyses, *Biotropica*, 14(2), 108, 1982.
210. **Kuo, J. and Pate, J. S.,** The extrafloral nectaries of cowpea *Vigna unguiculata*. I. Morphology anatomy and fine structure, *Planta (Berl.)*, 166(1), 15, 1985.
211. **Kupicha, F. K.,** The infrageneric structure of *Vicia*, *Notes R. Bot. Gard. Edinb.*, 34, 287, 1976.
212. **Lamont, B.,** The role of extrafloral nectaries in mulga, *Annu. Rep. Mulga Research Centre*, 9, 1978.
213. **Lamont, B.,** Extrafloral nectaries in Australian plants, with special reference to *Acacia*, *Annu. Rep. 2 Mulga Research Centre*, 15, 1978.
214. **Lanza, J. and Krauss, B. R.,** Detection of amino acids in artificial nectars by two tropical ants, *Leptothorax* and *Monomorium*, *Oecologia (Berl.)*, 63(3), 423, 1984.
215. **Lawton, J. H.,** The structure of the arthropod community on bracken (*Pteridium aquilinum* (L.)Kuhn), *Bot. J. Linn. Soc.*, 73, 187, 1976.
216. **Lawton, J. H.,** Plant architecture and the diversity of phytophagous insects, *Ann. Rev. Ent.*, 28, 23, 1983.
217. **Lawton, J. H., and Heads, P. A.,** Bracken, ants, and extrafloral nectaries. I. The components of the system, *J. Anim. Ecol.*, 53(3), 995, 1984.
218. **Leeper, J. R.,** Adult feeding behavior of *Lixophaga spenophori*, a tachinid parasite of the New Guinea sugarcane weevil, *Proc. Hawaiian Entomol. Soc.*, 21, 403, 1974.
219. **Leius, K.,** Attractiveness of different foods and flowers to the adults of some hymenopterous parasites, *Can. Entomol.*, 99, 444, 1960.
220. **Lersten, N. R. and Brubaker, C. L.,** Extrafloral nectaries in Leguminosae: review and original observations in *Erythrina* and *Mucuna* (Papilionoideae: Phaseoleae), *Bull. Torrey Bot. Club*, 114(4), 437, 1987.
221. **Lersten, N. R. and Pohl, R. W.,** Extrafloral nectaries in *Cipadessa* (Meliaceae), *Ann. Bot. (Lond.)*, 56(3), 363, 1985.
222. **Lersten, N. R. and Rugenstein, S. R.,** Foliar nectaries in mahogany (*Swietenia* Jacq.), *Ann. Bot.*, 49, 397, 1982.
223. **Letourneau, D. K.,** Passive aggression: an alternative hypothesis for the Piper-Pheidole association, *Oecologia (Berlin)*, 60, 122, 1983.
224. **Letourneau, D. K.,** The enemies hypothesis: tritrophic interactions and vegetational diversity in tropical agroecosystems, *Ecology*, 68, 1616, 1987.
225. **Letourneau, D. K.,** Two examples of natural enemy augmentation: a consequence of crop diversification, in *Research Approaches in Agricultural Ecology*, Gliessman, S. R. Ed., Springer-Verlag, New York, 1990, 11.
226. **Letourneau, D. K.,** Abundance patterns of leafhopper enemies in pure and mixed stands, *Environ. Entomol.*, 19, 505, 1990.
227. **Levings, S. C.,** Patterns of nest dispersion in a tropical ground ant community, *Ecology*, 63, 338, 1982.
228. **Levings, S. C.,** Seasonal, annual, and among-site variation in the ground ant community of a deciduous tropical forest: some causes of patchy species distributions, *Ecol. Monogr.*, 53(4), 435, 1983.
229. **Lingren, P. D. and Lukefahr, M. J.,** Effects of nectariless cotton on caged populations of *Campoletis sonorensis*, *Environ. Entomol.*, 6, 586, 1977.
230. **Lloyd, F. E.,** The extra-nuptial nectaries in the common brake, *Pteridium aquilinum*, *Science*, 8(336), 885, 1901.
231. **Lukefahr, M. J.,** Effects of nectariless cottons on populations of three lepidopterous insects, *J. Econ. Entomol.*, 53, 242, 1960.

232. **Lukefahr, M. J., Martin, D. F., and Meyer, J. R.,** Plant resistance to five Lepidoptera attacking cotton, *J. Econ. Entomol.*, 58, 516, 1965.
233. **Lüttge, U.,** Uber die zusammensetzung des nektars und den mechanismes seiner sekretion, I, *Planta*, 56, 189, 1961.
234. **Lüttge, U.,** Structure and function of plant glands, *Ann. Rev. Plant Physiol.*, 22, 23, 1971.
235. **Maheshwari, J. K.,** The structure and development of extrafloral nectaries in *Duranta plumeri* Jacq., *Phytomorphology*, 4, 208, 1954.
236. **Majer, J.,** The possible protective function of extrafloral nectaries of *Acacia saligna*, *Ann. Rep. 2 Mulga Research Centre*, 1978, 31.
237. **Marginson, R., Sedgley, M., Douglas, T. J., and Knox, R. B.,** Structure and secretion of the extrafloral nectaries of Australian acacias, *Isr. J. Bot.*, 34(2-4), 91, 1985
238. **Marginson, R., Sedgley, M., and Knox, R. B.,** Structure and histochemistry of the extrafloral nectary of *Acacia terminalis* Leguminosae Mimosoideae, *Protoplasma*, 127(1-2), 21, 1985.
239. **Mauseth, J. D.,** Development and ultrastructure of extrafloral nectaries in *Ancistrocactus scheeri* - Cactaceae, *Bot. Gaz.*, 143(3), 273, 1982.
240. **McKey, D.,** Interaction of the ant-plant *Leonardoxa africana* (Caesalpiniaceae) with its obligate inhabitants in a rainforest in Cameroon, Biotropica, 16, 81, 1984.
241. **McKey, D.,** Interactions between ants and leguminous plants, in *Advances in Legume Biology*, Stirton, C. H. and Zarucchi, J. L. Eds., Monogr. Syst. Bot., Missouri Botanical Garden, St. Louis, 1989, 673.
242. **Mclain, D. K.,** Ants, extrafloral nectaries and herbivory on the passion vine, *Passiflora incarnata*, *Am. Midl. Nat.*, 110, 433, 1983.
243. **Messina, F. J.,** Plant protection as a consequence of an ant-membracid mutualism: interactions on goldenrod (*Solidago* sp.), *Ecology*, 62(6), 1433, 1981.
244. **Metcalfe, C. R.,** Extrafloral nectaries in *Osmanthus* leaves, *Kew Bull.*, 6, 254, 1938.
245. **Metcalfe, C. R.,** The anatomical structure of the Dioncophyllaceae in relation to the taxonomic affinities of the family, *Kew Bulletin*, 3, 351, 1951.
246. **Metcalfe, C. R. and Chalk, L.,** Leaves, stem and wood in relation to taxonomy with notes on economic uses, in *Anatomy of the Dicotyledons*, Clarendon Press, Oxford, 1950.
247. **Miyakawa, Y., Fujishiro, N., Kifma, H. and Morita, H.,** Differences in feeding response to sugars between adult and larvae *Drosophila melanogaster*, *J. Insect Physiol.*, 26(10), 685, 1980.
248. **Mohan, J. S. S.,** Ultrastructure and secretion of extrafloral nectaries of *Plumeria rubra* L., *Ann. Bot.*, 57(3), 389, 1986.
249. **Monod, T.,** Nectaires extra-floraux et fleurs avortees chez les Pedaliacees (note preliminaire), *Bull. Mus. Natl. Hist. Nat. Sec B. Adansonia Bot. Phytochim*, 8(2), 103, 1986.
250. **Morellato, L. P. C. and Oliveira, P. S.,** Distribution of extrafloral nectaries in different vegetation types of Amazonian Brazil, *Flora (Jena)*, 185(1), 33, 1991.
251. **Mound, L. A.,** Extrafloral nectaries of cotton and their secretions, *Emp. Cott. Grow. Rev.*, 39, 254, 1962.
252. **Murphy, D. D.,** Nectar sources as constraints on the distribution of egg masses by the checkerspot butterfly, *Euphydryas chalcedona* (Lepidoptera: Nymphalidae), *Environ. Entomol.*, 12(2), 463, 1983.
253. **Murphy, D. D., Launer, A. E., and Ehrlich, P. R.,** The role of adult feeding in egg production and population dynamics of the checkerspot butterfly *Euphydryas editha*, *Oecologia (Berl.)*, 56, 257, 1983.
254. **Naganuma, K. and Hespenheide, H.,** Behavior of visitors at insect-produced analogs of extrafloral nectaries, *Southwest. Nat.*, 33(3), 275, 1988.
255. **Newbery, D. M. C.,** Interactions between the coccid, *Icerya seychellarum* (Westw.), and its host tree species on Aldabra Atoll. I. *Euphorbia pyrifolia* Lam., *Oecologia*, 46, 171, 1980.
256. **Newbery, D. M. C.,** Interactions between the coccid, *Icerya seychellarum* (Westw.) and its host tree species on Aldabra Atoll. II. *Scaevola taccada* (Gaertn.) Roxb., *Oecologia (Berl.)*, 46, 180, 1980.
257. **Newbery, D. M. C.,** Infestation of the coccid, *Icerya seychellarum* (Westw.), on the mangrove *Avicennia marina* (Forsk.) Vierh. on Aldabra Atoll, with special reference to tree age, *Oecologia*, 45, 325, 1980.
258. **Nieuwenhuis-Von Vexkull-Guildenband, M.,** Extrafloral zuckerausscheidungen und ameisenschutz., *Ann. Jard. Bot. Buitenzorg.*, 2 Ser., 6, 195, 1907.
259. **O'Dowd, D. J.,** Foliar nectar production and ant activity on a neotropical tree, *Ochroma pyramidale*, *Oecologia*, 43(2), 233, 1979.

260. **O'Dowd, D. J.,** Pearl bodies of a neotropical tree, *Ochroma pyrimidale*: ecological implications, *Am. J. Bot.*, 67(4), 543, 1980.
261. **O'Dowd, D. J. and Catchpole, E. A.,** Ants and extrafloral nectaries: no evidence for plant protection in *Helichrysum* spp.— ant interactions, *Oecologia*, 59(2-3), 145, 1983.
262. **Okoli, B. E. and Onofeghara, F. A.,** Distribution and morphology of extrafloral nectaries in some Cucurbitaceae, *Bot. J. Linn. Soc.*, 89(2), 153, 1984.
263. **Oliveira, P. S.,** Sobre a interacao de formigas com o pequi do cerrado, *Caryocar brasiliense* Camb. (Caryocaraceae): o significado ecologico de nectarios extraflorais, Doctor in Science thesis, Universidade Estadual de Campinas, Brazil, 1988.
264. **Oliveira, P. S. and Brandão, C. R. F.,** The ant community associated with extrafloral nectaries in Brazilian cerrados, and a review of ant assemblages at extrafloral nectaries, in *Ant-Plant Interactions*, Cutler, D. F. and Huxley, C. R., Eds., Oxford University Press, New York 1991, 198.
265. **Oliveira, P. S., Da Silva, A. F., and Martins, A. B.,** Ant foraging on extrafloral nectaries of *Qualea grandiflora* (Vochysiaceae) in cerrado vegetation: ants as potential antiherbivore agents, *Oecologia,* 74, 228, 1987.
266. **Oliveira, J. S. and Leitão-Filho, H. F.,** Extrafloral nectaries: their taxonomic distribution and abundance in the woody flora of cerrado vegetation in southeast Brazil, *Biotropica,* 19(2), 140, 1987.
267. **Oliveira, P. S., Oliveira-Filho, A. T., and Cintra, R.,** Ant foraging on ant-inhabited *Triplaris* (Polygonaceae) in western Brazil: a field experiment using live termite baits, *J. Trop. Ecol.*, 3, 193, 1987.
268. **Oliveira, P. S. and Oliveira-Filho, A. T.,** Distribution of extrafloral nectaries in the woody flora of tropical communities in Western Brazil, in *Evolutionary Ecology of Plant-Animal Interactions*, Price, P. W. et al. Eds., John Wiley & Sons, New York, 1991, 163.
269. **Orr, Y.,** The leaf glands of *Dioscorea maroura* Harms, *Notes R. Bot. Gard. Edinb.*, 14, 57, 1923.
270. **Orr, Y.,** On the secretory organs of Dioscoreaceae, *Notes R. Bot. Gard. Edinb.*, 15, 4, 1926.
271. **Page, C. N.,** Field observations on the nectaries of bracken, *Pteridium aquilinum*, in Britain, *Fern Gaz.*, 12(4), 233, 1982.
272. **Parkin, J.,** The extrafloral nectaries of *Hevea brasiliensis* Muell. Arg: an example of bud scales serving as nectaries, *Ann. Bot.*, 18, 217, 1904.
273. **Pate, J. S., Peoples, M. B., Storer, P. J., and Atkins, C. A.,** The extrafloral nectaries of cowpea *Vigna unguiculata* II. Nectar composition origin of nectar solutes and nectary functioning, *Planta (Berl.)*, 166(1), 28, 1985.
274. **Patetta, A. and Manino, A.,** I nettari extrafiorali del Lauraceraso, *Apic. Mod.*, 69(6), 193, 1978.
275. **Pausch, J. and Fraenkel, J. R.,** The nutrition of the larva of the oriental rat flea, *Xenopsylla cheopsis* (Rothschild), *Physiol. Zool.*, 39, 201, 1966.
276. **Pemberton, R. W.,** The abundance of plants bearing extrafloral nectaries in colorado and Mojave desert communities of southern California, *Madrono*, 35(3), 238, 1988.
277. **Pemberton, R. W.,** The occurrence of extrafloral nectaries in Korean plants, *Korean J. Ecol.*, 13, 251, 1990.
278. **Percival, M. S.,** *Floral Biology*, Pergamon Press, Oxford, 1965.
279. **Pickett, C. H. and Clark, W. D.,** The function of extrafloral nectaries in *Opuntia acanthocarpa* (Cactaceae), *Am. J. Bot.*, 66(6), 618, 1979.
280. **Pickles, W.,** Populations, territory and interrelations of the ants *Formica fusca*, *Acanthomyops niger* and *Myrmica scabrinodis* at Garforth (Yorkshire), *J. Anim. Ecol.*, 4, 22, 1935.
281. **Pickles, W.,** Populations and territories of the ants *Formica fusca*, *Acanthomyops flavus*, and *Myrmica ruginodis* at Thornhill (Yorks), *J. Anim. Ecol.*, 5, 262, 1936.
282. **Pierce, N. E.,** The Ecology and Evolution of Symbioses Between Lycaenid Butterflies and Ants, Ph.D. thesis, Harvard University, Cambridge, MA, 1983.
283. **Pierce, N. E. and Elgar, M. A.,** The influence of ants on host plant selection by *Jalmenus evagoras*, a myrmecophilous lycaenid butterfly, *Behav. Ecol. Sociobiol.*, 16, 209, 1985.
284. **Pierce, N. E. and Mead, P. S.,** Parasitoids as selective agents in the symbiosis between lycaenid butterfly larvae and ants, *Science*, 211, 1185, 1981.
285. **Polhill, R. M. and Raven, P. H., Eds.,** Advances in Legume Systematics Part 1. Royal Botanic Gardens, Kew, Surrey, England, 1981.
286. **Potter, C. F. and Bertin, R. I.,** Amino acids in artificial nectar: feeding preferences of the flesh fly *Sarcophaga bullata*, *Am. Midl. Nat.*, 120(1), 156, 1988.

287. **Poulsen, V. A.,** Plantanatomiske Bidrg II.III. Det extrafloral nektarium hos *Carapa guyanensis* Aub. Videnskabelige meddelelser fra Dansk naturhistorisk forening i kjobenhavn, 69, 338, 1918.
288. **Power, M. S. and Skog, J. E.,** Ultrastructure of the extrafloral nectaries of *Pteridium aquilinum*, *Am. Fern J.*, 77(1), 1, 1987.
289. **Price, P. W., Bouton, C. E., Gross, P., McPheron, B. A., Thompson, J. N., and Weis, A. E.,** Interactions among three trophic levels: influence of plants on interactions between insect herbivores and natural enemies, *Ann. Rev. Ecol. Syst.*, 11, 41, 1980.
290. **Proctor, M. and Yeo, P.,** *The Pollination of Flowers*, Taplinger Publishing Co., New York, 1972.
291. **Putman, W. L.,** Nectar of peach leaf glands as insect food, *Can. Entomol.*, 95, 108, 1963.
292. **Ramaswamy, S. B.,** Behavioural responses of *Heliothis virescens* (Lepidoptera: Noctuidae) to stimulation with sugars, *J. Insect Physiol.*, 33(10), 755, 1987.
293. **Rao, L. N.,** A short note on the extrafloral nectaries in *Spathodea stipulata*, *J. Indian Bot. Soc.*, 1925, 113, 1925.
294. **Rehr, S. S., Feeny, P. P., and Janzen, D. H.,** Chemical defence in central American non-ant-acacias, *J. Anim. Ecol.*, 42, 405, 1973.
295. **Ricks, B. L. and Vinson, S. B.,** Feeding acceptibility of certain insects and various water soluble compounds to two varieties of the imported fire ant, *J. Econ. Entomol.*, 63(1), 145, 1970.
296. **Rickson, F. R.,** Progressive loss of ant-related traits of *Cecropia peltata* on selected Caribbean Islands, *Am. J. Bot.*, 64(5), 585, 1977.
297. **Rico-Gray, V.,** The importance of floral and circum-floral nectar to ants inhabiting dry tropical lowlands, *Biol. J. Linn. Soc.*, 38, 173, 1989.
298. **Rico-Gray, V.,** Ant-mealybug interaction decreases reproductive fitness of *Schomburgkia tibicinis* (Orchidaceae) in Mexico, *J. Trop. Ecol.*, 5, 109, 1989.
299. **Rico-Gray, V., and Thien, L. B.,** Effect of different ant species on reproductive fitness of *Schomburgkia tibicinis* (Orchidaceae), *Oecologia*, 81, 487, 1989.
300. **Rico-Gray, V., Barber, J. T., Thien, L. B., Ellgaard, E. G., and Toney, J. J.,** An unusual animal-plant interaction: feeding of *Schomburgkia tibicinis* (Orchidaceae) by ants, *Am. J. Bot.*, 76(4), 603, 1989.
301. **Risch, S. J. and Rickson, F. R.,** Mutualism in which ants must be present before plants produce food bodies, *Nature*, 291, 149, 1981.
302. **Rogers, C. E.,** Extrafloral nectar: entomological implications. *Bull. Entomol. Soc. Am.* Fall, 15, 1985.
303. **Room, P. M.,** The fauna of the mistletoe *Tapinanthus bangwensis* growing on cocoa in Ghana: relationships between fauna and mistletoe, *J. Anim. Ecol.*, 41, 611, 1972.
304. **Ruffner, G. A. and Clark, W. D.,** Extrafloral nectar of *Ferocactus acanthodes* (Cactaceae): composition and its importance to ants, *Am. J. Bot.*, 73(2), 185, 1986.
305. **Salisbury, E. J.,** The extra-floral nectaries of the genus *Polygonum*, *Ann. Bot.*, 23, 229, 1909.
306. **Schaffner, G.,** Extraflorale nektarien bei *Cuscuta*, Beirichte der Deutschen Botanischen Gesellschaft, 92(2/3), 721, 1980.
307. **Schemske, D. W.,** The evolutionary significance of extrafloral nectar production by *Costus woodsonii* (Zingiberaceae): an experimental analysis of ant protection, *J. Ecol.*, 68, 959, 1980.
308. **Schemske, D. W.,** Ecological correlates of a neotropical mutualism: ant assemblages at *Costus* extrafloral nectaries, *Ecology*, 63, 932, 1982.
309. **Schemske, D. W. and Horvitz, C. C.,** Plant-animal interactions and fruit production in a neotropical herb: a path analysis, *Ecology*, 69(4), 1128, 1988.
310. **Schnell, R., Cussett, G., and Quenum, M.,** Contribution a l'etude des glandes extrafloral chez quelques groupes de plantes tropicales, *Rev. Gen. Bot.*, 70, 269, 1963.
311. **Schupp, E. W. and Feener, D. H., Jr.,** Phylogeny, lifeform, and habitat dependence of ant-defended plants in a Panamanian forest, in *Ant-Plant Interactions*, Cutler, D. F. and Huxley, C., Eds., Oxford University Press, Oxford, 1991, 175.
312. **Schuster, M. F., Lukefahr, M. J., and Maxwell, F. G.,** Impact of nectariless cotton on plant bugs and natural enemies, *J. Econ. Entomol.*, 69, 400, 1976.
313. **Scott, J. K.,** Extrafloral nectaries in *Alogyne hakeifolia* (Malvaceae) and their association with ants, *West. Aust. Nat.*, 15(1), 13, 1981.
314. **Seidel, J. L.,** The Monoterpenes of *Gutierrezia sarothrae*: Chemical Interactions between Ants and Plants in Neotropical Ant-gardens, Ph.D. dissertation, University of Utah, 1988.

315. **Shahjahan, M.,** Effect of diet on the longevity and fecundity of the adults of the tachinid parasite *Trichopoda pennipes pilipes*, *J. Econ. Entomol.*, 61, 1102, 1968.
316. **Sherbrooke, W. C. and Scheerens, J. C.,** Ant-visited extrafloral (calyx and foliar) nectaries and nectar sugars of *Erythrina flabelliformis* Kearney in Arizona, *Ann. MO. Bot. Gard.*, 66(3), 472, 1979.
317. **Shiraishi, A. and Kuwabara, M.,** The effect of amino acids on the labellar hair chemosensory cells of the fly, *J. Gen. Physiol.*, 56, 768, 1970.
318. **Skinner, G. J. and Whittaker, J. B.,** An experimental investigation of inter-relationships between the wood-ant (*Formica rufa*) and some tree-canopy herbivores, *J. Anim. Ecol.*, 50, 313, 1981.
319. **Smiley, J.,** Plant chemistry and the evolution of host specificity: new evidence for *Heliconius* and *Passiflora*, *Science*, 201, 745, 1978.
320. **Smiley, J.,** Heliconius caterpillar mortality during establishment on plants with and without attending ants, *Ecology*, 66:845, 1985
321. **Smiley, J.,** 1986. Ant constancy at *Passiflora* extrafloral nectaries: effects on caterpillar survival, *Ecology*, 67(2):516, 1986.
322. **Smiley, J. T., Atsatt, P. R., and Pierce, N. E.,** Local distribution of the Lycaenid butterfly, *Jalmenus evagoras*, in response to host ants and plants, *Oecologia*, 76, 416, 1988.
323. **Smith, L. L., Lanza, J., and Smith, G. C.,** Amino acid concentrations in the extrafloral nectar of *Impatiens sultani* increase after simulated herbivory, *Ecology*, 71(1), 107, 1990.
324. **Sprague, T. A.,** *Dioncophyllum*; with note on the anatomy by F. M. Scott, Bull. Misc. Inf., Kew, Surrey, England, 1916, 89.
325. **Starmer, W. T., Phaff, H. J., Bowles, J. M., and Lachance, M. A.,** Yeasts vectored by insects feeding on decaying saguaro cactus, *Southwest. Nat.*, 33(3), 362, 1988.
326. **Stephenson, A. G.,** The role of extrafloral nectaries of *Catalpa speciosa* in limiting herbivory and increasing fruit production, *Ecology*, 63, 663, 1982.
327. **Sterling, W. L.,** Fortuitous biological suppression of the boll weevil by the red imported fire ant, *Environ. Entomol.*, 7(4), 564, 1978.
328. **Stevens, J. A.,** Response of *Campsis radicans* (Bignoniaceae) to Simulated Herbivory and Ant Visitation, M.S. thesis, University of Missouri-St. Louis, 1990.
329. **Steward, J. L. and Keeler, K. H.,** Are there trade-offs among antiherbivore defenses in *Ipomoea* (Convolvulaceae)?, *Oikos*, 53, 79, 1988.
330. **Stout, J.,** An association of an ant, a mealy bug, and an understory tree from a Costa Rican rain forest, *Biotropica*, 11(4), 309, 1979.
331. **Subramanian, R. B. and Inamdar, J. A.,** Nectaries in *Bignonia illicum* ontogeny structure and functions, *Proc. Acad. Sci. Plant Sci.*, 96(2), 135, 1986.
332. **Sudd, J. H.,** Distribution of foraging woodants in relation to the distribution of aphids, *Insectes Soc.*, 30(3), 293, 1983.
333. **Sundby, R. A.,** Influence of food on the fecundity of *Chrysopa carnea* Stephens (Neuroptera: Chrysopidae), *Entomophaga*, 12, 475, 1967.
334. **Takeda, S., Kinomura, K., and Sakurai, H.,** Effects of ant attendance on the honeydew excretion and larviposition of the cowpea aphid, *Appl. Entomol. Zool.*, 17, 133, 1983.
335. **Tanowitz, B. D. and Koehler, D. L.,** Carbohydrate analysis of floral and extrafloral nectars in selected taxa of *Sansivieria* (Agavaceae), *Ann. Bot.*, 58(4), 541, 1986.
336. **Tempel, A. S.,** Bracken fern (*Pteridium aquilinum* Kuhn) and nectar-feeding ants: a non-mutualistic interaction, *Ecology*, 64(6), 1411, 1983.
337. **Terraciano, A.,** I nettarii estranuziali delle Bombaceae [sic.], in Borzi, *Contrib. alla biologia vegetale, Palerme*, 2, 139, 1899.
338. **Thorp, R. W. and Sugden, E. A.,** Extrafloral nectaries producing rewards for pollinator attraction in *Acacia longifolia* (Andr.) Willd, *Isr. J. Bot.*, 39, 177, 1990.
339. **Tilman, D.,** Cherries, ants and tent caterpillars: timing of nectar production in relation to susceptibility of caterpillars to ant predation, *Ecology*, 59(4), 686, 1978.
340. **Tomlinson, P. B.,** *The Biology of Trees Native to Tropical Florida*, Harvard University Printing Office, Allston, Mass., 1980.
341. **Topham, M. and Beardlsey, J. W., Jr.,** Influence of nectar source plants on the New Guinea sugarcane weevil parasite, *Lixophaga spehnophori* (Villenueve), *Proc. Hawaii Entomol. Soc.*, 21, 145, 1975.
342. **Trecul, A.,** Organisation des glandes pedicellees des feuilles du *Drosera rotundifolia*, *Ann. Sci. Nat.*, 4th series, 3(5), 1855.
343. **Trelease, W.,** The foliar nectar glands of *Populus*, *Bot. Gaz.*, 6, 284, 1881.
344. **Tyler, F. J.,** The nectaries of cotton, *Bull. Bur. Pl. Ind.*, U.S. Dept. Agric., 131, 45, 1908.

345. **Vansell, G. H.,** Nectar secretion in poinsettia blossoms, *J. Econ. Entomol.*, 33, 409, 1940.
346. **Von Tiechman, L. I. and Robbertse, P. J.,** Die anatomie en ultrastruktuur van die ekstraflorale nektarkliere van *Dioscorea sylvatica* Eckl. en die samestelling van die nektar, *J. South African Bot.*, 45(1), 63, 1979.
347. **Waller, G. D.,** Evaluating responses of honey bees to sugar solutions using an artificial-flower feeder, *Ann. Entomol. Soc. Am.*, 65, 857, 1972.
348. **Washburn, J. O.,** Mutualism between a cynipid gall wasp and ants, *Ecology*, 65, 654, 1984.
349. **Watanabe, M. and Kiyoshi, K.,** Nectar sucking behavior of *Tabanus iyoensis* (Diptera: Tabanidae), *Jpn. J. Sanit. Zool.*, 26(1), 41, 1975.
350. **Watt, W. B., Hoch, P. C., and Mills, S. G.,** Nectar resource use by *Colias* butterflies: chemical and visual aspects, *Oecologia (Berl.)*, 14(4), 353, 1974.
351. **Way, M. J.,** Mutualism between ants and honeydew-producing Homoptera, *Ann. Rev. Ent.*, 8, 307, 1963.
352. **Whalen, M. A. and Mackay, D. A.,** Patterns of ant and herbivore activity on five understory euphorbiaceous saplings in submontane Papua New Guinea, *Biotropica*, 20(4), 294, 1988.
353. **Willmer, P. G.,** The effects of insect visitors on nectar constituents in temperate plants, *Oecologia*, 47, 270, 1980.
354. **Wilson, R. L. and Wilson, F. D.,** Nectariless and glabrous cottons: effect on pink bollworm in Arizona, *J. Econ. Entomol.*, 69, 623, 1976.
355. **Wood, T. K.,** Ant-attended nymphal aggregations of *Echenopa binotata* complex (Homoptera: Membracidae), *Ann. Ent. Soc. Am.*, 75(6), 649, 1982.
356. **Yokoyama, V. Y.,** Relation of seasonal changes in extrafloral nectar and foliar protein and arthropod populations in cotton, *Environ. Entomol.*, 7, 799, 1978.
357. **Yuen, C. K. K. H. and Dehgan, D.,** Comparative morphology of the leaf epidermis in the genera *Codonanthe* and *Nematanthus* Gesneriaceae, *Bot. J. Linn. Soc.*, 85(4), 283, 1982.
358. **Zamora, P. M. and Vargas, N. S.,** Nectary-costule association in Phillippine drynarioid ferns, *Philipp. Agric.*, 57, 72, 1974.
359. **Zimmermann, M.,** Uber die extrafloralen nektarien der angiospermen, *Bot. Zentralblatt Beih.*, 49, 99, 1932.

5

The Role of Quinolizidine Alkaloids in Plant–Insect Interactions

Michael Wink
Universität Heidelberg
Institut für Pharmazeutische Biologie
Heidelberg, Germany

TABLE OF CONTENTS

I. INTRODUCTION

A. STRUCTURES OF QUINOLIZIDINE ALKALOIDS (QA)

More than 170 structures of quinolizidine alkaloids have been reported from plants.[42,50,60,73,98,99,136] According to the degree of substitution, at least seven different groups of QA can be distinguished (Figure 1): (1) the lupinine type, (2) the leontidine type, (3) the sparteine/lupanine/multiflorine type, (4) the α-pyridone type, (5) the matrine type, (6) the *Ormosia* type, and (7) piperidine and the dipiperidines, which often occur simultaneously with QA in the same plants and are biogenetically related.[156]

B. NATURAL DISTRIBUTION OF QA IN PLANTS

Quinolizidine alkaloids are mainly distributed within the Leguminosae, the third largest family of flowering plants after the Compositae and Orchidaceae, which consists of about 650 genera and 18,000 species. The QA-accumulating tribes — Sophoreae, Dalbergieae, Euchresteae, Thermopsidae, Genisteae, Bossiaeae, Brongniartieae, Podalyrieae, Liparieae, and Crotalarieae — are considered "primitive" within the Leguminosae.[60] However, restricted occurrences have been reported from a number of other families that are unrelated to the Leguminosae, such as the Berberidaceae, Ranunculaceae, Chenopodiaceae, and Rubiaceae (Table 1).

Parasitic or hemiparasitic plants tap phloem and/or xylem sap of their host plants and thus often acquire the respective secondary metabolities. QA were observed in a number of hemiparasitic species of the families Scrophulariaceae: *Orobanche*,[158] *Castilleja*,[109] and *Pedicularis*;[101,108] Loranthaceae: *Viscum*;[19] Cuscutaceae: *Cuscuta*;[136] and Santalaceae: *Osyris*.[67] These hemiparasitic plants are especially interesting for questions of plant–insect interactions, since the patterns of secondary metabolites may change within a population when the plants live on different hosts,[108,109] so these data are not informative for chemotaxonomic considerations.

Using plant cell suspension cultures of *Daucus carota, Conium maculatum, Atropa belladonna, Chenopodium rubrum, Spinacia oleracea,* and *Symphytum officinale* we observed that these cultures produce minute amounts of lupanine (Table 1), especially when challenged with elicitors.[152] Using a very sentive radioimmunoassay (RIA) for lupanine,[136] we found traces of this alkaloid in ca. 50 out of 100 nonleguminous plants studied.[132] These surprising findings suggest that the genes of QA biosynthesis might be widely distributed within the plant kingdom but are generally "silent". Only in the Leguminosae (and in a few other genera) are the genes for biosynthesis, transport, and storage concomitantly turned on.

The different structural types (Figure 1) of QA are not evenly distributed within the Leguminosae: common to all tribes are QA of the lupanine- (3a,3e,3f) and sparteine- (3b) type. Multiflorine (3c), aphylline (3d), and derivatives are only present in a few genera. α-Pyridones (4), such as cytisine or anagyrine, are also common but more restricted than types 3a and 3b. Lupinine and derivatives (1) have been detected in a comparably small number of species. Leontidine-type alkaloids (2) seem to be very specialized and have been found in *Camoensia, Melolobium* and *Maackia* so far (Table 1). Matrines (5) are common in only the genus *Sophora* and Ormosia-type alkaloids are especially found in the genus *Ormosia*. The chemotaxonomic implications cannot be discussed in this article but have been reviewed in Mears and Mabry,[73] Kinghorn and Balandrin,[60] Salatino and Gottlieb,[98] and Nowacki and Waller.[85]

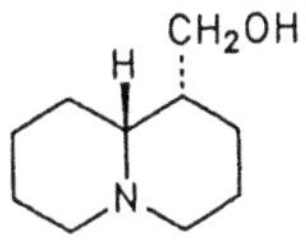

Lupinine (1)

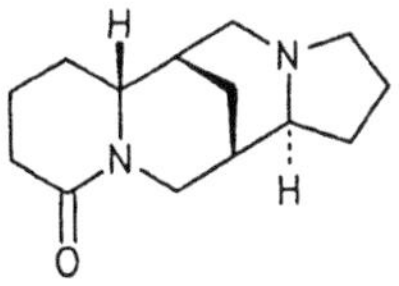

Camoensidine (2)

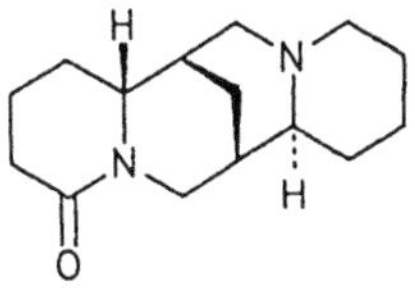

Lupanine (3a)

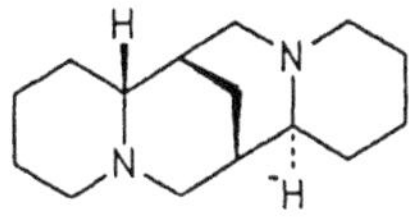

Sparteine (3b)

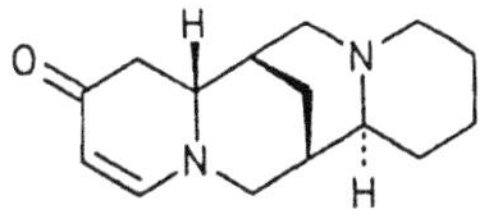

Multiflorine (3c)

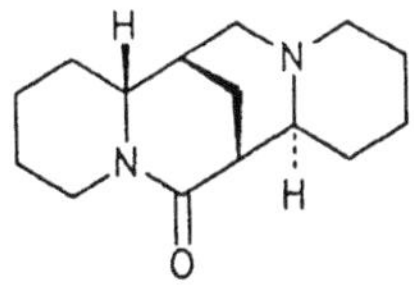

Aphylline (3d)

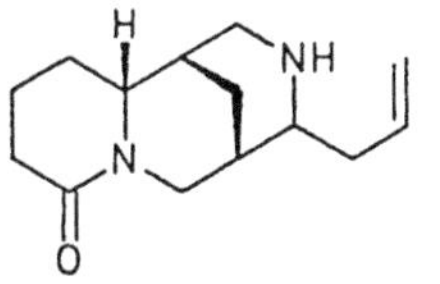

Angustifoline (3e)

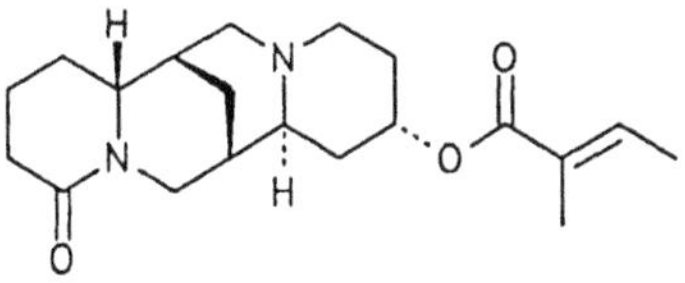

13α-Tigloyl-oxylupanine (3f)

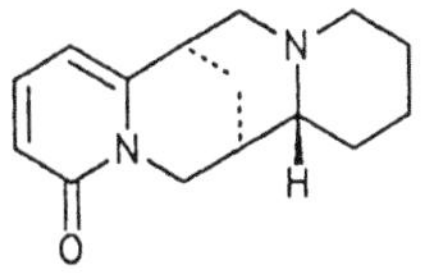

Anagyrine (4)

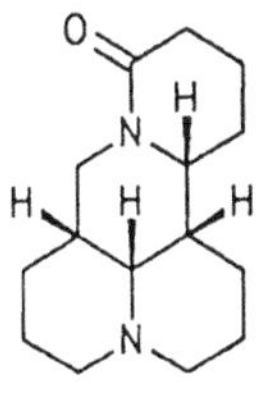

Matrine (5)

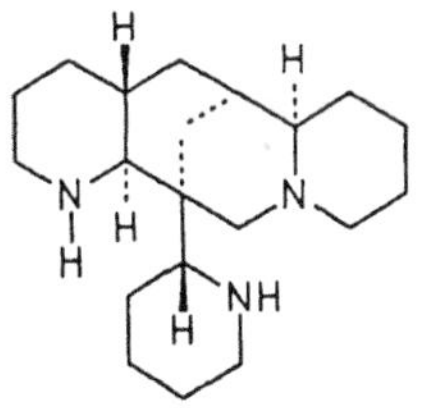

Piptanthine (6)

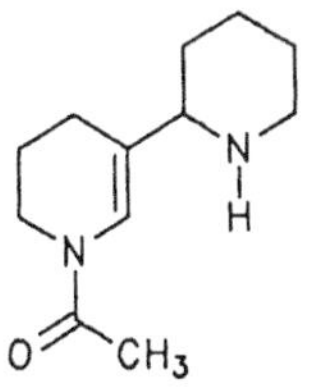

Ammodendrine (7)

FIGURE 1. Structures of some quinolizidine alkaloids. Numbers refer to alkaloidal types discussed in Section I.B.

Table 1
DISTRIBUTION OF QA IN THE PLANT KINGDOM

Family/tribe	Genus	1	2	3a	3b	3c	3d	3e	3f	4	5	6	7
		Alkaloidal types											
Leguminosae													
Sophoreae	Acosmium				*							*	
	Ammodendron			*	*					*			*
	Ammothamnus				*						*		
	Bolusanthus			*	*					*			
	Cadia	*		*	*				*				
	Calpurnia	*		*			*		*				
	Camoensia		*										
	Cladrastis	*								*			
	Clathrotropis			*						*			
	Diplotropis			*				*	*				
	Echinosophora			*						*			
	Goebelia	*								*	*		
	Haplormosia			*	*		*			*			*
	Keyserlingia				*					*	*		
	Maackia	*	*	*						*			*
	Nitraria											*	
	Ormosia			*	*		*	*		*		*	*
	Pericopsis			*	*					*			*
	Podopetalum											*	
	Sophora	*		*	*					*	*	*	*
Euchrestae	Euchresta	*								*	*		
Dalbergiae	Dalbergia			*					*				
Thermopsidae	Ammopiptanthus			*								*	
	Anagyris			*	*					*			
	Baptisia			*	*					*			*
	Piptanthus			*	*					*		*	
	Thermopsis	*		*	*					*			*
Genisteae	Adenocarpus				*								*
	Argyrocytisus									*			

	Argyrolobium	*		*	*		*			*			
	Calicotome									*			
	Chamaecytisus	*		*	*					*			
	Cytisusophyllum			*									*
	Cytisus			*	*			*	*	*			*
	Genista	*		*	*			*	*	*	*?		*
	Laburnum			*	*			*		*			*
	Lupinus	*		*	*	*	*	*	*	*			*
	Petteria			*						*			
	Retama			*	*					*	*?		*
	Spartium			*	*					*			*
	Teline						*			*			
	Ulex			*						*			
Podalyrieae	Virgilia	*		*			*			*			*
	Podalyria			*			*						
Liparieae	Liparia			*	*								*
Crotalarieae	Aspalathus			*	*								
	Dichilus			*						*			*
	Lebeckia			*	*								*
	Lotononis			*	*								*
	Melolobium		*	*	*					*			
	Pearsonia			*	*				*	*			*
	Polhillia			*	*					*			
	Rafnia			*	*								
	Rothia			*	*				*				*
	Spartidium				*								*
	Wiborgia			*	*								
Bossiaeeae	Lamprolobium	*								*			
	Templetonia	*		*	*					*		*	*
	Hovea			*						*		*	*
Brogniartieae	Harpalyce	*								*			
Berberidaceae	Caulophyllum	*		*	*					*			
	Leontice	*	*	*	*					*	*		
Boraginaceae	Symphytum*			*									
Chenopodiaceae	Anabasis	*					*						*
	Chenopodium*			*									
	Spinacia			*									

Table 1 (continued)
DISTRIBUTION OF QA IN THE PLANT KINGDOM

Family/tribe	Genus	Alkaloidal types											
		1	2	3a	3b	3c	3d	3e	3f	4	5	6	7
Cuscutaceae	Cuscuta**			*		*							
Loranthaceae	Viscum**			*	*					*			
Ranunculaceae	Cimicifuga									*?			
Rubiaceae	Readea								*				
Santalaceae	Osyris**								*	*			
Scrophulariaceae	Castilleja**	*		*			*	*		*		*	
	Orobanche**			*	*			*	*			*	
	Pedicularis**			*				*		*			
Umbelliferae	Conium*			*									

* Plant-cell suspension cultures; ** hemiparasitic plant living on a QA producing host.
For the identification of QA types, see Figure 1.

II. PHYSIOLOGY AND BIOCHEMISTRY OF QA PRODUCTION IN PLANTS

A. BIOSYNTHESIS

Lysine is the sole precursor for the alkaloid skeleton and is decarboxylated by lysine decarboxylase in the first step.[102] The resulting cadaverine is converted into lupanine and sparteine by oxosparteine synthase, a transaminating enzyme.[130,137,142] Many of the other QA are thought to be derived from lupanine or sparteine.[130,142] The biosynthesis of lupanine, sparteine, matrine, lupinine, and cytisine has also been studied by using radioactive isotopes, or more recently by ^{13}C- and ^{2}H-labeled precursors.[33,34,37,38,49,78,79,88,93] But many steps of QA biosynthesis are still unclear, and hardly anything is known on the pathways leading to the manyfold substitutions and variations of the quinolizidine ring skeletons, except for the formation of N-methylcytisine by S-adenosylmethionine cytisine N-methyltransferase,[120] and of esters of lupinine and of 13-hydroxylupanine by acyl-CoA: 13-hydroxylupanine O-acyltransferase[141] or acyl-CoA: lupinine O-acyltransferase80 (for a review, see Wink and Hartmann[142]).

QA biosynthesis takes place in the aerial green parts of legumes. Its intracellular site is the chloroplast,[139] where the biosynthesis of the precursor lysine also takes place. Apparently enzyme concentrations, lysine availability, hydrogen-ion concentrations, and the redox state of the enzymes vary diurnally,[124,140,142,153] thus QA concentrations display a diurnal rhythm, with a stimulated production period during the day.[140,153]

B. TRANSPORT AND ACCUMULATION

QA are not stored in the chloroplasts but are translocated by the phloem all over the plant.[140,153,157] For these transport studies, aphids were extremely useful (Table 4). QA accumulate predominantly in epidermal and subepidermal cell layers of stems and leaves.[127,131,144] The subcellular site of QA storage is the vacuole, into which the alkaloids are pumped by a carrier system.[74,134,145] Especially rich in alkaloids are the seeds (Table 3), which can store up to 8% of their dry weight as alkaloids.[123,124,126]

QA composition of a plant is most complex in the leaf (the site of biosynthesis) but is usually somewhat different in other organs, such as seeds, possibly reflecting selective transport and/or metabolic transformations. Furthermore, QA patterns change substantially, both qualitatively and quantitatively during ontogeny, e.g., at the stage of germination and growth of the seedling, at the time of flowering and fruit formation, and at the end of the vegetation period.[119,123,124,126,129,130,138,154,159] For a herbivore these changing patterns and concentrations are certainly demanding and difficult to cope with.

C. DEGRADATION

Seeds are especially rich in alkaloids. Most of the alkaloids are stored in the cotyledons. The testa contains between 1 and 7% alkaloids, depending on the species. During germination the alkaloids are translocated from the cotyledons to the newly formed tissues, where the alkaloids are partly degraded, obviously serving as a nitrogen source.[154] As soon as the new leaves have been formed, alkaloid formation sets in again. QA also disappear from senescing leaves, indicating that they are not end products but metabolically mobile compounds.[40,111,125,138,154]

III. PLANT–INSECT INTERACTIONS

Plants have to cope with many predators and pathogens. Especially abundant are insect herbivores, many of which are mono- or oligophagous.[5-7,28,110] In contrast, many vertebrate herbivores can be classified as polyphagous and close host-plant–herbivore interactions are uncommon.

Plants have, over time, evolved a number of antiherbivore strategies,[28,44,46,68,81,96,97,100,110,126,133] which in their complexity are only partly understood at present: physical barriers have been developed, such as impenetrable bark tissue, thick and impregnated cuticles, laticifers, and resin ducts full of latex or resin, respectively, which glue the mandibles of a chewing insect. Some plants have minimized the nutritional constituents of their tissues, so that they no longer provide optimal food to herbivores or microbes; examples are storage proteins lacking either the essential amino acids methionine or lysine (such as in legumes and in the Poaceae) or the presence of undigestible wood, lignins, or celluloses. Other plants produce antinutritive protease inhibitors, which inhibit protein-digesting proteases of herbivores; or lectins, which effectively inhibit cellular processes, such as translation. Furthermore, many plant tissues have an unfavorable ratio of carbon to nitrogen, which causes unproductive food intake. Some plants, such as grasses, accumulate silicates, which require special morphological adaptations by the herbivore.

An important feature in this context is the production of allelochemicals, the so-called secondary metabolites. All plant species make and store these compounds. Usually a wide variety of structurally different allelochemicals is present within one plant and are often distributed in a tissue-specific manner. Furthermore, the profile of secondary compounds and their concentrations often alter during the development of a plant.[79,128,134] The main purpose of these compounds seems to be chemical defense against herbivores, including insects, and against microbial pathogens, as well as against competing plants.[28,44,46,68,81,96,97,100,110,126,133] Since a single allelochemical is often involved in manyfold interactions, many allelochemicals must be considered as broad spectrum bioregulators. This statement does not contradict the observation that sometimes the same compound, which is antibiotic to some organisms (e.g., many monoterpenes and phenolics, flavones, anthocyanins), functions at the same time to attract pollinating insects or seed-dispersing animals.

In the following we shall concentrate on insect herbivores (which often behave very differently from vertebrate herbivores!) and ask in which way they may cope with allelochemicals, in this case, quinolizidine alkaloids (Tables 2 and 4). In general we see the following strategies:[5-7,27,148]

1. A herbivore detects the allelochemical present and avoids feeding. Insects are endowed with very competent olfactory and gustatory receptors (see the review in Kaissling[55]): these are involved in the initial decision.
2. A herbivore may feed on a plant with a certain set of allelochemicals. Those compounds that are passively absorbed are immediately "detoxified" by an elaborate system of enzymic reactions (including p450 cytochrome oxidases and glutathione peroxidase[1,11,12,163]). Others are eliminated by the Malpighian tubule system. If compounds are not taken up from the gut, they may pass out directly with the feces.
3. A herbivore may adapt to a certain set of allelochemicals and may develop absorption mechanisms in the gut, which allows the uptake of necessary nutrients but discriminates against the allelochemicals.
4. A herbivore may take advantage of the allelochemicals present, in that it actively absorbs, sequesters, and uses the allelochemicals for its own defense ("acquired defense").[27]

5. More elaborate adaptations may follow in that the acquired allelochemical is integrated in the physiology, biochemistry, or developmental program of the herbivore; examples are pheromones or morphogens in the Arctiidae.[13,83,100]

Strategies 1 and 2 are usually found in the group of polyphagous "generalists", whereas strategies 3 to 5 are more typical in mono- and oligophagous "specialists".

A. INSECTS THAT ARE NOT ADAPTED TO QA PLANTS ("GENERALISTS")

1. Alkaloid Elimination, Deterrence, and Toxicity

A mechanism to avoid toxins can be seen in a number of oligo- and polyphagous insects. For example, moth larvae, such as *Manduca sexta, Syntomis mogadorensis, Creatonotos transiens,* and beetle larvae, *Bruchidius villosus*, fed nontoxic doses of sparteine eliminate this QA with the feces, so that the body of these insects contains only trace amounts of the alkaloid[111,148] (Table 4). We assume that most of the sparteine molecules are not passively absorbed, because the molecules are charged under physiological conditions and are thus not able to pass biomembranes by simple diffusion. Instead, they are excreted unchanged with the feces. Those molecules that diffuse into the insect as free bases are probably either degraded by detoxifying enzymes or eliminated by the Malpighian tubules. Some experimental evidence indicates that both possibilities contribute to sparteine metabolism in *Manduca* and in *Syntomis*.[48,148]

QA were found to be feeding deterrents in a number of oligo- and polyphagous insects, including aphids, moth and butterfly larvae, beetles, grasshoppers, flies, bees, and ants, i.e., QA are deterrent over a wide range of insect orders (Table 2).

The few electrophysiological experiments performed with QA so far indicate that sparteine and other QA elicit responses from chemosensilla of *Entomoscelis, Phormia,* and *Pieris,* which lead to food rejection,[9,75,103] thus explaining the behavioral responses observed in the deterrency experiments (Table 2).

QA seem to be toxic to many insects, but it has to be considered that in relatively few instances have toxicity tests been performed, and only sparteine, lupanine, and 13-tigloyloxylupanine have been evaluated in some detail (Table 2). Most of the other QA have not been isolated in quantities and thus have not been available for testing yet, so that the full potential of the various QA derivatives is still unknown. Already the present data clearly show that QA have insecticidal properties, with LD_{100} values in the range between 0.4 and 50 m*M* (equivalent to 0.1 to 10% fresh weight). As compared to deterrent concentrations, the lethal concentrations are much higher. It should be remembered that alkaloid levels in legumes are usually in the same range or even much higher than the levels required for deterrence or toxicity (Table 3), i.e., QA contents are in the correct order of magnitude in nature and are especially abundant in those organs that are important for plant reproduction or survival. In this context, it should be recalled that QA are stored in epidermal and subepidermal cell layers, which are especially rich in alkaloids (up to 200 m*M*[127,144]) and are thus present in a strategically important position, since these tissues have to ward off an attack in the first instance.[127,144]

The mechanisms underlying the toxic effects observed in insects have not been elucidated. However, Twardowski and co-workers[63,114,150] have shown that QA can interfere with protein biosynthesis (Table 5). Since protein biosynthesis is an important and vulnerable process in most other organisms, we should expect that QA have toxic properties over a wide range of organisms that are indeed affected (Table 5). Furthermore, cytisine and probably other QA bind to acetylcholine (ACH) receptors with high affinity (K_d = 0.4 n*M*) and act as strong agonists.[14] This interaction is thought to be

Table 2
DETERRENT OR TOXIC EFFECTS OF QA AGAINST INSECT HERBIVORES

Organism	Alkaloid	Effect	Ref.
Homoptera			
Acyrthosiphon pisum[a]	Sparteine	Feeding deterrence ED_{50} 0.01%	26
	Lupinine	Feeding deterrence ED_{50} 0.08%	26
	Cytisine	Feeding deterrence ED_{50} 0.02%	26
Trialeurodes brassicae	QA extract	Insecticidal	91
Lepidoptera			
Spodoptera eridania	Sparteine	Reduction of growth and survivorship	54
	Lupanine	Reduction of growth and survivorship	54
Choristoneura fumiferana	13-Tigloyloxylupanine	Feeding deterrence (89% at 1.4 m*M*)	4
	13-trans-cinnamoyloxy-lupanine	Feeding deterrence (54% at 0.1 m*M*)	4
Syntomis mogadorensis	Sparteine	Feeding deterrence 70% at 1%	148
	Lupanine	Feeding deterrence 50% at 0.1%	148
	Cytisine	Feeding deterrence 100% at 0.1%	148
	17-Oxosparteine	Feeding deterrence 100% at 0.1%	148
Manduca sexta	Sparteine	Feeding deterrence ED_{50} 0.05%	48
Plutella maculipennis	Lupanine	Toxicity LD_{100} 12 m*M*	124,126
	Sparteine	Toxicity LD_{100} 50 m*M*	124,126
	13-Tigloyloxylupanine	Toxicity LD_{100} 6 m*M*	124,126
Pieris brassicae	Sparteine	Phagorepellent	103
Heteroptera			
Dysdercus	Lupanine	Toxicity LD_{100} 12 m*M*	124,126
	Sparteine	Toxicity LD_{100} 50 m*M*	124,126
	13-Tigloyloxylupanine	Toxicity LD_{100} 6 m*M*	124,126
Orthoptera			
Melanoplus bivittatus	Lupinine	Feeding reduction at 0.37% in diet	45
Coleoptera			
Entomoscelis americana	Sparteine	Feeding deterrence (10 m*M* 100%)	75

Callosobruchus maculatus	Sparteine	Lethal at 0.1% in diet	52
Leptinotarsa decemlineata	QA extract from *Lupinus*	Feeding deterrence, growth reduction	64
Phaedon spp.	Lupanine	Toxicity LD_{100} 12 m*M*	124,126
	13-Tigloyloxylupanine	Toxicity LD_{100} 6 m*M*	124,126
Diptera			
Phormia regina	Sparteine	Feeding deterrent	9
Ceratitis capitata	Lupanine	Toxicity LD_{100} 3 m*M*	124,126
	Sparteine	Toxicity LD_{100} 9 m*M*	124,126
	13-Tigloyloxylupanine	Toxicity LD_{100} 6 m*M*	124,126
Hymenoptera			
Apis melifera	Sparteine	Feeding deterrent, ED_{50} 0.03%	23
		Toxicity LD_{50} 0.05%	23
Formica rufa	Sparteine	Feeding deterrent, ED_{100} 1%	151
	Lupanine	Feeding deterrent, ED_{100} 1%	151
	Cytisine	Feeding deterrent, ED_{100} 0.1%	151

[a] Values of the original publication were wrong by a factor of 100 and are corrected here.

Table 3
ORGAN SPECIFIC QA CONCENTRATIONS IN A FEW SELECTED LEGUMES

Species	Organ/tissue	Total alkaloid (per g fresh weight)	Ref.
Cytisus scoparius	Stem epidermis	46 mg/g; 200 m*M*	144
	Shoots	2 mg	124,158
	Leaves	0.2–1 mg	124,158
	Seeds	2 mg[a]	124,158
	Roots	0.03 mg	124,158
Lupinus polyphyllus	Petiole epidermis	1.7–10 mg	124,127,144
	Stem epidermis	6.3 mg	127
	Leaves	1-4 mg	124,126,138
	Stems	1-2 mg	124,126,138
	Flower		
	Pollen	1.8 mg	124,126,138
	Carpels	1.3 mg	124,126,138
	Petals	0.4 mg	124,126,138
	Fruits	1.6 mg	124,126,138
	Seeds	30–40 mg[a]	124,126,138
	Roots	0.2 mg	124,126,138
L. luteus	Stem epidermis	0.6 mg	127
L. mutabilis	Stem epidermis	5.3 mg	127
L. albus	Stem epidermis	6.3 mg	127
	Phloem sap	0.5–1.2 mg/ml	153
	Leaves	2.8 mg	124,126,136,153
	Stem	0.7 mg	124,126,136,153
	Flower	4.1 mg	124,126,136,153
	Fruit	3.1 mg	124,126,136,153
	Seed	43.0 mg[a]	124,126,136,153
	Roots	0.5 mg	124,126,136,153
L. angustifolius	Phloem sap	0.8 mg/ml	143
	Xylem sap	0.05 mg/ml	143

L. consentinii	Phloem sap	5 mg/ml	143
	Xylem	0.05 mg/ml	143
Laburnum anagyroides	Leaves	0.3 mg	111,124,126,
	Twigs		
	Bark	11.1 mg	111,124,126
	Wood	0.5 mg	111,124,126
	Flower	0.4 mg	111,124,126
	Fruit	0.5 mg	111,124,126
	Seed	10–30 mg[a]	111,124,126
	Endosperm	21 mg	111,
	Testa	2 mg	111

[a] Dry weight.

Table 4
UPTAKE AND STORAGE OF QA IN RELATION TO THE ALKALOID PATTERNS OF THE HOST PLANT

Organism	Total QA	Alkaloid pattern (% of total QA)																				
	µg/g FW	N-MC	CYT	DHL	LUP	ANA	RHF	SPA	THR	ANG	HOL	AOL	TOL	3OH	MUL	OXO	DHS	NFC	TIT	MAT	APH	DAP	Ref.
HP: *Teline monspessulana*	2400–4145	5	+		+	+															41	33	160
I: *Uresiphita reversalis*[a]																							
Complete larvae	26–110																						
Larval exuviae	18	84	9																		+	7	160
Cocoon silk	89	81	9																		+	1	160
Feces	3590	+	-																		34	42	160
Pupae	2	96	+																				76, 160
Imagines	1	>97																					160
HP: *Spartium junceum*	n.d.	58	18	+	+	15	9	+										+	2				157
I: *Aphis genistae*	n.d.	12	65	+	+	19												2					157
HP: *Sophora davidii*	600		2	+	15	2					4									58			157
I: *Aphis genistae*	3200		+	+	26	2					5									48			157
HP: *Genista tinctoria*	500	22	58	+	5	14	+	+											4				157
I: *Aphis genistae*	1400	+	97	2	+	+																	157
HP: *Laburnum anagyroides*		10	85	2	4																		111
I: *Aphis cytisorum*	182–1012	3	94	3	0.4																		111
I: *Lasius niger*	45	3	95	1	1																		111
I: *Formica rufibarbis*/ F. cunicularia	n.d.	2	91	7	0.3																		111
I: *Bruchidius villosus*	5–13		100																				111
Feces	31276	1	98																				111
I: *Triaspis thoracius* (Braconidae)	3		100																				111

I: Other parasitoids (pteromalids)	1		100															111
HP: *Petteria ramentacea* (stems)	1200	17	32	6	2	43	+											155
I: *Aphis cytisorum*	4020	+	99															155
HP: *Petteria ramentacea*	400	5	94	+	+	+	+											111
I: *Aphis genistae*	3800	+	98	+	+													111
HP: *Lupinus polyphyllus* (stems)	3800				33			+	6	12	40	+	4					146
I: *Macrosiphum albifrons*	1300				44			+	7	11	25	+	3					146
HP: *L. polyphyllus*	2400				29			+	+	26	30	+	6	+	2	+		157
I: *Macrosiphum albifrons*	200				62			+	+	12	11	+	2	+	+			157
I: *Syntomis mogadorensis* (Lep.)	9																	148
Feces	1023																	148
HP: *L. angustifolius*	1500				28				2	14	55		+					146
I: *Macrosiphum albifrons*	1300				35					7	58							146
HP: *L. angustifolius*	1500				40				+	30	20	+	6					157
I: *M. albifrons*	800				70				+	2	19		2					157
HP: *L. arboreus*	1800				6			90	+	+	+							146
I: *Macrosiphum albifrons*	1500				7			87	+		5							146
HP: *L. mutabilis*	1200				26			+	10		9	7	25	6				146
I: *Macrosiphum albifrons*	700				9			+	1		11	9	40	7				146
Honeydew					35			5	5		25	1	2	20				146
HP: *L. albus*	2200				58						4		+	26				146
I: *Macrosiphum albifrons*	1300				49						15			23				146
HP: *L. albus*	2200				50			+			12	+	10	10				157
I. *M. albifrons*	1500				48			+			6	+	22	15				157
HP: *Cytisus scoparius*	2000				+			85	+		+	+	+			10	4	143
I: *Aphis cytisorum*	30–500				2			36								48	16	143
Pulvinaria vitis (Coc)	10															+		143

Table 4 (continued)
UPTAKE AND STORAGE OF QA IN RELATION TO THE ALKALOID PATTERNS OF THE HOST PLANT

	Total QA	Alkaloid pattern (% of total QA)																					
Organism	**μg/g FW**	**N-MC**	**CYT**	**DHL**	**LUP**	**ANA**	**RHF**	**SPA**	**THR**	**ANG**	**HOL**	**AOL**	**TOL**	**3OH**	**MUL**	**OXO**	**DHS**	**NFC**	**TIT**	**MAT**	**APH**	**DAP**	**Ref.**
Sitona tibialis (Col)	1							+															143
Apion striatum (Col.)	1															+							143
Phytodecta olivacea	0																						143
Sparteine feeding																							
I: *Creatonotos transiens* (Lep)	1																						148
Feces	548																						148
I: *Manduca sexta*	24																						48
Feces	1483																						48

[a] Alkaloid content per animal, d.w. in the case of plants.

HP = hostplant; I = insect; n.d. = not determined; N-MC = N-methylcytisine; CYT = cytisine; DHL = 5,6-dehydrolupanine; LUP = lupanine; ANA = anagyrine; RHF = rhombifoline; SPA = sparteine; THR = tetrahydrorhombifoline; ANG = angustidoline; HOL = 13-hydroxylupanine; AOL = 13-angeloyloxylupanine; TOL = 13-tigloyloxylupanine; 3OH = 3-hydroxylupanine (formerly 4-hydroxylupanine = nuttaline); MUL = multiflorine; OXO = 17-oxosparteine; DHS = 11,12-dehydrosparteine; NFC = N-formylcytisine; TIT = tinctorine; MAT = matrine; APH = epiaphylline/aphylline; DAP = dehydroaphylline; + = trace amounts.

Table 5
BIOLOGICAL ACTIVITIES OF QA

Organism/target	Alkaloid	Effect	Ref.
1. Allelopathy			
Poa spp.	QA mixture	Inhibition of seed germination; ED_{50} 1.5 m*M*	118
Lactuca sativa	QA mixture	Inhibition of seed germination; ED_{50} 1.5 m*M*	118
		Reduction of radicle length	118
	Sparteine	Inhibition of seed germination; 20% at 4 m*M*	118
		Inhibition of radicle growth in *Lactuca* and *Lepidium* ED_{50} 0.1%	150
	Lupanine	Inhibition of seed germination; 65% at 10 m*M*	118
	Cytisine	Inhibition of seed germination; 50% at 6 m*M*	118
		Inhibition of radicle growth of *Lepidium* seedlings (ED_{50} 0.1 %)	150
	13-Tigloyloxylupanine	Inhibition of seed germination; 90% at 6 m*M*	118
Lupinus albus	QA mixture	No inhibition at 8 m*M*	118
Rhaphanus	Sparteine	Inhibition of radicle growth; 40% at 0.01%	151
Sinapis alba	Sparteine	Inhibition of radicle growth; 72% at 1%	151
2. Antiviral activity			
Potato X-virus in *Nicotiana*	Sparteine	Inhibition of viral multiplication	129
3. Antibacterial activity			
Airborne bacteria	Sparteine	Growth inhibition, 60% at 5 mM	122
	Lupanine	Growth inhibition, 80% at 5 mM	122
	13-Tigloyloxylupanine	Growth inhibition, 70% at 5 m*M*	122
Streptococcus viridis (+)	Sparteine	Growth inhibition, $ED_{50} < 0.5$ m*M*	122
Micrococcus luteus (+)	Sparteine	Growth inhibition, ED_{50} 1.5 m*M*	122
Mycobacterium phlei (+)	Sparteine	Growth inhibition, ED_{50} 7.5 m*M*	122
Bacillus megaterium (+)	Sparteine	Growth inhibition, ED_{50} 3 m*M*	122
Bacillus subtilis (+)	Sparteine	Growth inhibition, ED_{50} 0.5 m*M*	122
Serratia marcescens (-)	Sparteine	Growth inhibition, 50% at 10 m*M*	122
Staphylococcus aureus (+)	13-Hydroxylupanine	Growth inhibition, MIC 50 m*M*	114
	Lupanine	Growth inhibition, MIC 50 m*M*	114
	Sparteine	Growth inhibition, MIC 50 m*M*	114
	Angustifoline	Growth inhibition, MIC 50 m*M*	114

Table 5 (continued)
BIOLOGICAL ACTIVITIES OF QA

Organism/target	Alkaloid	Effect	Ref.
Bacillus subtilis (+)	13-Hydroxylupanine	Growth inhibition, MIC 50 m*M*	114
	Lupanine	Growth inhibition, MIC 50 m*M*	114
	Sparteine	Growth inhibition, MIC 50 m*M*	114
	Angustifoline	Growth inhibition, MIC 50 m*M*	114
Bacillus thurigensis (+)	13-Hydroxylupanine	Growth inhibition, MIC 50 m*M*	114
	Lupanine	Growth inhibition, MIC 50 m*M*	114
	Sparteine	Growth inhibition, MIC 50 m*M*	114
	Angustifoline	Growth inhibition, MIC 50 m*M*	114
Escherichia coli (-)	13-Hydroxylupanine	Growth inhibition, MIC 50 m*M*	114
	Lupanine	Growth inhibition, MIC 50 m*M*	114
	Sparteine	Growth inhibition, MIC 50 m*M*	114
	Angustifoline	Growth inhibition, MIC 50 m*M*	114
Pseudomonas aeruginosa (-)	13-Hydroxylupanine	Growth inhibition, MIC 50 m*M*	114
	Lupanine	Growth inhibition, MIC 50 m*M*	114
	Sparteine	Growth inhibition, MIC 50 m*M*	114
	Angustifoline	Growth inhibition, MIC 50 m*M*	114
4. Antifungal activity			
Aspergillus oryzae	Sparteine	Growth inhibition, 42 % at 15 m*M*	122
Alternaria porri	Sparteine	Growth inhibition, 40 % at 15 m*M*	122
Piricularia orycae	Sparteine	Growth inhibition, 18 % at 15 m*M*	122
Helminthosporium carbonum	Sparteine	Growth inhibition, 33 % at 15 m*M*	122
Rhizoctonia solani	Sparteine	Growth inhibition, 15 % at 15 m*M*	122
Fusarium oxysporum	Sparteine	Growth inhibition, 5 % at 15 m*M*	122
Erysiphe graminis f.sp.hordei	Lupanine	Inhibition of conidia germination, ED_{50} 2 m*M*	161
	Sparteine	Inhibition of conidia germination, ED_{50} <2 m*M*	161
	13-Tigloyloxylupanine	Inhibition of mildew development	161
5. Antinematode and antihelminth activity			

Bursaphelenchus xylophilus	N-Methylcytisine	Nematicidal	70
	Anagyrine	Nematicidal	70
	Cytisine	Nematicidal	70
Angiostrongylus cantonens	N-Methylcytisine	Reduction of motility (spastical)(0.001–0.1 m*M*)	112
	Matrine	Reduction of motility (paralytical) (0.01–0.1 m*M*)	112
Dipylidium caninum	N-Methylcytisine	Reduction of motility (spastical) (0.1–1.2 m*M*)	112
	Matrine	Reduction of motility (paralytical) (0.1-1.2 m*M*)	112
Fasciola hepatica	N-Methylcytisine	Reduction of motility (spastical) (0.1–1.2 m*M*)	112
	Matrine	Reduction of motility (paralytical) (0.1–1.2 m*M*)	112
Tufifex	Sparteine	Lethal at 0.03%	151
6. Mollusc deterrence			
Helix pomatia	Sparteine	Feeding deterrent, ED_{50} 0.6–0.7 m*M*	121
	Lupanine	Feeding deterrent, ED_{50} 1–7 m*M*	121
	Cytisine	Feeding deterrent, ED_{50} 2.5 m*M*	121
Biomphalaria glabrata	2,3-Dehydro-O-(2-pyrrolylcarbonyl virgiline	Molluscicidal activity	65
7. Vertebrate toxicity			
K⁺ channels	Sparteine	Inhibition of K^+ channels	61
		Increase of insulin release in β-cells	87
Nerve cells	Matrine	Inhibition of glutamate action	51
Protein biosynthesis	Sparteine	Inhibition of aminoacyl tRNA synthetase	165
	13-Hydroxylupanine	Inhibition of Phe-tRNA binding and elongation, ED_{50} c 0.5 m*M*	63,114
		Inhibition of in vitro translation (wheat germ), 43% at 0.6 m*M*	150
	Lupanine	Inhibition of Phe-tRNA binding and elongation, ED_{50} c 0.5 m*M*	63,114
		Inhibition of in vitro translation (wheat germ), 38% at 0.6 m*M*	150
	Sparteine	Inhibition of Phe-tRNA binding and elongation, ED_{50} c 0.5 m*M*	63,114
		Inhibition of in vitro translation (wheat germ), 38% at 0.6 m*M*	150
	Angustifoline	Inhibition of Phe-tRNA binding and elongation, ED_{50} c 0.5 m*M*	63,114
	13-Tigloyloxylupanine	Inhibition of phe-tRNA binding 70% at 1.8 m*M*	150
		Inhibition of in vitro translation (wheat germ), 37% at 0.6 m*M*	150
	17-Oxosparteine	Inhibition of phe-tRNA binding 50% at 3.1 m*M*	150
		Inhibition of in vitro translation (wheat germ), 35% at 0.6 m*M*	150
	Cytisine	Inhibition of phe-tRNA binding 89% at 2.6 m*M*	150
		Inhibition of in vitro translation (wheat germ), 32% at 0.6 m*M*	150

Table 5 (continued)
BIOLOGICAL ACTIVITIES OF QA

Organism/target	Alkaloid	Effect	Ref.
Mutagenicity	Anagyrine	Congenital malformations in calves	56
	Cytisine	Teratogenic in chicks and rabbits	56
Cytotoxicity	Matrine	Antitumor activity in Ehrlich ascites tumor	62
		Antitumor activity in mouse sarcoma	180
	Matrine-N oxide	Antitumor activity in mouse sarcoma	180 62
General toxicity $LD_{50/100}$	Sparteine	LD_{100} i.p. guinea pig 23–30 mg/kg	20
		LD_{50} i.p. rat, 42–44 mg/kg	92
		s.c. rat, 68–75 mg/kg	92
		LD_{50} i.p. mouse, 55(m)–67(f) mg/kg	164
		i.v. mouse, 17(m)–20(f) mg/kg	164
		p.o. mouse, 350(m)–510(f) mg/kg	164
		200 mg/kg	162
		LD_{100} p.o. rabbit, 450 mg/kg	86
		Lethal dose, i.v. rabbit, 20–30 mg/kg	69
		Lethal dose i.v. dog, 50–70 mg/kg	69
		Lethal dose i.v. pigeon, 40–50 mg/ kg	69
	Lupanine	LD_{100}, i.p. guinea pig 22–25 mg/kg	20
		LD_{50}, i.p. mouse, 80 mg/kg	39
		i.p. rat, 180–192 mg /kg	39
		i.p . guinea pig, 210 mg/kg	39
		LD_{50} i.p. mouse, 175 mg/kg	162
		p.o. mouse, 410 mg /kg	162
		LD_{50} i.p. rat, 177 mg/kg	89
		p.o. rat, 1464 mg/kg	89
	Lupinine	LD_{100}, i.p., guinea pig, 28–30 mg/kg	20
	Epilupinine	LD_{50}, i.p., rat, 20-–400	89
	13-Hydroxylupanine	LD_{100}, i.p. guinea pig, 228 mg/kg	20
		s.c. guinea pig, 456 mg/kg	20
		LD_{50} , i.p. rat, 199 mg/kg	89
		LD_{50}, i.p. mouse, 172 mg/kg	71

	17-Oxolupanine	LD_{50}, i.p. mouse 690 mg/kg	72
	Cytisine	LD_{100}, s.c. cat, 3 mg/kg	36
		s.c. dog, 4 mg/kg	36
		s.c. goat, 109 mg/kg	36
	Matrine	LD_{50}, i.p. mouse, 150 mg/kg	62
	Matrine-N-oxide	LD_{50}, i.p. mouse, 750 mg/kg	62
		i.v. mouse, 150 mg/kg	62
Pharmacological effects	Matrine	Antipyretic (20–30 mg/kg)	18
		Amoebicidal	57
		Antiinflammatory (ED_{50} 12.6 mg/kg)	17
	Matrine-N-oxide	Amoebicidal	57
	Retama raetam with QA	Reduction of reproduction in rat	105
	Multiflorine	CNS depressant	60
	Lupinine	CNS depressant	60
	Sparteine	Oxytocic, uterotonic	60
		Antiarrhythmic	60,22
		Diuretic, hypoglycemic	60
		Respiratory depressant/stimulant	60
		CNS depressant	60
	Retamine	Uterotonic, hypotensive, diuretic	60
	Lupanine	Antiarrhythmic, hypotensive, hypoglycemic	60,22
	13-Hydroxylupanine	Antiarrhythmic, hypotensive	60
	Calpurnine	Ichthyotoxic, antiarrhythmic	60
	Cytisine	Respiratory stimulant (nicotinergic)	60
		Hallucinogenic	60
		Amoebicidal	57
		Hallucinogenic, oxytocic, uterotonic	60

Note: For insect related topics see Table 2.

responsible for the strong human toxicity (Table 5) of goldenrain (*Laburnum*) seeds, which contain cytisine in some quantities.[111,126] Since the ACH receptors are conservative proteins we can assume that insect ACH receptors have similar binding properties and are thus vulnerable to cytisine and to other QA (Table 1). Other targets could be K^+ channels, which are influenced by sparteine in vertebrates[61] or glutamate receptors, which can be modulated by matrine in cray fish neurons[51] (Table 5). The list is certainly far from complete and the topic urgently needs a thorough investigation.

2. Raison d'Être of Alkaloids in the Plant

Plants that produce QA, such as *Cytisus, Genista, Chamaecytisus, Laburnum, Lupinus, Baptisia,* and *Thermopsis,* are relatively abundant in temperate climatic zones and are often dominant members of plant associations, present in pastures, meadows, heath, and ruderal habitats. QA plants suffer severely from the attack of insects and other herbivores rather infrequently, and relatively few adapted insects have been described. (In the latter case, some damage can occur; see Section III.C). Plant breeders have selected lupins (i.e., *L. albus, L. luteus, L. angustifolius,* and more recently, *L. mutabilis*[24]) that produce only low levels of alkaloids (the so-called sweet lupins). Whereas the original alkaloid-rich "bitter" plants were more or less immune to the attack by generalist herbivores, the "sweet" varieties become highly susceptible to them. The reduced fitness of sweet lupins was shown experimentally involving some aphid species, Agromyzidae, thrips, and *Sitona lineatus* (Table 6), but probably includes species such as *Agrotis* spp.(Lepidoptera), *Phorbia platura* (Diptera), *Tipula* spp. (Diptera), *Agriotes* spp. (Coleoptera), *Smithurus viridis* (Collembola), *Ciampa arietaria* (Lepidoptera), *Heliothis punctigera, H. armigera,* (Lepidoptera), *Phytomyza horticola* (Diptera), *Sitona griseus* (Coleoptera), *Acyrthosiphon pisum* (Homoptera), *A. kondoi* (Homoptera), *Aphis craccivora* (Homoptera), *Myzus persicae* (Homoptera), *Calocoris norvegicus* (Heteroptera), *Adelphocoris lineolatus* (Heteroptera), and *Lygus rugulipennis* (Heteroptera).[35,90] It can be assumed that the sweet forms only differ in their alkaloid content from the bitter varieties, and not in other allelochemicals. These data clearly support the importance of alkaloid production for chemical defense against herbivores. Since other allelochemicals are present in both forms, such as flavonoids, isoflavones, anthocyanins, phenolics, saponins, stachyose, etc., some residual resistance is still maintained, even in the absence of QA. This is especially true for roots, which are a rich source of antifeedant and antifungal isoflavones.[66]

B. "SPECIALISTS": UPTAKE AND STORAGE OF QA

Whereas "generalists" and mono-/oligophagous insects, which are not specialized on QA plants, are deterred or intoxicated by QA, just the very opposite seems to be true for those relatively few species that are adapted to QA plants ("few", if we consider the huge number of potential herbivores that live in the same region).

1. Feeding Deterrents or Phagostimulants

Larvae of *Euphydryas editha* (Lepidoptera) is a specialist on *Pedicularis semibarbata,* a parasitic plant that taps the phloem sap of QA-producing species. This herbivore was found not to be influenced by QA and probably it is fully adapted to the alkaloid sequestration by its host plant.[108] Sparteine was reported to be a feeding stimulant for *Acyrthosiphon spartii,* an aphid of sparteine-rich broom (*Cytisus scoparius*).[106] Cytisine was phagostimulatory for larvae of the moth *Uresiphita reversalis,* which feed on *Teline monspessulana,* which produces α-pyridone alkaloids, besides aphylline-type QA (Figure 1)[8,76,77] (Table 4). The lupin aphid, *Macrosiphum albifrons*, lives on lupins that

Table 6
BITTER (ALKALOID-RICH) VERSUS SWEET (LOW-ALKALOID) LUPINS

Species	Lupin	Alkaloid content	Effect	Ref.
Non-Adapted Herbivores				
Vertebrates				
Sheep			Sweet lupins are preferred, bitter discriminated	3
Lepus europaeus			Sweet lupins are preferred, bitter discriminated	113,116
Orytolagus europaeus	*L. albus*	0.01 mg/g	Herbivory almost 100%	126,129,133
		2.0 mg/g	Herbivory <10%	
Insects				
Agromyzidae	*L. albus*	0.01 mg/g	Heavy infestation, 100% incidence	129,133
		2.0 mg/g	Infestation <1%	
		2.2 mg/g	Infestation <1%	
Sitona lineatus	*L. albus*	<0.02 mg/g	100% herbivory	16
		1500 mg/g	Low or no herbivory	
	L. mutabilis	2500 mg/g	Low or no herbivory	
Myzus spp.	*L. luteus*	0.01 mg/g	Infestation 100%	143
		>0.7 mg/g	Infestation <1%	
Acyrthosiphon pisum	*Lupinus*	Sweet	High infestation	116,117
		Bitter	No infestation	
Aphis fabae	*L. polyphyllus*	Sweet	Infestation	43
		Bitter	No infestation	
Frankliniella tritici	*Lupinus*	Sweet	Heavy infestation	32
		Bitter	No infestation	
F. bispinosa	*Lupinus*	Sweet	Heavy infestation	32
		Bitter	No infestation	
Adapted Herbivores				
Macrosiphum albifrons	*L. albus*	0.01 mg/g	Infestation <10%	146
		2.0 mg/g	Infestation 100%	
		2.2 mg/g	Infestation 100%	
	L. polyphyllus	>1 mg/g	Infestation 80%	146
	L. angustifolius	1.5 mg/g	Infestation 100%	146
	L. mutabilis	2.5 mg/g	Infestation 30%	146

contain lupanine as the major alkaloid.[146] Lupins with lupinine, such as *L. luteus*, are usually not accepted as host plants.[29,43,146] Plants storing other QA were not infested[146] — even those with chemically related sparteine as a major alkaloid. Sweet varieties of *L. albus* (the bitter forms are favored host plants) are also discriminated against (Table 6), if the aphids have the choice. Under artificial conditions, i.e., if the aphids were positioned on sweet lupins, normal growth and development occurred, indicating that QA are not a nutritionally or physiologically important factor,[29] but that they regulate behavior. When kept on *L. luteus*, aphids show increased mortality and growth retardation, which might be due to the QA lupinine that is present in this species.[29] All these data imply that the lupin aphid prefers the following situation in nature: first, plants must be rich in alkaloids, and second, plants should contain lupanine as the major alkaloid. Thus QA are critical factors determining host-plant choice in this and probably other species.[7,25]

2. Uptake, Storage, and Processing of QA

A number of studies can be evaluated in which both the alkaloid levels and patterns of host plants and insects have been determined with modern analytical methods (e.g., capillary GLC and GLC-MS[136,159]) (Table 4). Two general strategies can be observed:

1. Nonstorage (and probably non-uptake) of dietary alkaloids and their excretion via the feces (Table 4)

This feature was observed with several beetles that live on *Cytisus* and *Laburnum,* but also on other legumes. Examples include *Bruchidius villosus, Sitonia tibialis, Apion striatum, Phytodecta olivacea,* and *Pulvinaria vitis.* Only for *B. villosus* was the elimination of QA with the feces shown.[111] Three moth larvae that usually avoid QA-rich food showed a similar pattern in that most of the dietary QA was found in the frass[148] (Table 4). Even a species that stores QA, such as *Uresiphita reversalis,* excretes the majority of the ingested QA (Table 4).

2. Uptake and storage of dietary alkaloids

Well-documented examples for QA sequestration are the aphids *Macrosiphum albifrons, Aphis cytisorum,* and *A. genistae*, which exploit the alkaloid-rich phloem sap (Table 4) of QA plants. Alkaloid contents of the aphids were up to 4 mg/g f.w. (equivalent to 20 m*M*). Honeydew analyses showed that part of the QA, originally present in the phloem, was excreted (Table 4), whereas a significant part was stored in the insects. It is difficult to interpret differences in the alkaloid patterns of stored alkaloids vs. QA produced by the plants, since the exact QA composition of the phloem is often unknown; only for *L. albus, L. angustifolius*, and *L. consentinii* are analytical data available.[124,143,153] For *M. albifrons* feeding on *L. albus* and *L. angustifolius,* patterns of both phloem and insect were similar.[153] Patterns differed to some degree between *A. cytisorum* and shoots of the host plant *Cytisus scoparius*, in that 17-oxosparteine was more abundant in the aphid (Table 4). Aphid species from plants with α-pyridone alkaloids usually stored cytisine, being more than 90% of the total alkaloids present, whereas N-methylcytisine was more abundant in the plant. These data either imply that 17-oxosparteine and cytisine were also dominant in the phloem sap (data are not available) or that the aphids converted the original sparteine into 17-oxosparteine and N-methylcytisine into cytisine, respectively.[157]

So far, the only other example of QA sequestration comes from the pyralid moth, *Uresiphita reversalis*, which displays a very interesting adaptation towards QA:[8,76,77,160] in California, larvae are abundant on the scrub and QA plant, *Teline monspessulana.*

The aposematic larvae sequester some QA from their diet but eliminate the major part with the feces. Whereas alkaloid levels are reasonably high in larvae, the pupae and cryptically colored adults are virtually without QA (Table 4). QA are lost with larval exuviae and the cocoon silk that surrounds the pupae. The deposition into the cocoon silk seems to be a very ingenious defense strategy. QA distribution within the larvae suggests that the major proportion is stored in the larval head and in the integument.[160] This is a striking parallel to alkaloid storage in the plant, where also the epidermal tissues constitute the first "defense line".[127,144]

The alkaloid patterns of the larvae and their food plant differ remarkably in that the larvae mainly sequester N-methylcytisine and cytisine, whereas aphylline-type alkaloids dominate in the plant. These aphylline-type alkaloids are found in the feces (Table 4). How is this selectivity achieved? Experiments provide evidence for a carrier-protein mediated uptake of cytisine,[160] which could explain the selectivity observed. Similar carrier mechanisms have been described recently for the uptake of pyrrolizidine alkaloids by larvae of the arctiid moth, *Creatonotos transiens,* and for cardiac glycosides by larvae of the ctenuchid moth, *Syntomeida epilais.*[82,147] It should be recalled that QA are charged molecules under physiological conditions and do not pass biomembranes easily by simple diffusion. In lupins, where these alkaloids are stored in the vacuole, a carrier-mediated uptake mechanism across the vacuolar membrane was described,[74] which is an analogous situation to that in the gut of larval *Uresiphita*.

The aphid *M. albifrons* has its origin in North America, as do most lupins, such as *L. polyphyllus* or *L. arboreus*. Although *L. polyphyllus* was introduced to Europe about three centuries ago, no European aphid has adapted to this new host, which is abundant and naturalized in many parts of Europe (and no aphid is known from the European lupins, such as *L. albus, L. luteus, and L. angustifolius*). The only sucking insect that we have found in one local population of *L. polyphyllus* was *Palomena prasina* (Pentatomidae), which, however, does not sequester QA: QA contents of larvae and adults were between 1 and 10 μg/g FW (Figure 2) — only traces of alkaloids as compared to the plants. The situation changed in 1981 with the arrival of the North American lupin aphid (*Macrosiphum albifrons*) in Europe for the first time — in England. Since then it has spread over large parts of Great Britain and since 1984 also over Germany and other central European countries.[31,43] The lupin aphid may cause severe damage in cultivated lupins. If plants are introduced to new areas (islands, continents), then they often leave their enemies behind (at least temporarily).

C. ECOLOGICAL IMPLICATIONS OF ALKALOID STORAGE

The three aphid species (*M. albifrons, A. genistae,* and *A. cytisorum*) clearly sequester the dietary alkaloids present in their host plants. Total alkaloids were in the range of 0.6 to 1.8 mg/g FW (=2 to 7 mM lupanine) for *M. albifrons,*[146,155,157] (Table 4) and between 1 and 4 mg/g FW (=5 to 20 mM cytisine) for *A. cytisorum.*[146,155,157] It has been determined previously that QA concentrations between 1 and 5 m*M* are repellent and/or toxic for insects (Table 2) and vertebrates (Table 5), or inhibitory for microorganisms (Table 5). Considering the overall biological activities of QA, it is likely that QA-containing aphids are defended against predators, i.e., other insects or birds. For *M. albifrons* it could be demonstrated that carabids (*Carabus problematicus*) were paralyzed for nearly 48 h after the feeding of eight to nine alkaloid-rich aphids.[146] The mortality of larvae of *Coccinella septempunctata*, a known aphid predator, was 100% after 5 d when feeding on alkaloid-rich *M. albifrons* but was only 20% when feeding on alkaloid-free *Acyrthosiphon pisum.*[30,43] Motility of adult *C. septempunctata* was decreased by a factor of three to four when they had been feeding on alkaloid-rich *M. albifrons.*[43] In choice experiments, adult

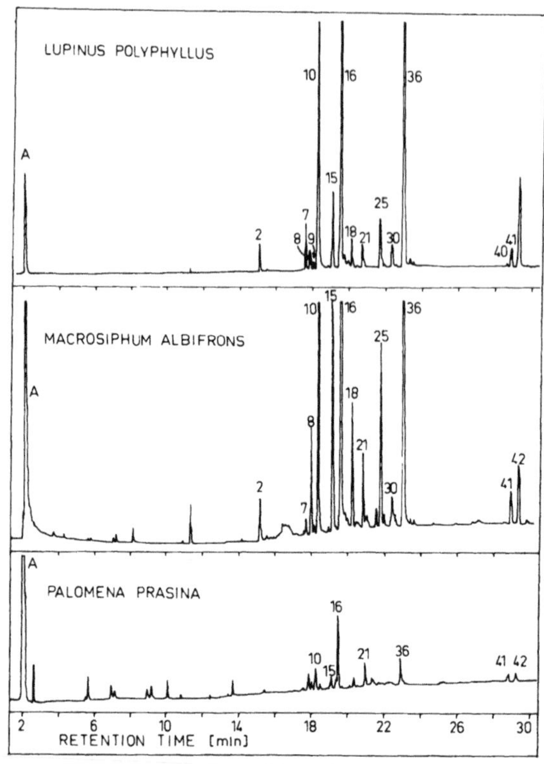

FIGURE 2. Separation of alkaloid extracts of the host plant *Lupinus polyphyllus* and its insect herbivores, *Macrosiphum albifrons* and *Palomena prasina*, by capillary GLC[157] on DB-1 columns. 2 = ammodendrine; 7 = Isoangustifoline; 8 = tetrahydrorhombifoline; 10 = angustifoline; 15 = dehydrolupanine; 16 = lupanine; 18 = 11,12-seco-12,13-didehydromultiflorine (formerly N-methylalbine); 21 = 3-hydroxylupanine (formerly nuttaline); 25 = multiflorine; 30 = A-1; 36 = 13-hydroxylupanine; 40 = 13-(2-methylbutyryl)-oxylupanine; 41 = 13-angeloyloxylupanine; 42 = 13-tigloyloxylupanine.

C. septempunctata killed significantly more *Acyrthosiphon pisum* than *M. albifrons*, indicating a deterrent effect. Syrphid larvae (*Episyrphus balteatus*) displayed growth retardation and an increased mortality when reared on *M. albifrons*.[30] These data provide good evidence for the protective function of QA, acquired from the host plants. No data exist for *A. cytisorum* and *A. genistae*, but since cytisine is considered to be even more toxic than sparteine or lupanine (Table 5), and since the alkaloid concentration is usually higher in *A. cytisorum* and *A. genistae* than in *M. albifrons*, it is very likely that QA storage also confers protection on *A. cytisorum* and *A. genistae*.

Aphis cytisorum and *A. genistae* colonies are regularly visited by ants, which collect the honeydew. It was shown by Szentesi and Wink[111] that *Lasius niger* that were collected from an *A. cytisorum* colony contained about 45 μg/g FW QA, mostly cytisine (Table 4). Thus QA pass through several trophic levels. However, it is not known what role, if any, QA play in ants. A related phenomenon was described for ants visiting *Aphis jacobaeae* on *Senecio jacobaea*, which exploited the pyrrolizidine alkaloids present. The ants defended their aphids and host plants against the attack of herbivorous larvae of the arctiid moth, *Tyria jacobaea*.[115] We suggest that the interaction between QA plants, their aphids, and ants could be of a similar nature, but this is not unusual and is not restricted to alkaloidal plants.

Uresiphita larvae obviously gain some advantage from QA storage: two hymenopteran predators, the Argentine ant, *Iridomyrmex humilis*, and the paper wasp,

Mischocyttarus flavitarsus, strongly prefer other, more palatable, lepidopteran larvae over *U. reversalis*. Extracts from *U. reversalis,* but also pure cytisine and sparteine solutions, that were applied on the surface of otherwise palatable prey, such as larvae of *Phthorimaea operculella,* were highly deterrent to both ants and wasps.[77] Thus it is likely that QA function as acquired defense compounds also in this context.

IV. QA AS CHEMICAL DEFENSE COMPOUNDS IN OTHER INTERACTIONS

It could be shown experimentally that pure QA or QA mixtures are also biologically active in interrelationships other (Table 5) than with plants and insects (for review of the older literature, see Kinghorn and Balandrin[60]).

A. ALLELOPATHY

Some evidence suggests activity against other competing plants.[118,150] During germination QA are exported from the roots into the soil. The main alkaloids detected were 13-tigloyloxylupanine and other esters that displayed the strongest herbicidal effects in vitro.[118,126] Since QA are rapidly degraded in soil,[41] it remains to be established whether and how an allelopathic effect can be obtained under field conditions.

B. ANTIVIRAL ACTIVITY

QA inhibited the multiplication of potato-X virus in laboratory experiments,[129] but under field conditions even alkaloid-rich lupins are infected by a number of viruses. They are transmitted by aphids that do not live on bitter lupins but on other legumes or other crops. They nevertheless probe bitter lupins and transmit the virus.[95]

C. ANTIMICROBIAL ACTIVITY

The growth of gram-negative and gram-positive bacteria[114,122] and of certain fungi[122,161] is significantly reduced by some QA in vitro (Table 5). However, even "bitter" lupins suffer from a number of fungal pathogens, such as *Rhizoctonia*, *Fusarium*, *Phoma*, *Ascochyta*, *Colletotrichum*, *Uromyces*, and *Erisyphe*,[35,90,95] which are obviously not very sensitive to QA. In the case of the fungus *Phomopsis leptostromiformis,* it is tempting to speculate that lupin–fungus interactions are not parasitic but symbiotic in reality: lupins are known to produce a disease in sheep and other herbivores, the so-called lupinosis. Originally, alkaloids were assumed to be the causative agents. Recently, a group of fungal toxins, which are hexapeptides called phomopsins, has been found to be responsible. Phomopsin levels up to 1.5 mg/kg seeds or foliage have been detected. Phomopsins are strong poisons, with LD_{50} values (p.o.) in rats and sheep of 1 to 2 mg/kg. Their molecular target is tubulin and the inhibition is as strong as observed for the known tubulin poison, vincristine. As a consequence, mitosis is blocked and long-term consequences are hepatocarcinomas.[21] Since the fungus does not kill the host plant, the production of a potent phytotoxin could be helpful for the plant in the defense against herbivores, in addition to chemical defense by QA.

D. ANTIHERBIVORE ACTIVITY

QA deter or repel feeding of a number of noninsect herbivores (and other animals), e.g., nematodes, snails, rabbits, and cows, or are directly toxic or mutagenic to them (Table 5). For example, anagyrine causes malformations, the so-called crooked calf disease, in young sheep and calfs, when their mothers feed on lupins or broom containing anagyrine (ingested at 7 to 11 mg/kg/d during days 40 and 75 of gestation).[56,59] Alkaloid extracts of *Lupinus angustifolius* that contain lupanine, angustifoline, and 13-hydroxylupanine did not show mutagenic activity in a *Salmonella* and Chinese hamster cell system.[21] This suggests that the mutagenic effect is limited to anagyrine and perhaps other α-pyridone alkaloids but is not caused by the lupanine-type QA.

QA are obviously bitter to humans and other mammals: mean detection levels in humans are 0.00085% for sparteine, 0.0021% for lupanine, and 0.017% for 13-hydroxylupanine.[10] In animals, the sensitivity to bitter lupins increases from sheep, over rabbit, guinea pig, to mouse and pig.[104] This indicates that mammalian herbivores are able to detect and to avoid harmful levels (compare Table 5) of dietary QA . Interestingly, young and adult geese (*Anser anser*, *A. indicus* and *Branta canadensis*) do not detect the bitter sparteine and feed deliberately on an artificial diet with up to 1% sparteine.[149] Data on other birds, such as insect predators, are not available.

Vertebrate toxicity was assessed for those QA that are abundant and are frequently found in plants (Table 5): alkaloids of the sparteine-lupanine type are relatively toxic when injected (i.v or i.p.) but less so when given orally. The α-pyridone alkaloids, such as cytisine, are approximately one order of magnitude more toxic.

The toxic effects may be due to inhibition of K^+ ion channels, to interactions with the acetylcholine receptor (like cytisine and nicotine), to inhibition of protein synthesis, and to other mechanisms that have not been elucidated.[14,36,61,63,87,150]

E. ECOLOGICAL FUNCTION

QA concentrations are sufficiently high in the plant to allow their inhibitory effects observed in vitro (Table 3). In addition, QA content can be increased by wounding: this short-term effect was highest under greenhouse conditions, but was also measurable in the field.[53,119] In this context QA localization in epidermal tissues can be interpreted as a strategically important adaptation, since this tissue has to ward off small herbivores and pathogens in the first instance. As has been discussed above, "sweet" lupins have a substantially reduced fitness under natural conditions as compared to their "bitter" wild forms (Table 6). They are preferentially eaten by rabbits and sheep (and insects, see Section III.B), and are vulnerable to other pathogens.[126,129,133,135] Thus it seems well established that QA function as chemical defense compounds against herbivores, but probably also against microorganisms, although saponins, lectins, protease inhibitors, stachyose and verbascose, phenolics, isoflavonoids, and anthocyanins[47] found in QA plants are certainly also important in the latter context.

An analogous defense system in vertebrates, the immune system, is closely integrated into the developmental program, the physiology and metabolism of an animal. Many thousand man years of research were needed and are certainly still demanded to fully understand its mechanism, function, and regulation. Similar to the plants' defense system we find specialists (parasites, microbial pathogens), which have overcome the normally functioning immune system.

V. CONCLUSIONS

Quinolizidine alkaloids constitute one group of allelochemicals that have been studied in enough detail already (although many facets are still missing!) that it is possible to obtain an inkling of the intricate complexity that underlies their biosynthesis, transport, storage, and degradation, and their integration in the ecology of a plant. All available data suggest that QA-producing plants exploit and depend on their alkaloids for chemical defense, mainly against insect and vertebrate herbivores, but additional activities at the antiviral, antibacterial, antifungal, and herbicidal level are likely. Herbivores, on the other hand, have developed several strategies to cope with this defense chemistry, and the present data indicate strong differences between generalists and the specialists, which are closely adapted to their QA-containing host plants. Specialists that have found biochemical means to sequester QA profit from them as acquired defense compounds against predators. Other groups of natural products that show a similar complexity of interactions and interdependencies include the pyrrolizidine alkaloids and the cardiac glycosides.[13,44,81,97,100]

Only when we understand how plants defend themselves can we devise ecologically compatible strategies to develop crops resistant to pests and disease. Our lupin example clearly shows that the traditional way of plant breeding does not consider arguments of fitness mediated by allelochemicals. Reducing alkaloids seems plausible only on first sight; sweet lupins can only be grown if their lost chemical defense is replaced by human-made pesticides and fences. We have proposed two alternatives:[135] (1) To select for lupin mutants that do not translocate the alkaloids to the seeds. In this case, the plant would retain its chemical resistance but would provide the valuable alkaloid-free seeds at the same time. (2) To grow alkaloid-rich plants but to process the seeds, which are important because of their high levels of proteins and oil (in the case of *L. mutabilis*). During such a technological processing, the original products can be diversified and refined. One component of this diversification would be an alkaloid fraction. Considering the biological activities of QA (Tables 2 and 5) it seems promising to exploit the alkaloids as natural plant protectants, similar to the use of other allelochemicals, such as pyrethroids or the constituents of the neem tree.

In a long-term perspective, it might be interesting to transfer the genes that are responsible for the production of allelochemicals from one species to another one in order to improve the fitness of the recipient species. However, much more basic knowledge is necessary before such a strategy could be realized.

Since the alkaloids and other natural products are not just random structures but have been shaped during evolution by natural selection, these compounds provide a valuable catalogue of biologically active compounds. However, their targets are not known in most instances. These active structures were and might be interesting in the future for medicine, because of their pharmacological activity, which is a "side effect" of their main biological function. They may also be important as biochemical probes for certain reactions and mechanisms in the way that cytisine is being used to localize nicotinergic receptors on neurons. Many different applications may be feasible (Table 5). Although a large body of knowledge exists on the traditional use of herbs for folk medicine, much awaits to be explored in the future. In this context it is interesting to recall that the Indians of Latin America have utilized bitter *L. mutabilis* plants for a long time.[2,15] They removed the alkaloids from the seeds by cooking and then by leaching them out in water. The extracted alkaloids were used for treatment of heart disease (sparteine is still in use as an antiarrhythmic drug in Europe), against skin diseases, rheumatism, against ectoparasites of sheep, llamas and alpacas, and as an antihelminth, insecticide, and fish poison.[2,15] But besides these applied aspects, the study of plant–herbivore and plant–pathogen interactions is rewarding and important in itself, since it provides interesting insight into the biology, ecology, and evolution of plants and animals.

An unexplored possibility is that, among insects that are adapted to QA and sequester them, protection is obtained not just from predators but from pathogens also and that the ecological complexity of QA is even greater.

ACKNOWLEDGMENTS

Our research was supported by grants of the Deutsche Forschungsgemeinschaft.

REFERENCES

1. **Ahmad, S.,** Mixed-function oxidase activity in a generalist herbivore in relation to its biology, food plants and feeding history, *Ecology*, 64, 235, 1983.
2. **Antunez de Mayolo, S.,** Tarwi in ancient Peru, in *Agricultural and Nutritional Aspects of Lupines*, Gross, R. and Bunting, E. S., Eds., GTZ, Eschborn, 1982, 1.
3. **Arnold, G. W. and Hill, J. I.,** Chemical factors affecting selection of food plants by ruminants, in *Phytochemical Ecology*, Harborne, J. B., Ed., Academic Press, New York, 1971, 92.
4. **Bentley, M. D., Leonard, D. E., Reynolds, E. K., Leach, S., Beck, A. B., and Murakoshi, I.,** Lupine alkaloids as larval feeding deterrents for spruce budworm, *Choristoneura fumiferana* (Lepidoptera: Tortricidae), *Ann. Entomol. Soc. Am.*, 77, 398, 1984.
5. **Bernays, E.,** The insect on a plant — a closer look, in *Insect-Plant Relationships*, Visser, J. H. and Minks, A. K., Eds., Wageningen, 1982, 3.
6. **Bernays, E. A. and Chapman, R.,** The evolution of deterrent responses in plant-feeding insects, in *Perspectives in Chemoreception and Behaviour*, Bernays, E., Chapman, M., and Stoffolano, J., Eds., Springer, New York, 1987.
7. **Bernays, E. A. and Graham, M.,** On the evolution of host specificity in phytophagous arthropods. *Ecology*, 69, 886, 1988.
8. **Bernays, C. B. and Montllor, J.,** Aposematism of *Uresiphita reversalis* larvae (Pyralidae), *J. Lepidopt. Soc.*, 43, 261, 1989.
9. **Blades, D. and Mitchell, B. K.,** Effects of alkaloids on feeding by *Phormia regina*, *Entomol. Exp. Appl.*, 41, 299, 1986.
10. **Bleitgen, R., Gross, R., and Gross, U.,** Die Lupine, ein Beitrag zur Nahrungsversorgung in den Anden. 5. Einige Beobachtungen zur traditionellen Entbitterung von Lupinen in Wasser, *Z. Ernährungswiss*, 18, 104, 1979.
11. **Brattsten, L. B.,** Enzymatic adaptations in leaf-feeding insects to host-plant allelochemicals, *J. Chem. Ecol.*, 14, 1919, 1988.
12. **Brattsten, L. B. and Ahmad, S.,** Molecular aspects of insect-plant associations, Plenum Press, New York, 1986.
13. **Boppré, M.,** Lepidoptera and pyrrolizidine alkaloids, *J. Chem. Ecol.*, 16, 165, 1990.
14. **Boska, P. and Quiron, R.,** *Eur. J. Pharmacol.* 137, 323, 1987.
15. **Brücher, H.,** Die genetischen Reserven Südamerikas für die Kulturpflanzenzüchtung, *Theor. Appl. Genet.*, 38, 9, 1968.
16. **Cantot, P. and Papineau, J.,** Discrimination des lupins basse teneur en alcaloides par les adultes de *Sitonia lineatus* L. (Col.; Curculionidae), *Agronomie*, 3, 937, 1983.
17. **Cho, C. H. and Chuang, C. Y.,** Study on the antiinflammatory action of matrine: an alkaloid isolated from *Sophora subprostrata*, *IRCS Med. Sci.*, 14, 441, 1986.
18. **Cho, C. H., Chuang, C. Y., and Chen, C. F.,** Study of the antipyretic activity of matrine. A lupin alkaloid isolated from *Sophora subprostrata*, *Planta Med.*, 343, 1986.
19. **Cordero, C. M., Ayuso, J., Richomme, P., and Bruneton, J.,** Quinolizidine alkaloids from *Viscum cruciatum*, hemiparsitic scrub of *Lygos sphaerocarpa*, *Planta Med.*, 55, 196, 1989.
20. **Couch, J. F.,** Relative toxicity of lupin alkaloids, *J. Agr. Res.*, 32, 51, 1926.
21. **Culvenor, C. C. J. and Petterson, D. S.,** Lupin toxins—alkaloids and phomopsins, in *Proc. 4th Intl. Lupin Conference*, Geralton, 1986, 188.
22. **Czarnecka, E., Kolinska-Marzec, A., and Szadowska, A.,** Effects of some lupine alkaloids on rhythm of the isolated heart, *Acta Polon. Pharm.*, 24, 545, 1967.
23. **Detzel, A. and Wink, M.** (in preparation).
24. **Dovrat, A.,** Aspects of grain lupin cultivation: an overview, in *Lupin Production and Bio-Processing for Feed, Food and Other By-Products*, Birk, Y., Dovrat, A., Waldman, M., and Uzureau, C., Eds., EEC, Luxembourg, 1990, 5.
25. **Dreyer, D. and Campbell, B. C.,** Chemical basis of host-plant resistance to aphids, *Plant Cell Environ.*, 10, 353, 1987.
26. **Dreyer, D., Jones, K. C., and Molyneux, R. J.,** Feeding deterrency of some pyrrolizidine, indolizidine, and quinolizidine alkaloids towards pea aphid (*Acyrthosiphon pisum*) and evidence for phloem transport of the indolizidine alkaloid swainsonine, *J. Chem. Ecol.*, 11, 1045, 1985.
27. **Duffey, J.,** Sequestration of plant natural products by insects, *Annu. Rev. Entomol.*, 25, 447, 1980.
28. **Ehrlich, P. R. and Raven, P. H.,** Butterflies and plants: a study of coevolution, *Evolution*, 18, 586, 1964.

29. **Emrich, B. H. and Schmutterer, H.,** Host-modulated biological characters of the lupin aphid, *Macrosiphum albifrons* Essig (Homoptera: Aphididae), *Med. Fac. Landbouww. Rijsuniv. Gent.,* 55, 463, 1990.
30. **Emrich, B. H.,** Erworbene Toxizität bei der Lupinenblattlaus *Macrosiphum albifrons* und ihr Einfluβ auf die aphidophagen Prädatoren *Coccinella septempunctata, Episyrphus balteatus* und *Chryoperla carnea, Z. Pflanzenkrankh. Pflanzensch.,* (in press).
31. **Eppler, A. and Hinz, U.,** Die Lupinenblattlaus *Macrosiphum albifrons* Essig, ein neuer Schaderreger und Virusvektor in Deutschland, *J. Appl. Entomol.,* 104, 510, 1987.
32. **Forbes, I. and Beck, E. W.,** A rapid biological technique for screening the blue lupin population for low alkaloid plants, *Agron. J.,* 46, 528, 1954.
33. **Fraser, A. M. and Robins, D. J.,** Incorporation of enantiomeric cadaverines1-d into the alkaloids (+)-sparteine and (-)-N-methylcytisine in *Baptisia australis, J. Chem. Soc. Chem. Comm.,* 545, 1986.
34. **Fraser, A. M. and Robins, D. J.,** Application of deuterium NMR spectroscopy to study the incorporation of enantiomeric deuterium-labelled cadaverines into alkaloids, *J. Chem. Soc. Perkin Trans.,* I, 105, 1987.
35. **Frey, F. and Yabar, E.,** *Enfermedades y Plagas de Lupinos en el Peru,* GTZ, Eschborn, 1983.
36. **Gessner, O.,** *Die Gift- und Arzneipflanzen in Mitteleuropa,* C. Winter, Heidelberg, 1953.
37. **Golebiewski, W. M. and Spenser, I. D.,** Deuteron NMR spectroscopy as a probe of the stereochemistry of biosynthetic reactions: the biosynthesis of lupanine and sparteine, *J. Am. Chem. Soc.,* 106, 7925, 1984.
38. **Golebiewski, W. M. and Spenser, I. D.,** Biosynthesis of the lupine alkaloids, *Can. J. Chem.,* 63, 2707, 1986.
39. **Gordon, W. C. and Henderson. J. H. M.,** The alkaloid content of blue lupine and its toxicity on small laboratory animals, *J. Agr. Sci.,* 41, 141, 1951.
40. **Greinwald, R., Wink, M., Witte, L., and Czygan, F.-C. ,** Das Alkaloidmuster der Pfropfchimäre *Laburnocytisus adamii* (Fabaceae), *Biochem. Physiol. Pflanzen.,* 87, 385, 1991.
41. **Gross, R. and Wink, M.,** Degradation of sparteine in soil, *Lupin Newsl.,* 9, 15, 1986.
42. **Grundon, M. F.,** Quinolizidine alkaloids, *Alkaloids* (London), 4, 104, 1974; 5, 93, 1975; 6, 90, 1976; 7, 69, 1977; 8, 66, 1978; 9, 69; 1979; 10, 66, 1980; 11, 63, 1981; 12, 73, 1982; 13, 87, 1983; *Nat. Prod. Rep.,* 1, 349, 1984; 2, 235, 1985; 4, 415, 1987; 6, 523, 1989.
43. **Gruppe, A. and Römer, P.,** The Lupin Aphid (*Macrosiphum albifrons* Essig, 1911) (Hom., Aphididae) in West Germany: its occurrence, host plants and natural enemies, *J. Appl. Entomol.,* 106, 135, 1988.
44. **Harborne, J. B.,** *Introduction to Ecological Biochemistry,* 2nd ed. Academic Press, New York, 1982.
45. **Harley, K. L. S. and Thorsteinson, A. J.,** The influence of plant chemicals on the feeding behaviour, development and survival of the two striped grasshopper *Melanoplus bivittatus,* Acrididae, Orthoptera, *Can. J. Zool.,* 45, 305, 1967.
46. **Hartmann, T.,** Prinzipien des pflanzlichen Sekundärstoffwechsels, *Plant Syst. Evol.,* 150, 15, 1985.
47. **Hatzold, T.,** Chemische und chemisch-technische Untersuchungen zur Beurteilung von Lupinen (*L. mutabilis*) als Nahrungsmittel für den Menschen. Schwerpunkt: Lipid- und Alkaloidfraktionen, Dissertation, Univ. Gieβen, 1982.
48. **Heidrich, P. and Wink, M.,** (in preparation).
49. **Herbert, R. B.,** Biosynthesis-I. General., *Alkaloids* (London), 1, 1, 1971; 2, 1, 1972; 3, 1, 1973; 4, 1, 1974; 5, 1 , 1975; 6, 90, 1976; 7, 1, 1977; 8, 1, 1986; 9, 1, 1979; 10, 1, 1980; 11, 1, 1981; 12, 1, 1982; 13, 1, 1983; *Nat. Prod. Rep.,* 2, 165, 1985; 3, 185, 1986; 4, 423, 1987.
50. **Howard, A. S. and Michael, J. P.,** Simple indolizidine and quinolizidine alkaloids, in *The Alkaloids,* 28, 183, 1986.
51. **Ishida, M. and Shinozaki, H.,** Glutamate inhibitory action of matrine at the crayfish neuromuscular junction, *Br. J. Pharmacol.,* 82, 523, 1984.
52. **Janzen, D., Inster, H. B. , and Bell, E. A.,** The toxicity of secondary compounds to the seed eating larvae of the Bruchid beetle, *Callosobruchus maculatus, Phytochemistry,* 16, 223, 1977.
53. **Johnson, N. D., Rigney, L., and Bentley, B. L.,** Short-term induction of alkaloid production in lupines. Differences between N_2-fixing and nitrogen-limited plants., *J. Chem. Ecol.,* 15, 2425, 1989.

54. **Johnson, N. D. and Bentley, B. B.,** Effects of dietary protein and lupine alkaloids on growth and survivorship of *Spodoptera eridania*, *J. Chem. Ecol.*, 14, 1391, 1988.
55. **Kaissling, R. H.** *Wright Lectures on Insect Olfaction,* Simon Fraser University, Burnaby, 1987.
56. **Keeler, R. F.,** Lupin alkaloids from teratogenic and nonteratogenic lupins: III. Identification of anagyrine as the probable teratogen by feeding trials, *J. Toxicol. Envir. Health*, 1, 887, 1976.
57. **Keene, A., Harris, A., Phillipson, J. D., and Warhurst, D. C.,** In vitro amoebicidal testing of natural products; Part 1. Methodology, *Planta Med.*, 278, 285, 1986.
58. **Keller, W. J.,** Alkaloids from *Sophora secundiflora*, *Phytochemistry*, 14, 2305, 1975.
59. **Kilgore, W. W., Crosby, D. G., Craigmill, A. L., and Pappen, N. K.,** Toxic plants as possible human teratogens, *Calif. Agric.*, 6, 1981.
60. **Kinghorn, A. D. and Balandrin, M. F.,** Quinolizidine alkaloids of the Leguminosae: structural types, analysis, chemotaxonomy, and biological activities, in *Alkaloids: Chemical and Biological Perspectives*, Vol. 2, Pelletier, W. S., Ed., Wiley, New York, 1984, 105.
61. **Kleinhaus, A. L.,** *J. Physiol.*, 299, 309, 1980.
62. **Kojima, R., Fukushima, S., Ueno, A., and Saiki, Y.,** Antitumour activity of Leguminosae plant constituents. I. Antitumour activity of constituents of *Sophora subprostrata*, *Chem. Pharm. Bull.*, 18, 2555, 1970.
63. **Korcz, A., Markiewicz, M., Pulikowska, J., and Twardowski, T.,** Species-specific inhibitory effects of lupine alkaloids on translation in plants, *J. Plant Physiol.*, 128, 433, 1987.
64. **Krzymanska, J., Waligora, D., Michalski, Z., Peretiatkowicz, B., and Gulewicz, K.,** Observations on the influence of spraying potatoes with lupine extract on the feeding and development of potato beetle populations (*Leptinotarsa decemlineata*), *Bull. Pol. Acad. Sci.*, Biol. Ser. 36, 45, 1988.
65. **Kubo, I., Matsumoto, T., Kozuka, M., and Chapya, A.,** The quinolizidine alkaloids from the African medicinal plant, *Calpurnea aurea:* molluscicidal activity and structural study by 2D-NMR, *Agric. Biol. Chem.*, 48, 2839, 1984.
66. **Lane, G. A., Sutherland, O. R. W., and Skipp, R. A.,** Isoflavonoids as insect feeding deterrents and antifungal compounds from roots of *Lupinus angustifolius*, *J. Chem. Ecol.*, 13, 771, 1987.
67. **Le Scao-Bogaert F., Faugeras, G., and Paris, R. R.,** Sur la presence d'alcaloides quinolizidinique chez *l'Osyris alba* (San talacees), *Plant. Medic. Phytother*, 12, 315, 1987.
68. **Levin, D. A.,** The chemical defences of plants to pathogens and herbivores, *Ann. Rev. Ecol. Syst.*, 7, 121, 1976.
69. **Lu, G. C. ,** The cause of death and resuscitation in sparteine poisoning, *Toxicol. Appl. Pharmacol.*, 6, 328, 1963.
70. **Matsuda, K., Kimura, M., Komai, K., and Hamada, M.,** Nematicidal activities of (-)-N-methylcytisine and (-)-anagyrine from *Sophora flavescens* against pine wood nematodes, *Agric. Biol. Chem.*, 53, 2287, 1989.
71. **Mazur, M., Polakowski, P., and Szadowska,** Pharmacology of lupinine and 13-hydroxylupanine, *Acta Physiol. Polonica*, 17, 299, 1966.
72. **Mazur, M., Polakowski, P., and Szadowska,** Pharmacology of 17-oxolupanine, lupanine-N-oxide and 17-hydroxylupanine, *Acta Physiol. Polonica*, 17, 311.
73. **Mears, J. A. and Mabry, T. J.,** Alkaloids in the Leguminosae in *Chemotaxonomy of the Leguminosae*, Harborne, J. B., Boulter, D., and Turner, B. L., Eds., Academic Press, London, 1971, 73.
74. **Mende, P. and Wink, M.,** Uptake of the quinolizidine alkaloid lupanine by protoplasts and vacuoles of *Lupinus polyphyllus* cell suspension cultures, *J. Plant Physiol.*, 129, 229, 1987.
75. **Mitchell, B. K. and Sutcliffe, J. F.,** Sensory inhibition as a mechanism of feeding deterrence: effects of three alkaloids on leaf beetle feeding, *Physiol. Entomol.*, 9, 57, 1984.
76. **Montllor, C. B., Bernays, E. A., and Barbehenn, R. V.,** Importance of quinolizidine alkaloids in the relationship between larvae of *Uresiphita reversalis* (Lepidoptera: Pyralidae) and a host plant, *Genista monspessulana*, *J. Chem. Ecol.*, 16, 1853, 1990.
77. **Montllor, C. B., Bernays, E. A. and Cornelius, M. L.** Responses of two hymenopteran predators to surface chemistry of their prey: significance for an alkaloid-sequestering caterpillar, *J. Chem. Ecol.*, 17, 391, 1991.
78. **Mothes, K. and Schütte, H. R.,** *Biosynthese der Alkaloide.* VEB, Berlin, 1969.
79. **Mothes, K., Schütte, H. R., and Luckner, M.,** *Biochemistry of Alkaloids*, Verlag Chemie, Weinheim, 1985.

80. **Murakoshi, I., Ogawa, Toriizuka, K., Haginiwa, J., Ohmiya, S., and Otomasu, H.,** The enzymatic conversion of lupinine to trans-(4'hydroxycinnamoyl)lupinine by extracts of Lupinus seedlings, *Chem. Pharm. Bull.*, 25, 527, 1977.
81. **Nahrstedt, A.,** Strukturelle Beziehungen zwischen pflanzlichen und tierischen Sekundärstoffen, *Planta Med.*, 44, 2, 1982.
82. **Nickisch-Rosenegk, Detzel, A., Schneider, D., and Wink, M.,** Carrier-mediated uptake of digoxin by larvae of the cardenolide sequestering moth, *Syntomeida epilais, Naturwissenschaften*, 77, 336, 1990.
83. **Nickisch-Rosenegk, E., Schneider, D., and Wink, M.,** Time-course of pyrrolizidine alkaloid processing in the alkaloid exploiting arctiid moth, *Creatonotos transiens, Z. Naturforsch.*, 45c, 881, 1990.
84. **Nishida, R. and Fukami, H.,** Host plant iridoid-based chemical defense of an aphid, *Acyrthosiphon nipponicus*, against ladybird beetles, *J. Chem. Ecol.*, 15, 1837, 1989.
85. **Nowacki, E. and Waller, G. R.,** Quinolizidine alkaloids from Leguminosae, *Rev. Latinoamer. Quim.*, 8, 49, 1977.
86. **Nowacki, E. and Wezyk, S.,** Preliminary research on lupin alkaloids. Toxicity for rabbit organism, *Rozniki Nauk Rolnikzych*, 75, 385, 1960.
87. **Paolisso, G., Nenquin, M., Schmeer, W., Mathot, F., Meissner, H. P., and Henquin, J. C.,** Sparteine increases insulin release by decreasing K^+ permeability of the B-cell membrane, *Biochem. Pharmacol.*, 34, 2355, 1985.
88. **Perrey, R. and Wink, M.,** On the role of piperideine and tripiperideine in the biosynthesis of quinolizidine alkaloids, *Z. Naturforsch.*, 43c, 363, 1988.
89. **Petterson, D. S., Ellis, Z. C., Harris, D. J., and Spadek, Z. E.,** Acute toxicity of the major alkaloids of cultivated *Lupinus angustifolius* seeds to rats, *J. Appl. Toxicol.*, 7, 51, 1987.
90. **Plancquaert, P.,** Protection integree des cultures de lupin, in *Proc. 3rd Intl. Lupin Conference*, ILA, 1984, 238.
91. **Plank, A.,** Die Lupine: Chancen und Risiken einer Produktionsalternative. Untersuchungen über den Anbau der Lupine in Österreich und ihre Verwertung als Eiweiss- und Alkaloidpflanze, Dissertation, Univ. Wien, 1989.
92. **Poe, C. F. and Johnson, C. C.,** Toxicity of hydrastine, hydrastinine and sparteine, *Acat Pharmacol. Toxicol.*, 10, 338, 1954.
93. **Rana, J. and Robins, D. J.,** Biosynthesis of quinolizidine alkaloids. Incorporation of cadaverine-1-amino-^{15}N,1-^{13}C into lupanine, 13-hydroxylupanine and angustifoline, *J. Chem. Soc., Synop.*, 196.197, 1985.
94. **Robinson, T.,** Metabolism and function of alkaloids in plants, *Science*, 184, 430, 1974.
95. **Römer, P.,** Genetische und physiologische Untersuchungen an *Lupinus mutabilis* Sweet., Dissertation, Univ. Gieβen, 1990.
96. **Rosenthal, J.,** Plant nonprotein amino acids and imino acids, Academic Press, New York, 1982.
97. **Rosenthal, J. and Janzen, D.,** *Herbivores: Their Interactions with Plant Secondary Metabolites*, Academic Press, New York, 1979.
98. **Salatino, A. and Gottlieb, O. R.,** Quinolizidine alkaloids as systematic markers of the Papilionoidea, *Biochem. Syst. Ecol.*, 8, 133, 1980.
99. **Saxton, J. E.,** Quinolizidine alkaloids, *Alkaloids* (London). 1, 83, 1971; 2, 79, 1972; 3, 95, 1973.
100. **Schneider, D.,** The strange fate of pyrrolizidine alkaloids, in *Perspectives in Chemoreception Behaviour*, Chapman, R. F., Bernays, E. A., and Stoffolano, J. G., Eds., Springer, Berlin, 1987, 123.
101. **Schneider, M. J. and Stermitz, F. R.,** Uptake of host plant alkaloids by root parasitic *Pedicularis* species, *Phytochemistry*, 29, 1811, 1990.
102. **Schoofs, G., Teichmann, S., Hartmann, T., and Wink, M.,** Lysine decarboxylase in plants and its integration in quinolizidine alkaloid biosynthesis, *Phytochemistry*, 22, 65, 1983.
103. **Schoonhoven, L. M.,** Secondary plant substances and insects, in *Structural and Functional Aspects of Phytochemistry*, Runekless, C. V. and Tso, T. C., Eds, *Rec. Adv. Phytochem.*, 5, 197, 1972.
104. **von Sengbusch, R.,** Die Prüfung des Geschmacks und der Giftigkeit von Lupinen und anderen Leguminosen durch Tierversuche unter besonderer Berücksichtigung der züchterisch brauchbaren Methoden, *Züchter*, 6, 62, 1934.
105. **Shappira, Z., Terkel, J., Egozi, J., Nyska, A., and Friedman, J.,** Reduction of rodent fertility by plant consumption, *J. Chem. Ecol.*, 16, 2019, 1990.
106. **Smith,B. C.,** Effect of plant alkaloid sparteine on the distribution of the aphid *Acyrthosiphon spartii, Nature*, 212, 213, 1966.

107. **Spenser, I. D.,** Stereochemical aspects of the biosynthetic routes leading to pyrrolizidine and quinolizidine alkaloids, *Pure Appl. Chem.*, 57, 454, 1985.
108. **Stermitz, F. R., Belofsky, G. N., Ng, D., and Singer, M. C.**, Quinolizidine alkaloids obtained by *Pedicularis semibarbata* from *Lupinus fulcratus* fail to influence the specialist herbivore *Euphydryas editah* (Lepidoptera), *J. Chem. Ecol.*, 15, 2521, 1985.
109. **Stermitz, F. R. and Harris, G. H.,** Chemistry of the Scrophulariaceae. 10. Transfer of pyrrolizidine and quinolizidine alkaloids to Castilleja hemiparasites from composite and legume host plants, *J. Chem. Ecol.*, 13, 1917, 1983.
110. **Swain, T.,** Secondary compounds as protective agents, *Ann. Rev. Plant Physiol.*, 28, 479, 1977.
111. **Szentesi, A. and Wink, M.,** Fate of quinolizidine alkaloids through three trophic levels: *Laburnum anagyroides* and associated organisms, *J. Chem. Ecol.*, 17, 1557, 1991.
112. **Terada, M., Sano, M., Ishii, A.I., Kino, H., Fukushima, S., and Nora, T.,** (Japanese title), *Folia Pharmacol. japon.*, 79, 105, 1982.
113. **Troll, H. J.,** Entwicklung und Probleme der Müncheberger Lupinenzüchtung, *Züchter*, 19, 153, 1948/49.
114. **Tyski, S., Markiewicz, M., Gulewicz, K., and Twardowski, T.,** The effect of lupin alkaloids and ethanol extracts from seeds of *Lupinus angustifolius* on selected bacterial strains, *J. Plant Physiol.*, 133, 240, 1988.
115. **Vrieling, K., Smit, W., and van der Meijden, E.,** Tritrophic interactions between aphids (*Aphis jacobaeae* Schrank), ant species, *Tyria jacobaeae* L. and *Senecio jacobaea* L. lead to maintenance of genetic variation in pyrrolizidine alkaloid concentration, *Oecologia*, 86, 177, 1991.
116. **Waller, G. R. and Nowacki, E.,** *Alkaloid Biology and Metabolism in Plants*, Plenum Press, New York, 1978.
117. **Wegorek, W. and Krzymanska, J.,** Further studies on the resistance of lupine to *Acyrthosiphon pisum, Pr. Nauk Inst. Ochr. Rosl.*, 13, 7, 1971.
118. **Wink, M.,** Inhibition of seed germination by quinolizidine alkaloids. Aspects of allelopathy in *Lupinus albus* L., *Planta*, 158, 365, 1983.
119. **Wink, M.,** Wounding-induced increase of quinolizidine alkaloid accumulation in lupin leaves, *Z. Naturforsch.*, 38c, 905, 1983.
120. **Wink, M.,** N-Methylation of quinolizidien alkaloids: An S- adenosyl-L-methionine: cytisine N-methyltransferase from *Laburnum anagyroides* plants and cell suspension cultures of *Cytisus canariensis*, *Planta*, 161, 339, 1984.
121. **Wink, M.,** Chemical defense of lupins. Mollusc-repellent properties of quinolizidine alkaloids, *Z. Naturforsch.*, 39c, 553, 1984.
122. **Wink, M.,** Chemical defense of Leguminosae. Are quinolizidine alkaloids part of the antimicrobial defense system of lupins?, *Z. Naturforsch.*, 39c, 548, 1984.
123. **Wink, M.,** Biochemistry and chemical ecology of lupin alkaloids, in *Proc. 3rd Intl. Lupin Conf.*, La Rochelle, 1984, 326.
124. **Wink, M.,** Stoffwechsel und Funktion der Chinolizidinalkaloide in Pflanzen und pflanzlichen Zellkulturen. Habilitation-thesis, Technische Universität Braunschweig, 1984.
125. **Wink, M.,** Metabolism of quinolizidine alkaloids in plants and cell suspension cultures, in *Primary and Secondary Metabolism of Plant Cell Cultures*, Neumann, K. H., Barz, W., and Reinhard, E., Eds., Springer, Heidelberg, 1985, 107.
126. **Wink, M.,** Chemische Verteidigung der Lupinen: Zur biologischen Bedeutung der Chinolizidinalkaloide, *Plant Syst. Evol.*, 150, 65, 1985.
127. **Wink, M.,** Storage of quinolizidine alkaloids in epidermal tissues, *Z. Naturforsch.*, 41c, 375, 1986.
128. **Wink, M.,** Physiology of the accumulation of secondary metabolites with special references to alkaloids, in *Cell Culture and Somatic Cell Genetics of Plants*, Constabel, F., Ed., Academic Press, London, 1987, 17.
129. **Wink, M.,** Chemical ecology of quinolizidine alkaloids, *ACS Symp. Ser.*, 330, 524, 1987.
130. **Wink, M.,** Quinolizidine alkaloids: biochemistry, metabolism, and function in plants and cell suspension cultures, *Planta Med.*, 53, 509, 1987.
131. **Wink, M.,** Site of lupanine and sparteine biosynthesis in intact plants and in vitro organ cultures, *Z. Naturforsch.*, 42c, 868, 1987.
132. **Wink, M.,** Abstract, Ann. Meeting of Gesellschaft f. Arzneipflanzenforschung, Leiden, 1987.
133. **Wink, M.,** Plant breeding: importance of plant secondary metabolites for protection against pathogens and herbivores, *Theor. Appl. Genet.*, 75, 225, 1988.

134. **Wink, M.,** Physiology of secondary product formation in plants, in *Secondary Products from Plant Tissue Culture,* Charlwood, B. V. and Rhodes, M. J. C., Eds., Clarendon Press, Oxford, 1990, 23.
135. **Wink, M.,** Plant breeding: High or low alkaloid levels?, in *Proc. 6th Intl. Lupin Conf. Temuco,* 1991, 326.
136. **Wink, M.,** Quinolizidine alkaloids, in *Methods of Plant Biochemistry,* (in press).
137. **Wink, M. and Hartmann, T.,** Cadaverine-pyruvate trans-amination: the principal step of enzymic quinolizidine alkaloid biosynthesis in *Lupinus polyphyllus* cell suspension cultures, *FEBS Lett.,* 101, 343, 1979.
138. **Wink, M. and Hartmann, T.,** Sites of enzymatic synthesis of quinolizidine alkaloids and their accumulation in *Lupinus polyphyllus, Z. Pflanzenphysiol.,* 102, 337, 1981.
139. **Wink, M. and Hartmann, T.,** Localization of the enzymes of quinolizidine alkaloid biosynthesis in leaf chloroplast of *Lupinus polyphyllus, Plant Physiol.,* 70, 74, 1982.
140. **Wink, M. and Hartmann, T.,** Diurnal fluctuation of quinolizidine alkaloid accumulation in legume plants and photomixotrophic cell suspension cultures, *Z. Naturforsch.,* 37c, 369, 1982.
141. **Wink, M. and Hartmann, T.,** Enzymatic synthesis of quinolizidine alkaloid esters: a tigloyl-CoA:13-hydroxylupanine O-tigloyltransferase from *Lupinus albus* L., *Planta,* 156, 560, 1982.
142. **Wink, M. and Hartmann, T.,** Enzymology of quinolizidine alkaloid biosynthesis, in *Natural Products Chemistry* 1984, Zalewski, R. I. and Skolik, J. J., Eds., Elsevier, Amsterdam, 1985, 511.
143. **Wink, M., Hartmann, T., Witte, L., and Rheinheimer, J.,** Interrelationship between quinolizidine alkaloid producing legumes and infesting insects: exploitation of the alkaloid-containing phloem sap of *Cytisus scoparius* by the Broom aphid (*Aphis cytisorum*), *Z. Naturforsch.,* 37c, 1081, 1982.
144. **Wink, M., Heinen, H. J., Vogt, H., and Schiebel, H. M.,** Cellular localization of quinolizidine alkaloids by laser desorption mass spectrometry (LAMMA 1000), *Plant Cell Rep.,* 3, 230, 1984.
145. **Wink, M. and Mende, P.,** Uptake of lupanine by alkaloid-storing epidermal cells of *Lupinus polyphyllus, Planta Med.,* 53, 465, 1987.
146. **Wink, M. and Römer, P.,** Acquired toxicity—the advantages of specializing on alkaloid-rich lupins to *Macrosiphum albifrons* (Aphidae), *Naturwissenschaften,* 73, 210, 1986.
147. **Wink, M. and Schneider, D.,** Carrier-mediated uptake of pyrrolizidine alkaloids in larvae of the aposematic and alkaloid-exploiting moth, Creatonotos, *Naturwissenschaften,* 75, 524, 1988.
148. **Wink, M. and Schneider, D.,** Fate of plant-derived secondary metabolites in three moth species (*Syntomis mogadorensis, Syntomeida epilais,* and *Creatonotos transiens*), *J. Comp. Physiol.* B, 160, 389, 1990.
149. **Wink, M. and Schneider, D.,** Geese and plant allelochemicals, (in preparation).
150. **Wink, M. and Twardowski, T.,** Allelochemical properties of alkaloids. Effects on plants, bacteria and protein biosynthesis, in *Allelopathy: Basic and Applied Aspects,* Chapmann & Hall 1992, 129.
151. **Wink, M.,** Allelochemical properties and the raison d'etre of alkaloids, in *The Alkaloids,* Brossi, A., Ed., (in press).
152. **Wink, M. and Witte, L.,** Evidence for a wide-spread occurrence of the genes of quinolizidine alkaloid biosynthesis. Induction of alkaloid accumulation in cell suspension cultures of alkaloid-"free" species, *FEBS Lett.,* 159, 196, 1983.
153. **Wink, M. and Witte, L.,** Turnover and transport of quinolizidine alkaloids: diurnal variation of lupanine in the phloem sap, leaves and fruits of *Lupinus albus* L., *Planta,* 161, 519, 1984.
154. **Wink, M. and Witte, L.,** Quinolizidine alkaloids as nitrogen source for lupin seedlings and cell suspension cultures, *Z. Naturforsch.,* 40c, 767, 1985.
155. **Wink, M. and Witte, L.,** Quinolizidine alkaloids in *Petteria ramentacea* and the infesting aphids, *Aphis cytisorum, Phytochemistry,* 24, 2567, 1985.
156. **Wink, M. and Witte, L.,** Cell-free synthesis of the alkaloids ammodendrine and smipine, *Z. Naturforsch.,* 42c, 197, 1987.
157. **Wink, M. and Witte, L.,** Storage of quinolizidine alkaloids in *Macrosiphum albifrons* and *Aphis genistae* (Homoptera: Aphididae), *Entomol. Gener.,* 15, 254, 1991.
158. **Wink, M., Witte, L., and Hartmann, T.,** Quinolizidine alkaloids of cell suspension cultures and plants of *Cytisus* (Sarothamnus) *scoparius* and of its root parasite *Orobanche rapum-genistae, Planta Med.,* 43, 342, 1981.

159. **Wink, M., Witte, L., Hartmann, T., Theuring, C., and Volz, V.,** Accumulation of quinolizidine alkaloids in plants and cell suspension cultures: genera Lupinus, Cytisus, Baptisia, Genista, Laburnum, and Sophora, *Planta Med.*, 48, 253, 1983.

160. **Wink, M., Montllor, C., Bernays, E. A., and Witte L.,** *Uresiphita reversalis* (Lep. Pyralidae): carrier-mediated uptake and sequestration of quinolizidine alkaloids obtained from the host plant *Teline monspessulana, Z. Naturforsch.*, 46c, 1080, 1991.

161. **Wippich, C. and Wink, M.,** Biological properties of alkaloids. Influence of quinolizidine alkaloids and gramine on the germination and development of powdery mildew, *Erysiphe graminis* f.sp.hordei, *Experientia*, 41, 1477, 1985.

162. **Yovo, K., Huguet, F., Pothier, J., Durand, M., Breteau, M., and Narcisse, G.,** Comparative pharmacological study of sparteine and its ketonic derivative lupanine from seeds of *Lupinus albus., Planta Med.*, 420, 1984.

163. **Yu, S. J.,** Consequences of induction of foreign compound-metabolizing enzymes in insects, in *Molecular Aspects of Insect Plant Associations,* Brattsten, L. B. and Ahmad, S., Eds., Plenum Press, New York, 153, 1986.

164. **Zetler, G. and Strubelt, O.,** Antifibrillatory, cardiovascular and toxic effects of sparteine, butylsparteine and pentyls parteine, *Arzneim.-Forsch.*, 30, 1497, 1980.

165. **Zwierzynski, T., Joachimak, A., Barciszewska, M., Kulinska, K., and Barciszewski, J.,** Interaction of alkaloids with plant transfer ribonucleic acids. Effect of sparteine on lupin arginyl-tRNA formation, *Chem. Biol. Interact.*, 42, 107, 1982.

The Impact of Plant Stress on Herbivore Population Dynamics

Gwendolyn L. Waring
Department of Biology
Museum of Northern Arizona
Flagstaff, Arizona

and

Neil S. Cobb
Department of Biological Sciences
Northern Arizona University
Flagstaff, Arizona

TABLE OF CONTENTS

I. INTRODUCTION

The effects of water and nutrient deficits on plant–insect interactions have been of interest to ecologists for some time. While environmental stress is known to exert strong effects on plants, we still do not have a clear understanding of how herbivores typically respond to stressed host plants, despite a considerable amount of research on this subject. Various studies have concluded that plant stress has clear and positive effects on herbivores,[195,196,262] while others suggest that stress effects are negative, or even negligible.[18,56,64,119,179,206,217,262] Despite a lack of consensus, the view that plant stress exerts strong positive effects on herbivore populations has acquired near paradigm status in ecology. This review is an effort to bring together a large number of studies to help resolve the controversy.

The fact that stress often does have strong effects on plants reinforces the view that the interaction between herbivores and plant stress will be significant, whether positive or negative. Different forms of nutrient nitrogen and starches often increase substantially in plants suffering water deficits.[56,89,97,139,140,167,262] Environmental stress also alters plant defensive chemistry.[74,139,140,195,196,226] The spectrum of effects on plants is broad, also including genetic changes and changes in growth, reproduction, leaf temperature, and water potential.[139,140] However, before we can appreciate the mechanistic importance of these factors in the herbivore–plant stress dynamic, we first need to elucidate patterns of herbivore responses to plant stress.

Here we present the first comprehensive review of studies of herbivore responses to two forms of environmental stress in host plants: water deficits and nutrient deficits. We examine (1) herbivore performance in nitrogen, phosphorus, and potassium fertilization studies, experimental water-stress studies, and studies that correlated herbivory with drought conditions. (2) We consider the effect of plant type (i.e., cultivated, wild, woody, herbaceous plants, and conifers and broadleaf trees) and herbivore feeding guild (i.e., woodborers, chewing, sucking, and gallforming insects and mites) on the herbivore–plant stress interaction. (3) We then summarize research in which both nutrient and water deficits were studied simultaneously, and studies on the influence of stress on the phenotypic expression of genetically based resistance traits. (4) We evaluate the relationship between measured plant responses to stress, such as changes in xylem water potential or leaf nitrogen, and corresponding herbivore responses, to look for patterns that might point to underlying mechanisms. (5) We compare herbivore responses to similar stress conditions when different herbivore life-history traits (survivorship, growth, and fecundity) or population densities are measured. (6) We discuss several biological realities that should be considered when studying the relationship between herbivory and plant stress.

II. METHODS

Definitions of plant stress are numerous and diverse,[103] and no one definition has been universally accepted. The stress gradient can be broad for many plant species, ranging between negligible and catastrophic reductions in plant productivity.[89] In light of this and the idiosyncratic responses of different plant species to stress, it seems that the definition of plant stress must remain general and operational on a relative scale. However, most definitions of plant stress do specify that productivity will decline in plants exposed to deficits such as water and nutrient deficits.[20,80,89] We adhered to this general definition in our review of the large and diverse literature on herbivore responses to plant stress. This enabled us to identify the patterns and problems emerging from herbivore–plant research to date, from which we can direct future research on this important subject.

We reviewed the biological literature through March 1991 to locate studies on herbivorous insect and mite responses to nutrient and water deficits in plants. This included a large and diverse collection of studies, more than 450 studies or multiple cases within studies on herbivore responses to plant stress (see the Appendix on page 187). We subdivided the studies into two groups based on study design to determine what effects this might have on conclusions. The first group included studies that presented quantitative measurements of plant and herbivore responses to deficits, and compared these conditions to herbivore responses to a control or nonstressed plant group, referred to as stress demonstrated studies. Studies in the second group, referred to as stress assumed studies, presented quantitative measurements of herbivore responses, but assumed rather than verified stress responses in plants. Examples of the latter included studies of forest insect outbreaks in relation to drought, in which control conditions were not available for comparison. We did not use studies that did not include quantitative measurements of plant stress or herbivore responses, or studies with plant results that appeared to be at all ambiguous. Because all studies were different in many aspects, these criteria were the only ones we could apply to assure scientific rigor and still utilize the extensive herbivore–plant stress literature.

We reviewed studies on herbivore responses to fertilization of host plants with nitrogen or total nutrients, phosphorus (P), or potassium (K). These elements are essential to plant growth and development; and while nitrogen is known to be critical to herbivore development, the effects of phosphorus and potassium on herbivores are not well understood.[56] We included studies that fertilized with total, multiple, or undefined (e.g., manure) fertilizers or with N in the "total nutrient or N fertilization" category, because some studies used fertilizers that emphasized nitrogen without using it exclusively, while others used nitrogen only.

We reviewed two types of water stress–herbivory studies: experimental water stress studies, in which stress was artificially induced, and drought or chronic stress studies, in which stress was naturally occurring, due to climatic drought conditions or to the occurrence of the plant in a naturally stressful habitat resulting in chronic stress. All experimental water stress studies measured one or more plant responses and were categorized as stress-demonstrated studies, whereas most drought or chronic stress studies did not measure a plant response and were categorized as stress-assumed studies.

Herbivore responses were categorized as positive (+), negative (-), nonlinear (*), or no response (=), and were based on measurements of herbivore responses at the population level (= density), or at the individual level, measuring survivorship, growth rate, and/or fecundity. Most responses were based on data with statistical comparisons. Exceptions included several drought studies in which control conditions were not available. Nonlinearity refers to a nonlinear response by a herbivore population to at least three treatment levels, where a combination of positive, negative, or no responses occurred. "No response" indicates that herbivores performed comparably on stressed and nonstressed plants. When an herbivore exhibited variable responses within a study (n = 8 cases), we estimated the strongest response and used that in our tabulations. For example, in the only study that found contradictory responses, Ohmart et al.[172] found that the leaf-chewing *Paropsis atomaria* had improved survivorship and fecundity, and decreased growth on fertilized *Eucalyptus* sp. We summarized this as an overall positive response.

In the fertilization studies, a positive herbivore response reflected a positive response to fertilization or nonstress, while in the water-stress studies, a positive herbivore response reflected a positive response to water stress. We attempt to bridge this unavoidable discrepancy, where possible, by extrapolating from fertilization to the nutrient-deficit condition.

We tabulated plant type and herbivore guild for each herbivore–plant stress interaction. Plants were categorized as wild or cultivated, woody or herbaceous, and resistant or susceptible. We also classified plants as being conifers, deciduous trees, shrubs or vines, or herbaceous. Herbivores were classified into one of four feeding guilds: (1) sucking insects, including mesophyll and phloem feeders; (2) chewers, including woodborers, stemborers, rootfeeders, leafminers, and leafchewers; (3) mites; and (4) gallforming insects. We separated woodborers from other chewing insects in one analysis, but there were not enough cases to consider stemborers, rootborers, or leafminers separately.

We compared proportions of positive and negative herbivore responses with chi-square goodness-of-fit tests and contingency table analyses, comparing responses according to plant type and herbivore guild.[272] We used these tests in a heuristic sense to indicate general trends and major differences. Where observed-expected values, were less than expected values we substituted chi-square tests with g-tests. The small sample size precluded inclusion of nonlinear responses or "no response" responses in these tests.

The appendix lists all studies that we included in this review. Each case is organized according to treatment, herbivore and plant species, plant response, herbivore response and the reference for each "data point".

III. RESULTS AND DISCUSSION

A. HERBIVORE RESPONSES TO PLANT NUTRIENTS

1. N or Total Nutrients:

Plant fertilization or nutrient enhancement has strong, positive effects on herbivore population dynamics: most herbivore responses to N fertilization were positive (Figure 1, Appendix). Nearly 60% of the 186 fertilization studies reported positive responses by herbivores. Nearly 25% of the fertilization studies reported no herbivore response to fertilized vs. unfertilized plants; that is, herbivore densities or performance were comparable on both treatments. There were very few negative responses to fertilization (11%; Figure 1).

a. Feeding Guilds

Despite widely different feeding modes, responses of the four herbivore guilds (sucking, chewing, and gallforming insects and mites) to fertilization were typically positive (Figure 1, Table 1), indicating that different guilds respond similarly to the same changes in plant nutrient status. The greater proportions of positive responses vs. negative responses were the same for all guilds ($\chi^{2,3} = 3.38$, $p > 0.25$; Table 1). The second most common response to fertilization among the feeding guilds was no response at all, except for mites, which more often responded negatively to fertilization (19%; Figure 1).

Most herbivores used in multiple studies exhibited consistent responses to fertilization across different plant species and different studies (e.g., negative responses of *Empoasca* spp. on three plant species;[95,123,258] negative responses of *Euphydryas*

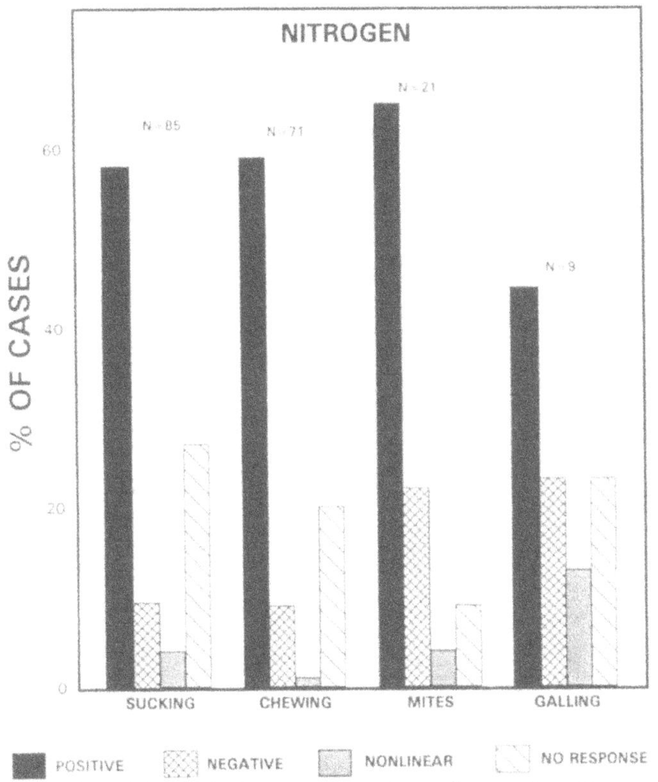

FIGURE 1. Responses of herbivores to nitrogen fertilization, measured as the percentage of studies. Data are presented for herbivore feeding guilds, including sucking, chewing, and gallforming insects and mites.

editha on three plant species;[63] and positive responses of *Cardiaspina albitextura* on two plant species[259] (see Appendix). However, several well-studied taxa, including *Adelges* spp. and *Neodiprion* spp., responded negatively and positively to nitrogen fertilization in different studies.[162] *Adelges picea* responded oppositely to seemingly slight differences in fertilizer types within the same study.[38]

b. Plant Type

Herbivores typically responded positively to fertilization in wild, cultivated, woody, and herbaceous plants (Figure 2, Table 1). The proportions of positive and negative herbivore responses to each plant type were significantly different (Table 1). However, proportionately more herbivores responded positively to cultivated plants than wild plants (χ^2,1 = 5.40, p <0.025, Table 1).

Coniferous trees were the only plants that did not elicit a strong positive herbivore response when fertilized. A comparable number of studies showed positive (n = 7) and negative (n = 8) responses to coniferous species (Figure 2). There were proportionately more negative responses by herbivores to conifers than broadleaf trees and herbs (Figure 2), suggesting that conifers are more likely than flowering plants to become more resistant and/or less attractive to herbivores when fertilized.

c. Herbivore Responses to Measured Plant Changes

Most herbivore responses were positively correlated with increases in plant growth and the nitrogen fraction (Table 2), indicating the predictive value that these factors can have in stress studies. In several cases, however, herbivore responses did not corre-

Table 1
GOODNESS OF FIT ANALYSES OF POSITIVE AND NEGATIVE RESPONSES OF HERBIVORE GUILDS TO PLANT FACTORS, IN RESPONSE TO FERTILIZATION WITH TOTAL NUTRIENTS OR N

	Total			Sucking			Chewing			Mites			Gallformer		
Factor	(+)	(–)	χ^2	(+)	(–)	χ^2	(+)	(–)	χ^2	(+)	(–)	χ^2	(+)	(–)	χ^2
All plants	111	23	**	50	9	**	43	8	**	14	4	*	4	2	
Cultivated plants	76	10	**	34	4	**	27	2	**	14	4	*	1	—	
Wild plants	35	13	**	16	5		16	6	*	—	—		3	2	
Woody plants	34	11	**	10	2	*	14	7		—	—		3	2	
Herbaceous plants	77	12	**	40	7	**	30	2	**	6	3		1	—	
Cultivated, woody plants	16	3	**	7	2		1	—		8	1		—	—	
Cultivated, herbaceous plants	60	7	**	27	2	**	26	2	**	6	3		1	—	
Wild, woody plants	18	8	ns	3	—		12	6	ns	—	—		3	2	
Wild, herbaceous plants	17	5	**	13	5	ns	4	—		—	—		—	—	

* $p < 0.05$; ** $p < 0.01$; ns = not statistically significant. Comparisons with expected values less than 5 were not analyzed.

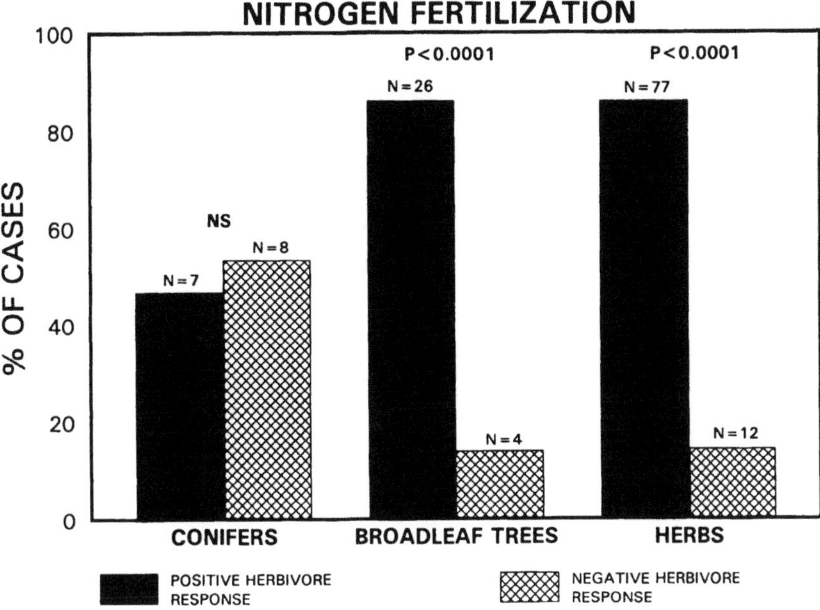

FIGURE 2. Percentage of studies in which herbivores responded positively or negatively to nitrogen fertilization in conifers, broadleaf trees, and herbaceous plants.

spond with higher nitrogen levels in plants (Appendix). This suggests that not all herbivores are nitrogen limited, that increased nitrogen may have deleterious effects on herbivores through direct effects or an increase in defensive chemistry, or that treatments were inadequate for eliciting herbivore responses that might otherwise have occurred. Where nitrates were used, these may have accumulated in some plant species, reducing their acceptability. Herbivores may also vary in the extent to which they can capitalize on increased plant nitrogen, with some species exhibiting a narrow or fixed response to varying nitrogen concentrations.[9] Herbivore responses to increases in amino acids were especially variable (Table 2), while some responses were negatively correlated with increased nitrate levels.[91,134]

Two studies found positive herbivore responses to decreases in total phenolics or tannins.[172,248] More studies are needed to better understand the relationship of plant defensive chemistry to herbivory and plant nutrient stress.

2. Phosphorus Fertilization

Although phosphorus is also limited in nature and is an essential element for plant development, the most common herbivore response to P-fertilized plants was no response at all (44% of 50 cases; Figure 3). The second most common response was positive (34%). When compared using contingency table analysis, there were no significant differences in the proportions of positive and negative responses to phosphorus. Other reviews have reported similar effects from fertilization with phosphorus.[56,217]

3. Potassium Fertilization

Most herbivores tested responded negatively (36%) or did not respond to plants fertilized with potassium (40%) (Figure 4). Fewer herbivores responded positively (24%). Contingency table analysis revealed no significant difference between the proportion of positive and negative herbivore responses. Among feeding guilds, more sucking insects responded negatively than positively to plants fertilized with K [(-) = 41%,

Table 2
A SUMMARY OF HERBIVORE RESPONSES TO INCREASED GROWTH OR INCREASED CONCENTRATIONS OF PLANT NUTRIENTS DUE TO FERTILIZATION WITH N OR TOTAL NUTRIENTS, P, OR K

	Herbivore response			
	Increase	Decrease	No response	Nonlinear
Plant Response				
+ Growth	46	1	14	3
+ Nitrogen	52	14	10	3
+ Nitrates	—	3	2	—
+ Protein	5	2	1	1
+ Amino acids	5	4	—	—
Phosphorus				
+ N	1	3	—	—
+ P	3	2	2	2
+ Potassium	1	1	—	—

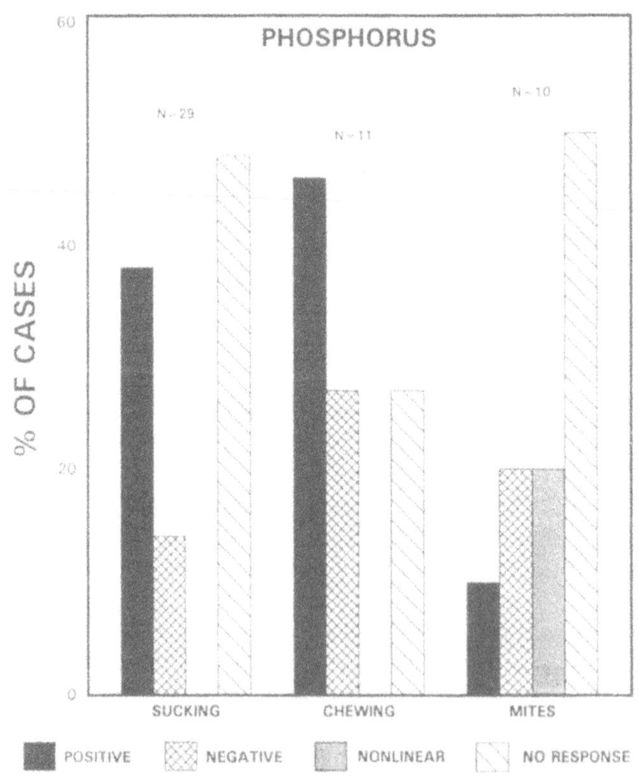

FIGURE 3. Responses of herbivores to phosphorus fertilization, measured as the percentage of studies. Data are presented for herbivore feeding guilds, including sucking and chewing insects and mites.

(+) = 7%; χ^2 = 10.57 p <0.01], while more chewing insects (62%, n = 5) responded positively to K fertilization. These results may relate to the fact that potassium deficiency blocks protein synthesis and results in the accumulation of amino acids, as well as soluble and reducing sugars, while fertilization with potassium causes increased levels of protein and starch formation in plants.[56]

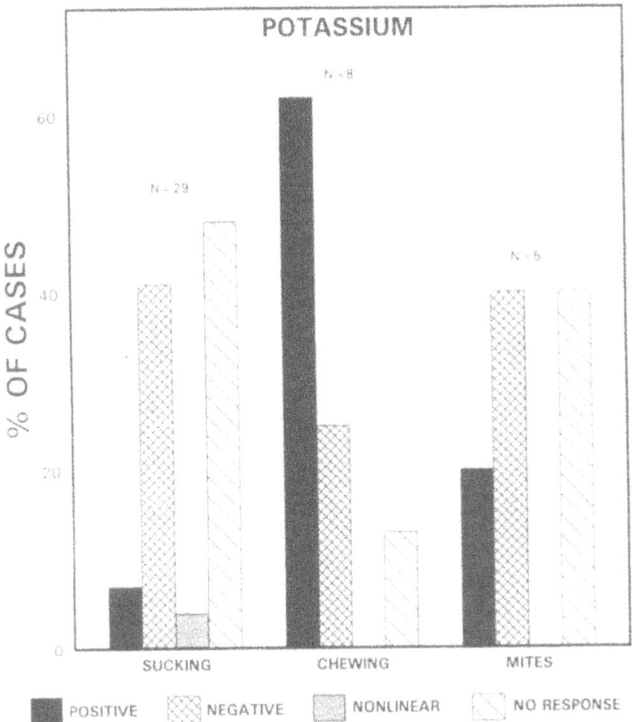

FIGURE 4. Responses of herbivores to potassium fertilization, measured as the percentage of studies. Data are presented for herbivore feeding guilds, including sucking and chewing insects and mites.

Other reviews have reported that herbivores typically respond negatively to plants fertilized with potassium.[18,179,217] In the only study of K effects on gallformers, Israel and Prakasa Rao[102] reported no relationship between densities of the gall midge, *Orseolia oryzae,* and level of K in rice. Increased K can also make flowers less attractive to pollinators.[247] Despite its effects on herbivores, the effect of potassium on the herbivore–plant stress interaction is not well studied; we found only 42 studies to review. It is difficult to generalize about our findings, since most studies involved cultivated, herbaceous plants (69%) and sucking insects (69%).

4. Conclusions

The positive responses of most herbivores to fertilized plants reinforce the view that nitrogen is both limiting and valuable to herbivores in nature, and a lack of it can have serious consequences. Herbivore performance on nutrient-poor plants was generally lower, leading to reduced survivorship, growth, fecundity, and densities, and suggesting that many herbivores are not able to compensate for poor nutrient quality with quantity, as has been predicted.[210]

However, it is not clear whether the positive responses of most herbivores to fertilization were due to increased nutrient levels, decreased defenses, or some combination of both in host plants. Few studies measured defensive chemistry. Perhaps low nutrient quality in and of itself represents a defense against herbivores.[9,149,165] Alternatively, a positive herbivore response to fertilization may reflect a shift in a plant's carbon-nutrient balance away from the production of phenolic carbon-based compounds to nitrogen-based compounds and growth. This shift by plants in response to fertilization is well documented.[28,138,149,209,248] Ultimately, lower nutrient status in plants does appear to limit herbivore success, although the underlying mechanism is not yet clear.

The positive responses of herbivores to unfertilized conifers and the negative responses of herbivores to unfertilized flowering plants suggests a major difference in the physiology of these two plant groups. Nutrient-stressed flowering plants should often be protected from herbivore infestations, because nutrient limitations are probably common in nature. By contrast, conifers, which often occur in nutrient-poor sites, stand to be more susceptible to their herbivores under these same conditions.

The positive response of herbivores to increased nitrogen or total nutrients seems to be finite in some instances. Brewer et al.[24] reported that forest herbivores, including the spruce budworm, ceased to respond or responded nonlinearly when large amounts of fertilizers were applied to plants. This may relate to a toxic effect that is tied to high nitrogen levels,[43] or to herbivore "indifference" to high nutrient levels.[9,158,159]

B. HERBIVORE RESPONSES TO PLANT WATER STRESS

Herbivore responses to host water stress depended on whether plants were experimentally water stressed or naturally stressed by drought or chronic water stress. Most herbivores responded negatively to experimentally or artificially induced plant water stress (51% of 99 cases), while a majority of herbivores responded positively to naturally water stressed plants (68% of 75 cases). This discrepancy raises a concern about how stress is studied, including how it is induced, and suggests that a better understanding of plant stress is needed. Despite these differences, patterns of herbivore responses based on feeding guilds and plant type did emerge from both data sets. Other than the preponderance of negative responses to experimental water stress, herbivore responses were highly variable, leading to no clear secondary pattern (Figure 5, top). This was not true for drought or chronic studies. The next most common response of herbivores to drought or chronic stress was negative (27%). There were few "no responses" (4%) to drought stress, which suggests that these systems may be selected for study based on an *a priori* sense of how they operate.

1. Feeding Guilds

Generally, the different feeding guilds responded similarly to one another, regardless of the type of water stress. Of all guilds, only chewing insects were well represented in both types of studies, and they clearly show opposite responses to the two forms of stress (Figure 5).

All guilds, except mites, responded negatively most often to experimental water stress (Figure 5, Table 3), although proportions of positive and negative responses among the guilds were not significantly different ($\chi^2 = 7.60$, $p > 0.05$). Approximately half of all sucking and chewing insect responses were negative, while 73% of gallformers responded negatively to experimentally water-stressed plants. By contrast, mite studies reported more positive (42%) than negative (26%) responses to stress, although the number of studies was limited (Figure 5). Mites were studied only on cultivated plants, which tend to elicit negative responses from herbivores (see below). The rapid generation time and potential population growth rates of mites[31,42] may enable them to capitalize on the early beneficial phases of water stress in a way that more slowly developing herbivores cannot. Although proportions of positive and negative responses of guilds were not statistically different, the proportions of positive responses varied as follows: mites (61%) > galling insects (33%) = chewing insects (33%) > sucking insects (18%).

Both sucking and chewing guilds responded positively to plants experiencing drought or chronic water deficits (70% and 73%, respectively; Figure 5, Table 4). The ratio of negative and positive responses of herbivores to this type of stress was independent of

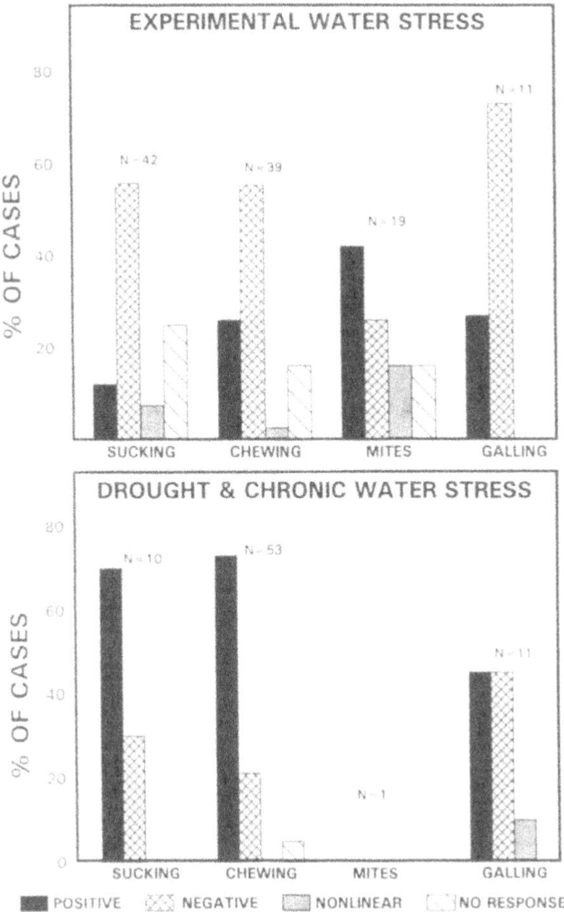

FIGURE 5. Responses of herbivore feeding guilds to experimental water stress (above), and drought and chronic water stress (below), measured as the percentage of studies.

guild type (χ^2 = 3.31, p >0.25). Among the chewing insects, all woodborers responded positively to drought stress, and most were associated with conifers (Table 4). The proportion of positive vs. negative responses of guilds to drought or chronic water stress were ranked as follows: woodborers (100%) > sucking insects (70%) = chewing insects (68%) > galling insects (50%; Table 4).

Gallforming insects exhibited equal numbers of positive and negative responses to natural stress (Figure 5), and this may relate to differing resource requirements. Among the gallforming cecidomyiids (*Asphondylia* spp.) on *Larrea tridentata*, the flower gallformer, *A. florea*, was most abundant on vigorous nonstressed plants with high flower production, while several leaf gallformers were most abundant on water-stressed plants that produced a more diffuse canopy with more oviposition sites.[249] This suggests that resource requirements of herbivores must be accounted for in plant stress studies.

Many herbivore genera and species that were observed in multiple studies exhibited consistent responses to water stress (i.e., *Aphis fabae*, *Cardiaspina* spp., *Choristoneura* spp., *Euphydryas* spp., and *Lymantria dispar*), while others, such as *Tetranychus urticae*, exhibited variable responses between studies (see Appendix). Variable *T. urticae* responses to stress resulted from testing it on different plant species and cultivars,[76] and measuring different life history parameters.[64] *Tetranychus urticae* re-

Table 3
GOODNESS OF FIT ANALYSES OF POSITIVE AND NEGATIVE RESPONSES OF HERBIVORE GUILDS TO PLANT FACTORS, IN RESPONSE TO EXPERIMENTAL WATER STRESS

	Total			Sucking			Chewing			Mites			Gallformers		
Plant type	(+)	(–)	χ^2	(+)	(–)	χ^2	(+)	(–)	χ^2	(+)	(–)	χ^2	(+)	(–)	χ^2
All plants	25	54	**	5	23	**	10	20	*	8	5ns		3	6	
Cultivated plants	18	23	ns	5	16	*	5	2		8	5ns		—	—	
Wild plants	7	31	**	—	7		5	18	**	—	—		2	6	
Woody plants	8	28	**	—	6		5	15	*	1	1		2	6	
Herbaceous plants	17	26	ns	5	17	*	5	5		7	4	ns	1		
Cultivated, woody plants	3	2		—	—		2	—		1	1		—	—	
Cultivated, herbaceous plants	15	21	ns	5	15	*	3	2		7	4	ns	—	—	
Wild, woody plants	5	26	**	—	5	3	15	*		—	—		2	6	
Wild, herbaceous plants	3	5		—	2	2	3			—	—		1	—	

* $p < 0.05$, ** $p < 0.01$.

Table 4
GOODNESS OF FIT ANALYSES OF POSITIVE AND NEGATIVE RESPONSES OF HERBIVORE GUILDS TO PLANT FACTORS, IN RESPONSE TO DROUGHT OR CHRONIC WATER DEFICITS

	Total			Sucking			Chewing			Mites			Gallformers		
Plant type	(+)	(−)	χ^2	(+)	(−)	χ^2	(+)	(−)	χ^2	(+)	(−)	χ^2	(+)	(−)	χ^2
All plants	51	20	**	7	3		39	11	**	—	1		5	5	
	(36)	(20)	*				(24)	(11)	*						
Cultivated plants	1	6		—	2		1	2		—	1		—	—	
Wild plants	50	14	**	7	1		38	9	**	—	—		5	5	
	(35)	(14)	**				(23)	(9)	*						
Woody plants	47	9	**	7	1		35	2	**	—	1		5	5	
	(32)	(9)	**				(20)	(2)	**						
Herbaceous plants	4	11		—	2		4	9		—	—		—	—	
Cultivated, woody plants	1	—		—	—		—	—		—	1		—	—	
Cultivated, herbaceous plants	1	5		—	2		1	2		—	—		—	—	
Wild, woody plants	47	8	**	7	1		35	2	**	—	—		5	5	
	(32)	(8)	**				(20)	(2)	**						
Wild, herbaceous plants	3	6		—	—		3	7		—	—		—	—	

* $p < 0.05$, ** $p < 0.01$. Numbers in parentheses are tallies without woodboring beetles.

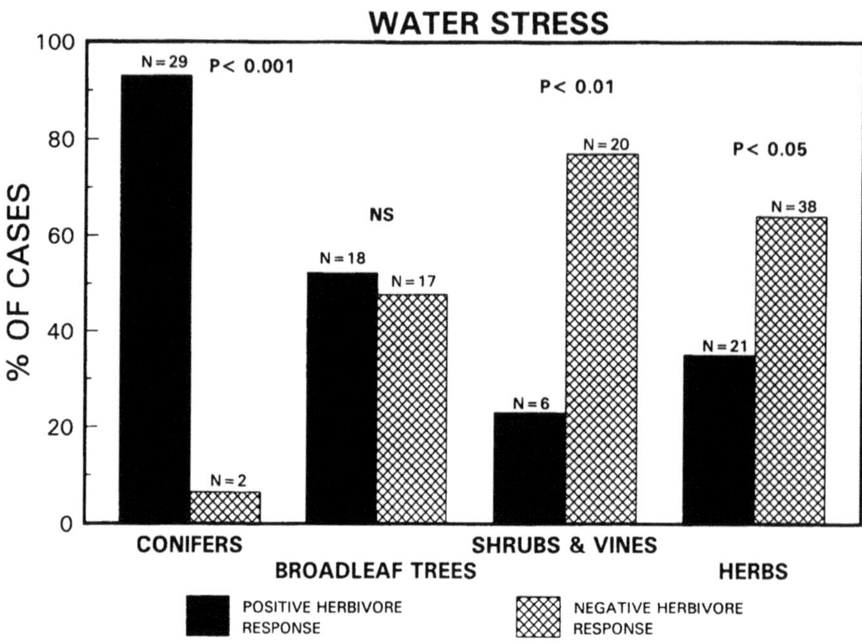

FIGURE 6. Percentage of studies in which herbivores responded positively or negatively to all forms of water stress in conifers, broadleaf trees, shrubs and vines, and herbaceous plants.

sponded positively to water-stressed peppermint, radish, and chrysanthemum,[96,154,191] and negatively to stressed soybean and alfalfa.[30,153] Bernays and Lewis[15] also found that the response of *Schistocerca gregaria* to water stress depended on which plant species it was offered, in a study in which they offered *S. gregaria* stressed and nonstressed forms of 41 plant species.

2. *Plant Type*

Plant type has a strong effect on the herbivore–plant stress interaction (Figure 6). Herbivores exhibited different responses to architecturally different plants in both experimental water stress, and drought, and chronic stress studies. Among tree species, nearly all responses to conifers were positive (93%), while there were comparable numbers of positive and negative responses to broadleaf trees, such as *Quercus* spp. and *Eucalyptus* spp. (Figure 6). Most herbivores responded negatively to stress in smaller plants, including shrubs and vines (77%), and herbaceous plants (67%; Figure 6). It seems likely that stress effects will be manifested more slowly in trees vs. herbaceous plants. Many herbaceous plants, as drought avoiders, are barely able to compensate for or cope with water stress.[167] Nutrient loss may also play a role in herbivore responses to different plant types, as suggested by Scriber,[209] who found that leaf nitrogen levels drop substantially with small changes in leaf water content in herbaceous plants, while they fall less in broadleaf trees, and not at all in conifers.

Most herbivores responded negatively to experimentally water-stressed wild plants (72%), while only 37% of herbivore responses to experimentally stressed cultivated plants were negative ($\chi^2 = 4.8$, $p < 0.05$; Table 3), suggesting that water stress has very different effects on wild and cultivated plants. Perhaps defensive chemistry becomes more pronounced in stressed wild plants,[195] while being largely absent in cultivated plants.

Table 5
SUMMARY OF MEASURED PLANT RESPONSES TO WATER STRESS AND HERBIVORE RESPONSES TO THOSE RESPONSES.

	Herbivore Response			
	Increase	Decrease	No Response	Nonlinear
Plant Responses				
Water deficits	13 (2)*	42 (1)	15	3
Decreased growth	8 (14)	12 (4)	2	3
Increased nitrogen	1	1	—	—
Increased protein	3	4	1	—
Increased amino acids	2	2	3	1
Decreased resins	3 (6)	2 (1)	—	—

* Experimental studies are presented in columns and drought or chronic stress studies are presented adjacently in parentheses.

In drought or chronic stress studies, herbivores typically responded positively to stressed woody plants (Table 4). Most positive herbivore responses were towards drought-stressed conifers, including *Pinus* spp., *Abies* spp., and *Pseudotsuga mensiezii* (Appendix). As with the nutrient studies, these results indicate that herbivores often respond differently to conifers vs. flowering plants. The few negative responses to water stressed conifers occurred in experimental water stress studies. The performance of *Neodiprion autumnalis* was depressed,[144] while *N. fulviceps* showed no response to *Pinus ponderosae* stressed by trenching.[51]

Negative responses to herbaceous plants were consistent across data sets, suggesting that this plant type tends not to accumulate beneficial nutrients such as amino acids when stressed.

3. Herbivore Responses to Measured Plant Changes

Herbivore responses correlated with water stress-related reductions in plant water potential, growth, and resin levels (Table 5). Herbivores most often responded negatively to plants with water deficits in experimental studies (χ^2 = 15.3, p <0.001). Herbivores have been shown to respond directly to plant water status.[266] Herbivore responses to decreased plant growth were most often positive in drought and chronic stress studies (χ^2 = 5.55, p <0.05), and though not significant, tended to be negative in most experimental studies that measured decreased plant growth (Table 5). Unlike nutrient studies, herbivores showed no clear response to water stress-related increases in the nitrogen fraction (Table 5). The lack of positive correlation between herbivore response and increased amino acid concentrations (Table 5) runs counter to Broadbeck and Strong's proposition[26] that increased amino acids lead to improved herbivore performance in stressed plants. In both experimental and drought and chronic studies, most herbivore responses to decreased resin levels were positive, although data were limited (Table 5). Resin levels often decrease in the bark, wood, and needles of water-stressed conifers.[120,139,140,226]

Only seven studies measured the response of secondary chemistry, other than resins, to water stress (Appendix). Concentrations of phenolic compounds increased, decreased, or stayed the same in stressed plants. This variable pattern has been reported elsewhere.[74] Herbivore responses did not suggest any relation to changes in the concentrations of these compounds. Cates et al.[40] found that increased growth and survivorship of the spruce budworm correlated positively with changes in terpene composition. Bernays et al.[14] found that increased survivorship of a cassava insect correlated with decreased HCN release rates in stressed cassava leaves. More effort is needed to understand the role of plant defensive chemistry in the herbivore–plant stress interaction.

4. Conclusions

Water-stressed plants become more suitable or less suitable for herbivores depending on how they are stressed and what type of plants they are. A review of 150 studies confirmed that herbivores typically respond negatively to experimental water stress in plants and positively to natural stress. We do not know which herbivore response is more representative because we do not understand how experimental stress relates to natural stress, and because the selection of most drought studies may be biased towards herbivore species that respond positively to stress.

The intensity of plant water stress may depend on the level, timing, tempo, or duration of water deficits. We are only beginning to understand the details of stress and its complexity.[42,76,169,170,220,237,246,271] Youngman et al.[271] found that herbivores responded positively to intermittent experimental water stress and negatively to continuous water stress. Nitrogen levels in this experiment were higher in intermittent treatment plants than in continuous treatment plants, contrary to conventional wisdom about nitrogen and water stress.[139,140,271] Drought stress may begin with a dry winter and persist for months, while experimental water stress may be initiated during a plant's growing season and be studied only briefly. Thomson and Shrimpton[237] found that not all drought conditions lead to mountain pine-beetle outbreaks, although all outbreaks are tied to drought. They determined that the timing of drought best predicted outbreaks. Mechanisms of experimentally inducing water stress, such as root trenching or xylem girdling, may also cause negative herbivore responses by reducing nutrient uptake or for other unknown reasons (Richard Tinus, U.S.F.S, personal communication).

The large proportion of positive herbivore responses and the lack of nonresponses to drought could relate to the nonrandom way in which many of these systems were selected for study, with an *a priori* expectation that a positive relationship exists. The pest status of many of the herbivores on drought-stressed plants, such as the spruce budworm and woodboring beetles, has provided much of the impetus for these studies, so the focus of this research is often different. An understanding of how less conspicuous or economically unimportant forest herbivores respond to drought is essential to validate the patterns reported here.

Few studies have compared experimental and natural stress. Among them, the chewing insect, *Choristoneura fumiferana*, exhibited no response to experimental water stress,[141] and a positive response to drought and chronic plant water stress.[110,214] Mattson et al.[141] regarded the water stress level in the experimental study to have been relatively weak. Conversely, Cobb et al. (unpublished) found that stress in pinyon pine improves herbivore performance in both observational and experimental studies. A better understanding of the details of naturally induced water stress will improve our understanding of its effects on herbivore–plant interactions. Future experimental studies should incorporate natural stress levels in a broad gradient of stress conditions.

The varied responses of herbivores to stress in different plant types indicate that the stress hypothesis endorsed by White,[262] Rhoades,[195,196] and others lacks broad applicability. Positive responses to conifers, negative responses to herbaceous plants, and varied responses to other plant types indicate that the herbivore–plant stress interaction is complicated. The hypothesis that plant stress benefits herbivores may be valid for conifers, although there is little experimental evidence to support this. Defensive chemicals may increase in stressed herbaceous plants and possibly contribute to their negative influence on herbivory so commonly encountered in this review. Concentrations of defensive compounds, including cyanogenic glycosides, glucosinolates, and alkaloids, often increase in water stressed plants.[74] Herbaceous plants may be less tolerant of water stress than other plant types, rendering them poorer hosts when water stressed.[148] The different feeding guilds responded similarly to water stress, as was true in the nutrient studies. While many herbivores, even closely related ones,[249] respond variably to water stress, feeding habits do not appear to influence response patterns. With the exception of woodboring beetles, the water stress data do not support Larsson's contention[120] that herbivore feeding type accounts for different herbivore responses to stress.

C. WATER AND NUTRIENT DEFICITS IN CONCERT

All studies that compared herbivore responses to water *and* nutrient deficits found that herbivores responded similarly to both, whether they responded negatively, positively, or not at all. There is no doubt that the two are inextricably connected,[226] and together they exert complementary effects on the dynamics of herbivore populations and communities.[115,122,124]

Two studies reported positive herbivore responses to plants receiving the highest levels of both nutrients and water in the same experiment,[45,124] while two others found positive herbivore responses on plants receiving the lowest water and nutrient levels.[8,122] Larsson and Tenow[122] found that drought-associated outbreaks of *Neodiprion sertifer* were greatest on plants growing on nutrient poor sites, and fertilization reduced outbreak levels.

Four studies examined the responses of herbivores to water stress and nutrient stress in separate experiments within the same studies and found that herbivores typically responded similarly to water stress and nutrient stress.[128,148,156,248]

Considered separately, a plant's water status and nutrient status also influence one another. Nutrient enhancement can have positive effects on a plant's water status by increasing water efficiency and reducing the transpiration rate,[122] while water stress can affect a plant's nutrient status by limiting nutrient uptake.[25] Several studies reported that fertilization resulted in increased leaf water content and higher osmotic pressure or turgor pressure of plant sap.[77,199,265,266] Wolfson[266] found that well-fertilized plants had a higher leaf water content and that ovipositing *Pieris rapae* were most attracted to high leaf water content. Leaf water content and nitrogen concentration can also be positively correlated in untreated plants.[142] Interactions between nutrients and water in plants influence the herbivore–plant stress interaction, and should be considered in future stress studies.

D. STRESS EFFECTS ON GENETIC RESISTANCE

Is the expression of genetically based resistance and susceptibility traits in plants affected by environmental stress? This subject is not well studied.[238] This review determined that the expression of genetically based resistance and susceptibility traits were maintained in four of five water stress studies compared to three of nine fertilization studies, suggesting that a plant's nutrient status is more likely to alter the expression of its genetically based resistance and susceptibility traits.[36,59,114,129,148,230,231,248] In three studies resistance traits were diminished or susceptibility traits were enhanced, resulting in a decreased difference in herbivore responses to resistant vs. susceptible plants.[114,148]

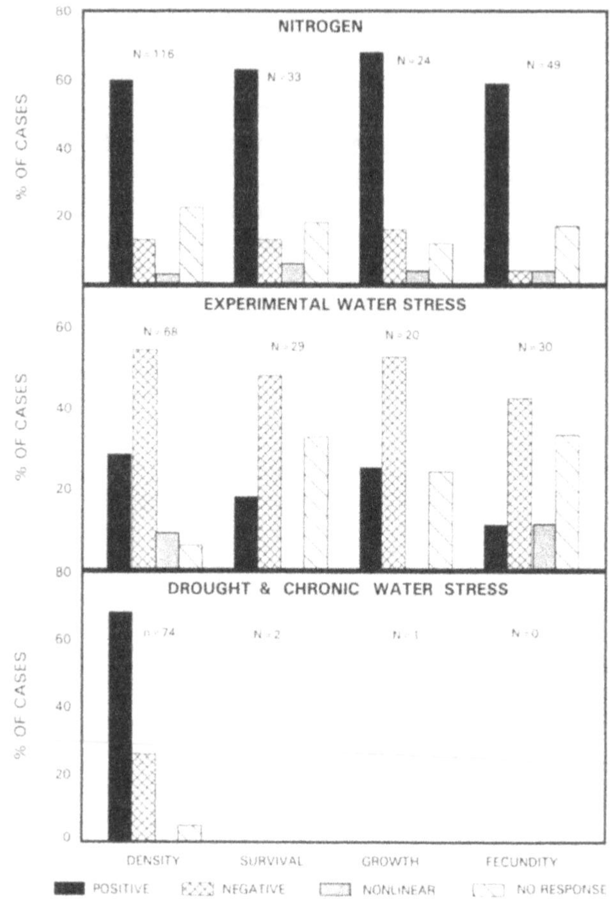

FIGURE 7. Herbivore responses, based on measurements of density, survival, growth, and fecundity, to nitrogen fertilization, experimental water stress, drought, and chronic water stress. Data are presented as the percentage of studies for all herbivores.

In four studies differences in herbivore responses to resistant and susceptible plant lines increased.[35,114,148]

In one study, in which resistance or susceptibility traits were not specifically considered, water stress eliminated all variation in herbivore densities on different plant genotypes. Kimberling et al.[113] reported that gall densities were highly variable on nonstressed grape clones, but gall densities were uniformly low on the same clones when water stressed.

E. STRESS AND HERBIVORE LIFE-HISTORY PARAMETERS

The results of this survey do not support the contention that there is strong differential sensitivity of different herbivore life-history parameters to plant stress.[108] There were no differences in responses to nitrogen fertilization or experimental water stress due to the different life history parameters measured, including survivorship, growth rate, and fecundity (Figure 7). These patterns also agreed with density or population level responses. This suggests that all of these measurements will typically provide a comparable estimate of the influence of stress on herbivore population dynamics.

F. NONLINEAR HERBIVORE RESPONSES TO PLANT STRESS

Nonlinear responses of herbivore populations to plant stress were rarely reported in the studies we reviewed (n = 12). Nearly all nonlinear responses to fertilization or water stress had a convex shape, indicating that herbivore performance increased until reaching some stress threshold, after which it declined. This response occurred in several studies that applied particularly high fertilizer treatments[24,248] or severe water stress treatments.[64] Probably all herbivores that respond positively to plant stress will exhibit negative responses as some stress threshold is surpassed, and studies can detect this by measuring herbivore performance across a wide stress gradient.[64,120,139,140] English-Loeb[64] measured the response of *Tetranychus urticae* to up to six water-stress treatments. By comparing the results of any two treatment levels, he found negative or positive responses, depending on which pair of treatment levels he compared. English-Loeb proposed that many contradictory responses of mites to water stress may result from researchers making measurements at different points along the stress gradient.

Nonlinear responses may arise for reasons other than extreme treatments. Rodriguez[199] found that *Tetranychus urticae* responded nonlinearly to P fertilization in tomato and linearly (negatively) to P fertilization in apple, because tomato absorbs more phosphorus than does apple. Leigh et al.[124] found that herbivore responses to fertilization in cotton varied according to plant water status and the time of sampling during the growing season. They suggested that interactions between nutrients and water availability, and plant growth and reproduction all affected herbivore responses. Several studies that examined the responses of two herbivore species to the same treatments reported that only one of the two responded nonlinearly.[88,243]

IV. CONDUCTING PLANT STRESS–HERBIVORY EXPERIMENTS

In addition to finding some important patterns of herbivore responses on stressed plants, this review has revealed other factors that may influence this interaction.

A. OTHER STRESS EFFECTS ON HERBIVORES

Most studies in this review assumed that herbivore responses were based on plant stress. However, stress can exert direct effects on herbivores and on their other ecological interactions. Water or nutrient stress can alter predator population dynamics.[10,145,229] Strauss[229] found that chrysomelid populations declined in fertilized fields due to negative interactions, with increased numbers of ants tending higher densities of aphids. Turchin et al.[241] refuted earlier claims that *Dendroctonus frontalis* is regulated by drought and contended that southern pine beetles are governed by delayed density-dependent factors, possibly including natural enemies. Increased leaf temperatures due to drought or experimental water stress can increase herbivore development rates.[173,258,271]

Drought stress can affect all stages of herbivore development. It causes higher levels of egg mortality in some tettigonids, scarabaeids, and other species.[101,184,185] Harris and Ring [85] reported that densities of soil-pupating pecan weevils declined during drought years because adult emergence was impeded by drier soils. Herbivore responses to water stress can be strongly influenced by instar and by leaf age.[15,234,252] Herbivore species vary in their behavioral and physiological abilities to tolerate water stress, with xeric species being better adapted to drought.[48,135]

B. STUDY DESIGN

The design of a study can have a strong influence on the results in stress research. The different herbivore responses to experimental water stress and drought stress may have been influenced by differing methods. Levels of stress, as well as timing and duration, influence herbivore responses to stress and indicate that the details of naturally induced plant stress need to be better understood.

A comparison of the herbivore–plant interaction between different systems can best be made if response curves are generated for each system that reflect herbivore responses to a broad gradient of stress and nonstress conditions in plants. The idiosyncratic responses of different plant species to comparable stress conditions (e.g., different absorption rates of nutrients[199] and different osmoregulatory abilities[71,167]) make such an approach essential. Such stress studies should include treatments that mimic natural stress conditions. Stark,[226] Shaw et al.,[213] Carrow and Betts,[38] and others have shown that even slight differences in fertilizers strongly affected herbivore responses.

It is imperative to make multiple and careful quantitative measurements of plant responses to stress. Storms[227] found that while fecundity in *Tetranychus urticae* was positively correlated with leaf N in apples, leaf N was not necessarily correlated with nutrient treatment. An increase in available plant feeding sites can lead to an increase in herbivore population size that need not relate at all to qualitative plant changes. Okigbo and Gyrisco[174] proposed that tillering in fertilized wheat, rather than qualitative plant changes, led to increased densities of *Mayetiola destructor* galls. Another measurement of herbivore performance would help to define this kind of relationship. Some of the many "no responses" to fertilization may have resulted from an erroneous assumption of plant response, rather than herbivore "indifference".

Finally, it is important to have a clear understanding of the natural history of the herbivore, and especially its specific resource needs and how they are affected by stress (e.g., flowers, certain types of leaves, stem diameter).

V. CONCLUSIONS

Stress has strong effects on herbivore population dynamics: more than 75% of the more than 450 studies we reviewed reported significant herbivore responses — whether positive, negative, or nonlinear — to plant water and/or nutrient deficits. In general, experimental nutrient stress and water stress appear to render plants poorer resources for herbivores. However, drought and chronic water stress appear to have the opposite effect. Differing methodologies of experimental vs. observational studies may account for these contradictory results. Herbivore responses to stress are also conditional on whether host plants are flowering plants or conifers, and perhaps on the ratios of nutrients and defenses in host plants. It may be that water- and nutrient-stressed conifers are attacked more often due to decreased resins and terpenes, while water-stressed conifers have the added attraction of increased nitrogen levels. The opposite may be true for stressed flowering plants; that is, defensive chemistry may increase and nutrients decrease, making them less palatable for herbivores. Perhaps flowering plants are typically more resistant than conifers to herbivores during drought and when living in poor microsites, by virtue of a different biochemical physiology.

Although there are many parallels between water stress and nutrient stress effects on plants and herbivores, the water stress studies revealed differences in herbivore responses to woody vs. herbaceous plants that were not evident in the nutrient studies. It may be that of the two stress forms, water stress has a more debilitating effect on herbaceous plants, and consequently on herbivores, because its effects are expressed more quickly.

Larsson[120] proposed that different herbivore feeding guilds respond differently to plant stress, however, this review found little variation in the responses of four different feeding guilds — sucking, chewing and gallforming insects and mites — to plant stress and fertilization. Instead, variation in herbivore responses to stress, even among closely related species, stemmed from different resource requirements, where resources were affected differently by stress, and by differences in treatment design, such as the use of different plant genotypes and species.

Exceptions exist to all of these patterns, and yet the patterns offer a reference point for assessing our current understanding of the influence of stress on herbivore–plant interactions. Perhaps the strongest message to emerge from this review is that in order to obtain reliable results in herbivore–plant stress studies, the nature of stress itself must be better understood. There is no doubt that stress exerts a strong influence over herbivore–plant interactions, perhaps especially through its extensive effects on plant chemistry, and a better understanding of how nutritional and defensive chemistry vary with plant stress is needed.

ACKNOWLEDGMENTS

We wish to thank K. Larson, S. Mopper, P. Price, L. Stevens, T. Whitham, and an anonymous reviewer for valuable discussions and comments on the manuscript. This project was funded in part by NSF grant BSR-8705347 and USDA grant GAM-8700709 to T. G. Whitham.

APPENDIX

Source data are organized in this appendix (beginning on the following page) according to treatment and plant type. Each case lists treatment; insect species; insect feeding guild (1 = sucking insect, 2 = chewing insect, 3 = mites, 4 = gallforming insects); plant species; plant response (NG = response information not given); herbivore response according to density (= #), survivorship, growth, and fecundity measurements); herbivore responses are measured as positive (+), negative (–), nonlinear (⋆) or no response (=); and the reference for each case. Fertilization studies and then water stress studies are presented. Each section is subdivided according to plant type (herbaceous, woody, wild, cultivated) and within each section cases are organized according to herbivore feeding guild and species.

TREATMENT	INSECT (GUILD)	PLANT	PLANT RESPONSE	HERBIVORE RESPONSE #	SURV	GROW	FECUND	REFERENCE
NITROGEN OR TOTAL NUTRIENTS:								
HERBACEOUS CULTIVATED PLANTS:								
FERTILIZATION	ACYRTHOSIPHON PISUM (1)	PEA	(+) GROWTH				(=)	TAYLOR ET AL. 1952 (235)
FERTILIZATION	ACYRTHOSIPHON PISUM (1)	PEA	(+) GROWTH				(=)	TAYLOR ET AL. 1952 (235)
FERTILIZATION	ACYRTHOSIPHON PISUM (1)	BROADBEAN	NG	(=)				MARKKULA & TIITTANEN 1969 (136)
FERTILIZATION	APHIS FABAE (1)	SUGAR BEET	(+) GROWTH, LEAF N	(*)				HEATHCOTE 1974 (88)
FERTILIZATION	APHIS GOSSYPII (1)	COTTON	NG	(+)				MCGARR 1942 (146)
FERTILIZATION	APHIS RHAMNI (1)	POTATO	(+) GROWTH	(+)				BROADBENT ET AL. 1952 (27)
FERTILIZATION	AULACORTHUM (1) SOLANI	POTATO	(+) GROWTH	(+)				BROADBENT ET AL. 1952 (27)
FERTILIZATION	BREVICORYNE (1) BRASSICAE	BRUSSELS SPROUTS	(+) LEAF N				(*)	VAN EMDEN 1966 (243)
FERTILIZATION	BREVICORYNE (1) BRASSICAE	BRUSSELS SPROUTS	(+) N				(+)	VAN EMDEN & BASHFORD 1969 (244)
FERTILIZATION	MACROSIPHUM (1) EUPHORBIAE	POTATO	(+) GROWTH	(+)				BROADBENT ET AL. 1952 (27)
FERTILIZATION	MACROSIPHUM PISI (1)	PEA	(+) GROWTH				(+)	BARKER & TAUBER 1951 (11)
FERTILIZATION	MACROSIPHUM PISI (1)	PEA	(+) GROWTH				(+)	BARKER & TAUBER 1951 (11)
FERTILIZATION	MACROSIPHUM (1) EUPHORBIAE	POTATO	(+) GROWTH				(=)	TAYLOR ET AL. 1952 (235)
FERTILIZATION	MACROSIPHUM (1) EUPHORBIAE	POTATO	(+) GROWTH				(=)	TAYLOR ET AL. 1952 (235)
FERTILIZATION	METOPOLOPHIUM (1) DIRHODUM	WHEAT	(–) NITRATE IN SAP	(+)				HENDERSON & PERRY 1978 (91)
FERTILIZATION	METOPOLOPHIUM (1) DIRHODUM	WHEAT	(+) YIELD	(+)				PREW ET AL. 1983 (190)

TREATMENT	INSECT (GUILD)	PLANT	PLANT RESPONSE	HERBIVORE RESPONSE				REFERENCE
				#	SURV	GROW	FECUND	
FERTILIZATION	MYZUS PERSICAE (1)	POTATO	(+) GROWTH	(+)				BROADBENT ET AL. 1952 (27)
FERTILIZATION	MYZUS PERSICAE (1)	POTATO	(+) LEAF NITRATES	(=)				JANSSON & SMILOWITZ 1986 (106)
FERTILIZATION	MYZUS PERSICAE (1)	BRUSSELS SPROUTS	(+) LEAF N				(+)	VAN EMDEN 1966 (243)
FERTILIZATION	MYZUS PERSICAE (1)	BRUSSELS SPROUTS	(+) N				(+)	VAN EMDEN & BASHFORD 1969 (244)
FERTILIZATION	MYZUS PERSICAE (1)	SUGAR BEET	(+) GROWTH, LEAF N	(+)				HEATHCOTE 1974 (88)
FERTILIZATION	MYZUS PERSICAE (1)	SUGAR BEET	NG	(+)				MARKKULA & TIITTANEN 1969 (136)
FERTILIZATION	MYZUS PERSICAE (1)	TOBACCO	(+) LEAF N	(+)	(+)	(+)		WOOLRIDGE & HARRISON 1969 (267)
FERTILIZATION	MYZUS PERSICAE (1)	BROADBEAN	NG	(=)				MARKKULA & TIITTANEN 1969 (136)
FERTILIZATION	MYZUS PERSICAE (1)	CHRYSANTHEMUM	NG	(=)				MARKKULA & TIITTANEN 1969 (136)
FERTILIZATION	MYZUS PERSICAE (1)	CUCUMBER	NG	(=)				MARKKULA & TIITTANEN 1969 (136)
FERTILIZATION	RHOPALOSIPHUM (1) FITCHII	OATS	(+) GROWTH	(+)				COON 1959 (50)
FERTILIZATION	RHOPALOSIPHUM MAIDIS (1)	SORGHUM	NG	(+)				BRANSON & SIMPSON 1966 (21)
FERTILIZATION	RHOPALOSIPHUM (1) PADI	WHEAT	(–) NITRATE IN SAP	(+)				HENDERSON & PERRY 1978 (91)
FERTILIZATION	SCHIZAPHIS GRAMINUM (1)	SORGHUM	(+) GROWTH	(+)				ARCHER ET AL. 1982 (7)
FERTILIZATION	SCHIZAPHIS GRAMINUM (1)	WHEAT	(+) GROWTH	(–)				DANIELS 1957 (57)
FERTILIZATION	SCHIZAPHIS GRAMINUM (1)	OATS	NG	(–)				ARANT & JONES 1951 (6)
FERTILIZATION	SITOBION AVENAE (1)	WHEAT	(–) NITRATE IN SAP	(+)				HENDERSON & PERRY 1978 (91)
FERTILIZATION	THERIOAPHIS MACULATA (1)	ALFALFA (RESISTANT)	(+) GROWTH, LEAF N				(=)	MCMURTRY 1962 (148)
FERTILIZATION	THERIOAPHIS MACULATA (1)	ALFALFA (SUSCEPTIBLE)	(+) GROWTH, LEAF N				(=)	MCMURTRY 1962 (148)
FERTILIZATION	THERIOAPHIS MACULATA (1)	ALFALFA (RESISTANT)	(+) GROWTH		(+)			KINDLER & STAPLES 1970 (114)

TREATMENT	INSECT (GUILD)	PLANT	PLANT RESPONSE	HERBIVORE RESPONSE				REFERENCE
				#	SURV	GROW	FECUND	
FERTILIZATION	THERIOAPHIS MACULATA (1)	ALFALFA (SUSCEPTIBLE)	(+) GROWTH		(=)			KINDLER & STAPLES 1970 (114)
FERTILIZATION	SACCHAROSYDNE (1) SACCHARIVORA	SUGARCANE	(+) N				(+)	METCALFE 1970 (156)
FERTILIZATION	DALBULUS MAIDIS (1)	CORN	(+) GROWTH	(=)				POWER 1989 (187)
FERTILIZATION	DALBULUS MAIDUS (1)	CORN	(+) GROWTH	(+)	(+)			POWER 1987 (186)
FERTILIZATION	EMPOASCA LIBYCA (1)	COTTON	(+) LEAF N	(+)				JOYCE 1958 (109)
FERTILIZATION	BLISSUS LEUCOPTERUS (1)	SORGHUM	(+) LEAF N		(+)		(+)	DAHMS 1947 (55)
FERTILIZATION	LYGUS HESPERUS (1)	COTTON	(+) YIELD	(*)				LEIGH ET AL. 1970 (124)
FERTILIZATION	BEMISIA TABACI (1)	COTTON	(+) LEAF N	(+)				JOYCE 1958 (109)
FERTILIZATION	HERCOTHRIPS (1) FUMIPENNIS	COTTON	(+) LEAF N	(+)				JOYCE 1958 (109)
FERTILIZATION	LEPTINOTARSA (2) DECEMLINEATA	POTATO	(+) LEAF NITRATES	(=)				JANSSON & SMILOWITZ 1985 (105)
FERTILIZATION	EPILACHNA (2) VARIESTRIS	SOYBEAN	(+) LEAF PROTEIN	(+)				TODD ET AL. 1972 (240)
FERTILIZATION	LISSORHOPTRUS (2) ORYZOPHILUS	RICE	(+) YIELD	(+)				BOWLING 1963 (19)
FERTILIZATION	BRACHYRHINUS (2) SULCATUS	STRAWBERRY	(=) GROWTH				(=)	CRAM 1965 (53)
FERTILIZATION	SCIOPITHES (2) OBSCURUS	STRAWBERRY	(=) GROWTH				(=)	CRAM 1965 (53)
FERTILIZATION	BRYOBIA PRAETIOSA (3)	BEAN	(+) GROWTH, LEAF N	(+)	(+)		(+)	MORRIS 1961 (168)
FERTILIZATION	SPODOPTERA (2) FRUGIPERDA	MILLET	(+) GROWTH		(+)	(+)		LEUCK 1972 (125)
FERTILIZATION	SPODOPTERA (2) FRUGIPERDA	MILLET	(=) GROWTH		(–)	(–)		LEUCK 1972 (125)

TREATMENT	INSECT (GUILD)	PLANT	PLANT RESPONSE	HERBIVORE RESPONSE				REFERENCE
				#	SURV	GROW	FECUND	
FERTILIZATION	ELDANA SACCHARINA (2)	SUGARCANE	(+) N, AA'S	(+)				ATKINSON & NUSS 1989 (8)
FERTILIZATION	MELANOPLUS MEXICANA (2)	WHEAT	(+) LEAF N		(+)	(+)	(+)	SMITH & NORTHCOTT 1951 (222)
FERTILIZATION	ANTHONOMUS GRANDIS (2)	COTTON	(+) GROWTH, REPROD.	(+)				MISTRIC 1968 (161)
FERTILIZATION	HELIOTHIS ARMIGERA (2)	CORN	(+) PROTEIN, REPROD.	(=)				DOUGLAS & ECKHART 1953 (60)
FERTILIZATION	HELIOTHIS ZEA (2)	CORN	(+) BIOMASS				(+)	WISEMAN ET AL. 1973 (264)
FERTILIZATION	HELIOTHIS ZEA (2)	CORN	(+) YIELD	(=)				KLOSTERMEYER 1950 (116)
FERTILIZATION	HELIOTHIS ZEA (2)	CORN	(+) GROWTH				(=)	TAYLOR ET AL. 1952 (235)
FERTILIZATION	HELIOTHIS ZEA (2)	CORN	(+) GROWTH	(+)				ADKISSON 1958 (2)
FERTILIZATION	SPODOPTERA ERIDANIA (2)	BLACK MUSTARD	(+) LEAF H_20, PROTEIN		(+)	(+)	(+)	WOLFSON 1982 (266)
FERTILIZATION	SPODOPTERA EXIGUA (2)	SUGAR BEET	(+) LEAF N		(+)	(+)	(+)	AL-ZUBAIDI & CAPINERA 1984 (5)
FERTILIZATION	SPODOPTERA (2) FRUGIPERDA	CORN	(+) BIOMASS		(+)	(+)	(+)	WISEMAN ET AL. 1973 (264)
FERTILIZATION	SPODOPTERA (2) FRUGIPERDA	BERMUDA GRASS	NG		(+)	(+)		LYNCH 1984 (130)
FERTILIZATION	PIERIS BRASSICAE (2)	CABBAGE	NG			(+)		ALLEN & SELMAN 1957 (4)
FERTILIZATION	PIERIS RAPAE (2)	CABBAGE	(+) N,P	(+)		(+)	(+)	MYERS 1985 (171)
FERTILIZATION	PIERIS RAPAE (2)	BLACK MUSTARD	(+) LEAF H_20, PROTEIN			(+)	(+)	WOLFSON 1982 (266)
FERTILIZATION	PIERIS RAPAE (2)	BLACK MUSTARD	(+) GROWTH	(+)				WOLFSON 1980 (265)
FERTIILIZATION	PIERIS RAPAE (2)	CABBAGE	NG			(+)		SLANSKY & FEENY 1977 (219)
FERTILIZATION	GEOMYZA TRIPUNCTATA (2)	RYE	NG	(+)				MOORE & CLEMENTS 1984 (163)
FERTILIZATION	ANTHONOMUS GRANDIS (2)	COTTON	NG	(=)				MCGARR 1942 (146)

TREATMENT	INSECT (GUILD)	PLANT	PLANT RESPONSE	HERBIVORE RESPONSE #	SURV	GROW	FECUND	REFERENCE
FERTILIZATION	PHAEDON COCHLEARIAE (2)	WATERCRESS	NG				(+)	ALLEN & SELMAN 1955 (3)
FERTILIZATION	ATHERIGONA (2) BITUBERCULATA	WHEAT	NG	(+)				RAWAT & SAHU 1969 (193)
FERTILIZATION	EPITRIX HIRTIPENNIS (2)	TOBACCO	NG	(=)				SEMTNER ET AL. 1980 (193)
FERTILIZATION	MANDUCA SEXTA (2)	TOBACCO	NG	(+)				SEMTNER ET AL. 1980 (211)
FERTILIZATION	MANDUCA SEXTA (2)	TOBACCO	NG	(+)				REAGAN ET AL. 1978 (194)
FERTILIZATION	OSCINELLA VASTATOR (2)	RYE	NG	(–)				MOORE & CLEMENTS 1984 (163)
FERTILIZATION	OSCINELLA FRIT (2)	RYE	NG	(+)				MOORE & CLEMENTS 1984 (163)
FERTILIZATION	OSCINELLA FRIT (2)	OATS	NG	(=)				CUNLIFFE 1928 (54)
FERTILIZATION	OSTRINIA NUBILIALIS (2)	CORN	(+) YIELD		(+)			PATCH 1947 (177)
FERTILIZATION	OSTRINIA NUBILALIS (2)	CORN (RESISTANT)	NG		(=)			CANNON & ORTEGA 1966 (36)
FERTILIZATION	OSTRINIA NUBILALIS (2)	CORN (SUSCEPTIBLE)	NG		(+)			CANNON & ORTEGA 1966 (36)
FERTILIZATION	TETRANYCHUS (3) ATLANTICUS	SOYBEAN	NG	(+)				CANNON & CONNELL 1965 (35)
FERTILIZATION	TETRANYCHUS (3) BIMACULATUS	TOMATO	(+) N, P	(–)				RODRIGUEZ 1951 (198)
FERTILIZATION	TETRANYCHUS (3) BIMACULATUS	COTTON	(+) YIELD	(*)				LEIGH ET AL. 1970 (124)
FERTILIZATION	TETRANYCHUS URTICAE (3)	RADISH	(=) GROWTH	(–)				MELLORS & PROPTS 1983 (154)
FERTILIZATION	TETRANYCHUS (3) URTICAE	TOMATO	(+) LEAF N, P, VITAMINS	(–)				RODRIGUEZ & RODRIGUEZ 1952 (200)
FERTILIZATION	TETRANYCHUS URTICAE (3)	BEAN	(+) LEAF N	(+)			(+)	HENNEBERRY 1962 (92), HENNEBERRY & SHRIVER 1964 (93)
FERTILIZATION	TETRANYCHUS URTICAE (3)	BROADBEAN	NG	(=)				MARKKULA & TIITTANEN 1969 (136)
FERTILIZATION	TETRANYCHUS URTICAE (3)	CHRYSANTHEMUM	NG	(+)				MARKKULA & TIITTANEN 1969 (136)

TREATMENT	INSECT (GUILD)	PLANT	PLANT RESPONSE	HERBIVORE RESPONSE #	SURV	GROW	FECUND	REFERENCE
FERTILIZATION	TETRANYCHUS URTICAE (3)	SUGAR BEET	NG	(+)				MARKKULA & TIITTANEN 1969 (136)
FERTILIZATION	TETRANYCHUS URTICAE (3)	TOMATO	NG	(+)				MARKKULA & TIITTANEN 1969 (136)
FERTILIZATION	TETRANYCHUS URTICAE (3)	TOMATO	NG	(=)				MARKKULA & TIITTANEN 1969 (136)
FERTILIZATION	TETRANYCHUS URTICAE (3)	BEAN	NG	(+)				GARMAN & KENNEDY 1949 (73)
FERTILIZATION	MAYETIOLA DESTRUCTOR (4)	WHEAT	(+) TILLERING	(+)				OKIGBO & GYRISCO 1962 (174)
HERBACEOUS WILD PLANTS:								
FERTILIZATION	APHIDIDAE SPP. (1)	ARTEMESIA LUDOVICIANA	(+) N, BIOMASS	(+)				STRAUSS 1987 (229)
FERTILIZATION	MYZUS PERSICAE (1)	SOLANUM SP.	(+) N				(+)	HARREWIJN 1970 (83)
FERTILIZATION	MEMBRACIDAE SPP. (1)	ARTEMESIA LUDOVICIANA	(+) N, BIOMASS	(+)				STRAUSS 1987 (229)
FERTILIZATION	CICADELLIDAE SPP. (1)	SPARTINA SPP.	(+) BIOMASS, N	(+)				VINCE & VALIELA 1981 (245)
FERTILIZATION	ZYGINIDIA SCUTELLARIS (1)	HOLCUS LANATUS	(+) LEAF N				(=)	PRESTIDGE 1982a (188)
FERTILIZATION	ZYGIDINIA SCUTELLARIS (1)	DACTYLIS GLOMERATA HOLCUS SPP.	(+) LEAF N	(–)				PRESTIDGE 1982b (189)
FERTILIZATION	EUCELIS INCISUS (1)	HOLCUS LANATUS	(+) LEAF N				(=)	PRESTIDGE 1982a (188)
FERTILIZATION	ELYMANA SULPHERELLA (1)	HOLCUS LANATUS	(+) LEAF N				(=)	PRESTIDGE 1982a (1988)
FERTILIZATION	ELYMANA SULPHURELLA (1)	HOLCUS SPP.	(+) LEAF N	(–)				PRESTIDGE 1982b (189)
FERTILIZATION	DICRANOTROPIS HAMATA (1)	HOLCUS LANATUS	(+) LEAF N		(=)		(=)	PRESTIDGE 1982a (188)
FERTILIZATION	DICRANOTROPIS HAMATA (1)	HOLCUS SPP.	(+) LEAF N	(+)				PRESTIDGE 1982b (189)
FERTILIZATION	APHRODES ALBIFRONS (1)	GRASSES	(+) LEAF N	(+)				PRESTIDGE 1982b (189)
FERTILIZATION	ADARRUS OCELLARIS (1)	HOLCUS SPP.	(+) LEAF N	(+)				PRESTIDGE 1982b (189)
FERTILIZATION	DIPLOCOLENUS (1) ABDOMINALIS	HOLCUS SPP.	(+) LEAF N	(–)				PRESTIDGE 1982b (189)
FERTILIZATION	MACROSTELES LAEVIS (1)	AGROSTIS TENUIS	(+) LEAF N	(+)				PRESTIDGE 1982b (189)

TREATMENT	INSECT (GUILD)	PLANT	PLANT RESPONSE	HERBIVORE RESPONSE				REFERENCE
				#	SURV	GROW	FECUND	
FERTILIZATION	MOCYDIOPSIS (1) PARVICAUDA	GRASS	(+) LEAF N	(–)				PRESTIDGE 1982b (189)
FERTILIZATION	PSAMMOTETTIX CONFINIS (1)	AGROSTIS TENUIS	(+) LEAF N	(+)				PRESTIDGE 1982b (189)
FERTILIZATION	RECILIA CORONIFERA (1)	HOLCUS SPP.	(+) LEAF N	(–)				PRESTIDGE 1982b (189)
FERTILIZATION	JAVESELLA PELLUCIDA (1)	NUMEROUS	(+) LEAF N	(+)				PRESTIDGE 1982b (189)
FERTILIZATION	PROKELISIA MARGINATA (1)	SPARTINA SP.	(+) LEAF SOLUBLE PROTEIN	(+)				DENNO ET AL. 1985 (58)
FERTILIZATION	MIRIDAE SPP. (1)	ARTEMESIA LUDOVICIANA	(+) N, BIOMASS	(+)				STRAUSS 1987 (229)
FERTILIZATION	MIRIDAE SPP. (1)	SPARTINA SPP.	(+) N, BIOMASS	(+)				VINCE & VALIELA 1981 (245)
FERTILIZATION	CYRTOBAGOUS (2) SALVINIAE	FERN	(+) GROWTH	(+)	(+)			ROOM ET AL. 1989 (201)
FERTILIZATION	SAMEA (2) MULTIPLICALIS	FERN	(+) GROWTH	(+)	(+)			ROOM ET AL. 1989 (201)
FERTILIZATION	GRASSHOPPERS (2)	SPARTINA SPP.	(+) BIOMASS, N	(+)				VINCE & VALIELA 1981 (245)
FERTILIZATION	CHELOBASIS PERPLEXA (2)	HELICONIA SP.	(+) LEAF N			(=)		AUERBACH & STRONG 1981 (9)
FERTILIZATION	CEPALOLEIA (2) CONSANGUINEA	HELICONIA SP.	(+) LEAF N			(=)		AUERBACK & STRONG 1981 (9)
FERTILIZATION	SPODOPTERA EXIGUA (2)	CHENOPODIUM ALBUM	(+) LEAF N		(+)	(+)	(+)	AL-ZUBAIDI & CAPINERA 1984 (5)
FERTILIZATION	OSCINELLA FRIT (2)	GRASSLAND	NG	(=)				HENDERSON & CLEMENTS 1977 (90)
FERTILIZATION	CHLOROPIDAE (2)	GRASSLAND	NG	(=)				HENDERSON & CLEMENTS 1977 (90)
FERTILIZATION	OPOMYZIDAE (2)	GRASSLAND	NG	(=)				HENDERSON & CLEMENTS 1977 (90)
FERTILIZATION	EUROSTA SOLIDAGINIS (4)	SOLIDAGO ALTISSIMA	(+) GROWTH	(=)		(=)		ABRAHAMSON ET AL. 1988 (1)
WOODY CULTIVATED PLANTS:								
FERTILIZATION	AONIDIELLA AURANTII (1)	CITRUS	(=)	(–)				SALAMA ET AL. 1972 (204)

TREATMENT	INSECT (GUILD)	PLANT	PLANT RESPONSE	HERBIVORE RESPONSE #	SURV	GROW	FECUND	REFERENCE
FERTILIZATION	LEPIDOSAPHES BECKII (1)	CITRUS	(=)	(+)				SALAMA ET AL. 1972 (204)
FERTILIZATION	SCIRTOTHRIPS CITRI (1)	CITRUS	(+) TOTAL N, AA'S, SOLUBLE PROTEINS	(+)				HARE ET AL. 1989 (82)
FERTILIZATION	PARLATORIA ZIZYPHUS (1)	CITRUS	NG	(+)				SALAMA ET AL. 1985 (205)
FERTILIZATION	ICERYA PURCHASI (1)	CITRUS	NG	(+)				SALAMA ET AL. 1985 (205)
FERTILIZATION	PSYLLA PYRICOLA (1)	PEAR	(+) N, AA'S				(–)	PFEIFFER & BURTS 1984 (182)
FERTILIZATION	PSYLLA PYRICOLA (1)	PEAR	(+) N, AA'S	(+)				PFEIFFER & BURTS 1983 (181)
FERTILIZATION	PHENACOCCUS (1) HARGREAVESI	COCOA	NG	(=)				CAMPBELL 1984 (33)
FERTILIZATION	PSEUDOCOCCOS (1) CONCAVOCERARII	COCOA	NG	(=)				CAMPBELL 1984 (33)
FERTILIZATION	MACONELLICOCCUS (1) UGANDAE	COCOA	NG	(=)				CAMPBELL 1984 (33)
FERTILIZATION	MESOHOMOTOMA (1) TESSMANNI	COCOA	NG	(=)				CAMPBELL 1984 (33)
FERTILIZATION	PLANOCCOIDES (1) NJALENSIS	COCOA	NG	(=)				CAMPBELL 1984 (33)
FERTILIZATION	PLANOCOCCUS (1) CITRI	COCOA	NG	(+)				CAMPBELL 1984 (33)
FERTILIZATION	GASCARDIA SP. NR. (1) ZONATA	COCOA	NG	(=)				CAMPBELL 1984 (33)
FERTILIZATION	PSEUDOCOCCUS (1) COMSTOCKI	APPLE	(+) GROWTH, REPROD.	(+)				SCHOENE 1941 (206)
FERTILIZATION	SYNANTHEDON EXITIOSA (2)	PEACH	(+) YIELD	(+)				SMITH & HARRIS 1952 (223)
FERTILIZATION	EOTETRANYCHUS (3) HICORIAE	PECAN	(+) GROWTH	(+)				JACKSON & HUNTER 1983 (104)
FERTILIZATION	METATETRANYCHUS (3) ULMI	APPLE	(+) GROWTH, LEAF N	(+)				BREUKEL & POST 1959 (23)

TREATMENT	INSECT (GUILD)	PLANT	PLANT RESPONSE	HERBIVORE RESPONSE				REFERENCE
				#	SURV	GROW	FECUND	
FERTILIZATION	PANONYCHUS CITRI (3)	CITRUS	(+) TOTAL N, AA'S, SOLUBLE PROTEINS	(–)				HARE ET AL. 1989 (82)
FERTILIZATION	PANONYCHUS ULMI (3)	APPLE	(+) GROWTH	(+)				HAMSTEAD & GOULD 1957 (81)
FERTILIZATION	PANONYCHUS ULMI (3)	APPLE	(+) LEAF N	(+)				RODRIGUEZ 1958 (199)
FERTILIZATION	TETRANYCHUS URTICAE (3)	APPLE	(+) N			(+)	(+)	WERMELINGER ET AL. 1985 (265)
FERTILIZATION	TETRANYCHUS URTICAE (3)	APPLE	(+) LEAF N				(+)	STORMS 1969 (227)
FERTILIZATION	TETRANYCHUS URTICAE (3)	APPLE	(+) LEAF N	(+)				RODRIGUEZ 1958 (199)
FERTILIZATION	TETRANYCHUS URTICAE (3)	PEACH	NG	(+)				GARMAN & KENNEDY 1949 (73)
WOODY WILD PLANTS:								
FERTILIZATION	FIORINA EXTERNA (1)	TSUGA CANADENSIS	(+) N		(+)	(+)	(+)	MCCLURE 1980, 1977 (142,143)
FERTILIZATION	SAP SUCKERS (1)	LARREA TRIDENTATA	(+) GROWTH, N	(+)				LIGHTFOOT & WHITFORD 1987 (128)
FERTILIZATION	ALEUROTRACHELUS (1) JELINEKI	VIBURNUM SP.	(+) AA'S	(+)			(+)	SOUTHWOOD & READER 1976 (224)
FERTILIZATION	GENERAL	EUCALYPTUS	(+) LEAF N			(+)		LANDSBERG ET AL. 1990 (118)
FERTILIZATION	CHEWERS (2)	LARREA TRIDENTATA	(+) GROWTH,N	(=)				LIGHTFOOT & WHITFORD 1987 (128)
FERTILIZATION	CHEWERS (2)	CERIOPS TAGAL V. TAGAL	NG	(=)				JOHNSTONE 1981 (107)
FERTILIZATION	HYLOBIUS (2) RHIZOPHAGOUS	PINUS BANKSIANA	(+) N, H_2O		(–)	(–)		GOYER & BENJAMIN 1972 (77)
FERTILIZATION	HYLOBIUS (2) RHIZOPHAGOUS	PINUS BANKSIANA	(+) N, P, H_2O		(–)	(–)		GOYER & BENJAMIN 1972 (77)
FERTILIZATION	PISSODES STROBU (2)	PINUS STROBUS	NG	(–)				XYDIUS & LEAF 1964 (268)
FERTILIZATION	PISSODES STROBU (2)	PINUS STROBUS	NG	(=)				XYDIUS & LEAF 1964 (268)
FERTILIZATION	DENDROCTONUS SP. (2)	PINUS CONTORTA	(+) LEAF,INNER BARK N	(–)				WARING & PITMAN 1985 (250)

TREATMENT	INSECT (GUILD)	PLANT	PLANT RESPONSE	HERBIVORE RESPONSE #	SURV	GROW	FECUND	REFERENCE
FERTILIZATION	PAROPSIS ATOMARIA (2)	EUCALYPTUS BLAKELYI	(+) N, (–) TANNINS		(+)	(–)	(+)	OHMART ET AL. 1985 (172)
FERTILIZATION	CHORISTONEURA (2) FUMIFERANA	ABIES BALSAMEA	(+) GROWTH, N (–) STARCH		(+)		(+)	SHAW ET AL. 1978 (213)
FERTILIZATION	CHORISTONEURA (2) FUMIFERANA	ABIES BALSAMEA	(+) GROWTH, N (–) STARCH		(=)		(=)	SHAW ET AL. 1978 (213)
FERTILIZATION	CHORISTONEURA (2) FUMIFERANA	ABIES BALSAMEA	NG		(+)	(+)		SHAW & LITTLE 1972 (212)
FERTILIZATION	CHORISTONEURA (2) OCCIDENTALIS	PSEUDOTSUGA MENZIESII	(+) LEAF N		(*)	(*)	(*)	BREWER ET AL. 1985 (212)
FERTILIZATION	ECDYTOLOPHA SP. (2)	RHIZOPHORA MANGLE	(+) N, GROWTH	(+)				ONUF ET AL. 1977 (24)
FERTILIZATION	PHOCIDES SP. (2)	RHIZOPHORA MANGLE	(+) N, GROWTH			(+)		ONUF ET AL. 1977 (176)
FERTILIZATION	ALARODIA SLOSSONIAE (2)	RHIZOPHORA MANGLE	(+) N, GROWTH	(+)				ONUF ET AL. 1977 (176)
FERTILIZATION	AUTOMERIS IO (2)	RHIZOPHORA MANGLE	(+) N, GROWTH	(+)				ONUF ET AL. 1977 (176)
FERTILIZATION	MEGALOPYGE (2) OPERCULARIS	RHIZOPHORA MANGLE	(+) N, GROWTH	(+)				ONUF ET AL. 1977 (176)
FERTILIZATION	POECILIPS (2) RHIZOPHORAE	RHIZOPHORA MANGLE	(+) N, GROWTH	(+)				ONUF ET AL. 1977 (176)
FERTILIZATION	EUURA LASIOLEPIS (4)	SALIX LASIOLEPIS	(+) PROTEIN, GROWTH (–) PHENOLS	(*)	(*)			WARING & PRICE 1988 (248)
FERTILIZATION	NEODIPRION (2) EDULICOLIS	PINUS EDULIS	(+) LEAF N, GROWTH				(+)	MOPPER & WHITHAM 1992 (164)
FERTILIZATION	NEODIPRION LECONTEI (2)	PINUS BANKSIANA	(+) GROWTH, N	(+)				POSEY & BENJAMIN 1969 (183)
FERTILIZATION	NEODIPRION SERTIFER (2)	PINUS SYLVESTRIS	NG	(–)	(=)			LARSSON & TENOW 1984 (122)
FERTILIZATION	NEODIPRION SWAINEI (2)	PINUS BANKSIANA	(+) N (–) CA, MG, MN & ZN	(–)				SMIRNOFF & BERNIER 1973 (221)
FERTILIZATION	ADELGES COOLEYI (4)	PSEUDOTSUGA MENZIESII	(+) N, NEEDLE GROWTH		(+)		(+)	MITCHELL & PAUL 1974 (162)

TREATMENT	INSECT (GUILD)	PLANT	PLANT RESPONSE	HERBIVORE RESPONSE #	SURV	GROW	FECUND	REFERENCE
FERTILIZATION	ADELGES PICEA (4)	ABIES AMABILIS	(+) AA'S, PROTEIN	(–)				CARROW & GRAHAM 1968 (39)
FERTILIZATION	ADELGES PICEA (4)	ABIES GRANDIS	(+) ARGININE, N	(+)	(=)		(+)	CARROW & BETTS 1973 (38)
FERTILIZATION	ADELGES PICEA (4)	ABIES GRANDIS	(+) ARGININE, N	(–)	(–)		(–)	CARROW & BETTS 1973 (38)
FERTILIZATION	PEMPHIGUS POPULIVENAE (4)	POPULUS FREMONTII	NG				(+)	WHITHAM 1978 (263)
FERTILIZATION	PEMPHIGUS BETAE (4)	POPULUS ANGUSTIFOLIA	NG				(=)	WHITHAM 1978 (263)
<u>PHOSPHORUS:</u>								
HERBACEOUS CULTIVATED PLANTS:								
FERTILIZATION	ACYRTHOSIPHON (1) PISUM	BROADBEAN	NG	(=)				MARKKULA & TIITTANEN 1969 (136)
FERTILIZATION	ACYRTHOSIPHON PISUM (1)	PEA	(+) GROWTH				(=)	TAYLOR ET AL. 1952 (235)
FERTILIZATION	APHIS RHAMNI (1)	POTATO	(+) GROWTH	(+)				BROADBENT ET AL. 1952 (27)
FERTILIZATION	AULACORTHUM (1) SOLANI	POTATO	(+) GROWTH	(+)				BROADBENT ET AL. 1952 (27)
FERTILIZATION	MACROSIPHUM (1) EUPHORBIAE	POTATO	(+) GROWTH	(+)				BROADBENT ET AL. 1952 (27)
FERTILIZATION	MACROSIPHUM (1) EUPHORBIAE	POTATO	(+) GROWTH				(=)	TAYLOR ET AL. 1952 (235)
FERTILIZATION	MACROSIPHUM PISI (1)	PEA	(+) GROWTH				(+)	BARKER & TAUBER 1951 (11)
FERTILIZATION	MYZUS PERSICAE (1)	POTATO	(+) GROWTH	(+)				BROADBENT ET AL. 1952 (27)
FERTILIZATION	MYZUS PERSICAE (1)	TOBACCO	(+) LEAF P	(+)	(+)	(+)		WOOLRIDGE & HARRISON 1968 (267)
FERTILIZATION	MYZUS PERSICAE (1)	BROADBEAN	NG	(=)				MARKKULA & TIITTANEN 1969 (136)
FERTILIZATION	MYZUS PERSICAE (1)	CHRYSANTHEMUM	NG	(=)				MARKKULA & TIITTANEN 1969 (136)
FERTILIZATION	MYZUS PERSICAE (1)	CUCUMBER	NG	(=)				MARKKULA & TIITTANEN 1969 (136)

TREATMENT	INSECT (GUILD)	PLANT	PLANT RESPONSE	HERBIVORE RESPONSE				REFERENCE
				#	SURV	GROW	FECUND	
FERTILIZATION	MYZUS PERSICAE (1)	SUGAR BEET	NG	(–)				MARKKULA & TIITTANEN 1969 (136)
FERTILIZATION	THERIOAPHIS MACULATA (1)	ALFALFA (RESISTANT)	(+) GROWTH, LEAF P				(+)	MCMURTRY 1962 (148)
FERTILIZATION	THERIOAPHIS MACULATA (1)	ALFALFA (SUSCEPTIBLE)	(+) GROWTH, LEAF P				(=)	MCMURTRY 1962 (148)
FERTILIZATION	THERIOAPHIS MACULATA (1)	ALFALFA (RESISTANT)	(+) GROWTH		(+)			KINDLER & STAPLES 1970 (114)
FERTILIZATION	THERIOAPHIS MACULATA (1)	ALFALFA (SUSCEPTIBLE)	(+) GROWTH		(=)			KINDLER & STAPLES 1970 (114)
FERTILIZATION	BLISSUS LEUCOPTERUS (1)	SORGHUM	(+) LEAF P		(–)		(–)	DAHMS 1947 (55)
FERTILIZATION	SPODOPTERA (2) FRUGIPERDA	MILLET	(–) GROWTH		(–)	(–)		LEUCK 1972 (125)
FERTILIZATION	HELIOTHIS ZEA (2)	CORN	(+) GROWTH				(=)	TAYLOR ET AL. 1952 (235)
FERTILIZATION	AGASICLES (2) HYGROPHILA	ALLIGATORWEED	NG				(+)	MADDOX & RHYNE 1975 (131)
FERTILIZATION	EPITRIX HIRTIPENNIS (2)	TOBACCO	NG	(–)				SEMTNER ET AL. 1980 (211)
FERTILIZATION	MANDUCA SEXTA (2)	TOBACCO	NG	(–)				SEMTNER ET AL. 1980 (211)
FERTILIZATION	PHAEDON COCHLEARIAE (2)	WATERCRESS	NG				(+)	ALLEN & SELMAN 1955 (3)
FERTILIZATION	PIERIS BRASSICAE (2)	CABBAGE	NG			(+)		ALLEN & SELMAN 1957 (4)
FERTILIZATION	OSCINELLA FRIT (2)	OATS	NG	(=)				CUNLIFFE 1928 (54)
FERTILIZATION	OSTRINIA NUBILALIS (2)	CORN (RESISTANT)	NG		(+)			CANNON & ORTEGA 1966 (36)
FERTILIZATION	OSTRINIA NUBILALIS (2)	CORN (SUSCEPTIBLE)	NG		(+)			CANNON & ORTEGA 1966 (36)
FERTILIZATION	BRYOBIA PRAETIOSA (3)	BEAN	(+) LEAF P	(+)	(+)		(+)	MORRIS 1961 (168)
FERTILIZATION	TETRANYCHUS URTICAE(3)	TOMATO	(+) N, P	(*)				RODRIGUEZ 1951 (198)
FERTILIZATION	TETRANYCHUS URTICAE (3)	CHRYSANTHEMUM	NG	(=)				MARKKULA & TIITTANEN 1969 (136)
FERTILIZATION	TETRANYCHUS URTICAE (3)	SUGAR BEET	NG	(=)				MARKKULA & TIITTANEN 1969 (136)
FERTILIZATION	TETRANYCHUS URTICAE (3)	TOMATO	NG	(=)				MARKKULA & TIITTANEN 1969 (136)

TREATMENT	INSECT (GUILD)	PLANT	PLANT RESPONSE	HERBIVORE RESPONSE				REFERENCE
				#	SURV	GROW	FECUND	
FERTILIZATION	TETRANYCHUS URTICAE (3)	BROADBEAN	NG	(=)				MARKKULA & TIITTANEN 1969 (136)
FERTILIZATION	TETRANYCHUS URTICAE (3)	CUCUMBER	NG	(=)				MARKKULA & TIITTANEN 1969 (136)
FERTILIZATION	TETRANYCHUS URTICAE (3)	BEAN	NG				(–)	HENNEBERRY 1962 (92), HENNEBERRY & SHRIVER 1964 (93)
WOODY CULTIVATED PLANTS:								
FERTILIZATION	AONIDIELLA AURANTII (1)	CITRUS	(=) NPK	(–)				SALAMA ET AL. 1972 (204)
FERTILIZATION	LEPIDOSAPHES BECKII (1)	CITRUS	(=) NPK	(+)				SALAMA ET AL. 1972 (204)
FERTILIZATION	PARLATORIA ZIZYPHUS (1)	CITRUS	NG	(+)				SALAMA ET AL. 1985 (205)
FERTILIZATION	ICERYA PURCHASI (1)	CITRUS	NG	(+)				SALAMA ET AL. 1985 (205)
FERTILIZATION	PHENACOCCUS (1) HARGREAVESI	COCOA	NG	(=)				CAMPBELL 1984 (33)
FERTILIZATION	PSEUDOCOCCOS (1) CONCAVOCERARII	COCOA	NG	(=)				CAMPBELL 1984 (33)
FERTILIZATION	MACONELLICOCCUS (1) UGANDAE	COCOA	NG	(=)				CAMPBELL 1984 (33)
FERTILIZATION	MESOHOMOTOMA (1) TESSMANNI	COCOA	NG	(–)				CAMPBELL 1984 (33)
FERTILIZATION	PLANOCCOIDES (1) NJALENSIS	COCOA	NG	(=)				CAMPBELL 1984 (33)
FERTILIZATION	PLANOCOCCUS (1) CITRI	COCOA	NG	(=)				CAMPBELL 1984 (33)
FERTILIZATION	GASCARDIA SP. NR. (1) ZONATA	COCOA	NG	(=)				CAMPBELL 1984 (33)
FERTILIZATION	PANONYCHUS ULMI (3)	APPLE	(+) LEAF P	(*)				RODRIGUEZ 1958 (199)
FERTILIZATION	TETRANYCHUS URTICAE (3)	APPLE	(+) LEAF P	(–)				RODRIGUEZ 1958 (199)

TREATMENT	INSECT (GUILD)	PLANT	PLANT RESPONSE	HERBIVORE RESPONSE #	SURV	GROW	FECUND	REFERENCE
WOODY WILD PLANTS:								
FERTILIZATION	HYLOBIUS (2) RHIZOPHAGOUS	PINUS BANKSIANA	(+) P		(=)	(=)		GOYER & BENJAMIN 1972 (77)
<u>POTASSIUM:</u>								
HERBACEOUS CULTIVATED PLANTS:								
FERTILIZATION	ACYRTHOSIPHON PISUM (1)	PEA	(+) GROWTH				(=)	TAYLOR ET AL. 1952 (235)
FERTILIZATION	APHIS RHAMNI (1)	POTATO	(+) GROWTH	(–)				BROADBENT ET AL. 1952 (27)
FERTILIZATION	AULACORTHUM (1) SOLANI	POTATO	(+) GROWTH	(–)				BROADBENT ET AL. 1952 (27)
FERTILIZATION	BREVICORYNE BRASSICAE(1)	BRUSSEL SPROUTS	(+) LEAF N				(*)	VAN EMDEN 1966 (243)
FERTILIZATION	MACROSIPHUM (1) EUPHORBIAE	POTATO	(+) GROWTH	(–)				BROADBENT ET AL. 1952 (27)
FERTILIZATION	MACROSIPHUM (1) EUPHORBIAE	POTATO	(+) GROWTH				(=)	TAYLOR ET AL. 1952 (235)
FERTILIZATION	MACROSIPHUM PISI (1)	PEA	(+) GROWTH				(–)	BARKER & TAUBER 1951 (11)
FERTILIZATION	MYZUS PERSICAE (1)	TOBACCO	(+) LEAF K	(+)	(+)	(+)		WOOLRIDGE & HARRISON 1968 (267)
FERTILIZATION	MYZUS PERSICAE (1)	BRUSSEL SPROUTS	(+) LEAF N				(–)	VAN EMDEN 1966 (243)
FERTILIZATION	MYZUS PERSICAE (1)	POTATO	(+) GROWTH	(–)				BROADBENT ET AL. 1952 (27)
FERTILIZATION	THERIOAPHIS MACULATA (1)	ALFALFA (SUSCEPTIBLE)	(+) GROWTH		(=)			KINDLER & STAPLES 1970 (114)
FERTILIZATION	THERIOAPHIS MACULATA (1)	ALFALFA (RESISTANT)	(+) GROWTH		(–)			KINDLER & STAPLES 1970 (114)
FERTILIZATION	THERIOAPHIS MACULATA (1)	ALFALFA (RESISTANT)	(+) GROWTH				(–)	MCMURTRY 1962 (148)
FERTILIZATION	THERIOAPHIS MACULATA (1)	ALFALFA (SUSCEPTIBLE)	(+) GROWTH				(=)	MCMURTRY 1962 (148)
FERTILIZATION	ACYRTHOSIPHON PISUM (1)	BROADBEAN	NG	(=)				MARKKULA & TIITTANEN 1969 (136)
FERTILIZATION	MYZUS PERSICAE (1)	CHRYSANTHEMUM	NG	(–)				MARKKULA & TIITTANEN 1969 (136)

TREATMENT	INSECT (GUILD)	PLANT	PLANT RESPONSE	HERBIVORE RESPONSE #	SURV	GROW	FECUND	REFERENCE
FERTILIZATION	MYZUS PERSICAE (1)	CUCUMBER	NG	(=)				MARKKULA & TIITTANEN 1969 (136)
FERTILIZATION	MYZUS PERSICAE (1)	SUGAR BEET	NG	(=)				MARKKULA & TIITTANEN 1969 (136)
FERTILIZATION	SPODOPTERA (2) FRUGIPERDA	MILLET	(–) GROWTH		(=)	(–)		LEUCK 1972 (125)
FERTILIZATION	HELIOTHIS ZEA (2)	CORN	(+) GROWTH				(=)	TAYLOR ET AL. 1952 (235)
FERTILIZATION	BRYOBIA PRAETIOSA (3)	BEAN	(+) LEAF K	(–)	(–)		(–)	MORRIS 1961 (168)
FERTILIZATION	EPITRIX HIRTIPENNIS (2)	TOBACCO	NG	(+)				SEMTNER ET AL. 1980 (211)
FERTILIZATION	MANDUCA SEXTA (2)	TOBACCO	NG	(+)				SEMTNER ET AL. 1980 (211)
FERTILIZATION	PHAEDON COCHLEARIAE (2)	WATERCRESS	NG				(+)	ALLEN & SELMAN 1955 (3)
FERTILIZATION	PIERIS BRASSICAE (2)	CABBAGE	NG			(+)		ALLEN & SELMAN 1957 (4)
FERTILIZATION	TETRANYCHUS URTICAE (3)	CHRYSANTHEMUM	NG	(–)				MARKKULA & TIITTANEN 1969 (136)
FERTILIZATION	TETRANYCHUS URTICAE (3)	SUGAR BEET	NG	(=)				MARKKULA & TIITTANEN 1969 (136)
FERTILIZATION	TETRANYCHUS URTICAE (3)	CUCUMBER	NG	(+)				MARKKULA & TIITTANEN 1969 (136)
FERTILIZATION	TETRANYCHUS URTICAE (3)	BROADBEAN	NG	(=)				MARKKULA & TIITTANEN 1969 (136)
WOODY CULTIVATED PLANTS:								
FERTILIZATION	AONIDIELLA AURANTII (1)	CITRUS	(=) NPK	(–)				SALAMA ET AL. 1972 (204)
FERTILIZATION	LEPIDOSAPHES BECKII (1)	CITRUS	(=) NPK	(+)				SALAMA ET AL. 1972 (204)
FERTILIZATION	PARLATORIA ZIZYPHUS (1)	CITRUS	NG	(=)				SALAMA ET AL. 1985 (205)
FERTILIZATION	ICERYA PURCHASI (1)	CITRUS	NG	(=)				SALAMA ET AL. 1985 (205)
FERTILIZATION	PSEUDOCOCCOS (1) CONCAVOCERARII	COCOA	NG	(=)				CAMPBELL 1984 (33)
FERTILIZATION	MACONELLICOCCUS (1) UGANDAE	COCOA	NG	(=)				CAMPBELL 1984 (33)

TREATMENT	INSECT (GUILD)	PLANT	PLANT RESPONSE	HERBIVORE RESPONSE #	SURV	GROW	FECUND	REFERENCE
FERTILIZATION	MESOHOMOTOMA (1) TESSMANNI	COCOA	NG	(−)				CAMPBELL 1984 (33)
FERTILIZATION	PLANOCCOIDES (1) NJALENSIS	COCOA	NG	(=)				CAMPBELL 1984 (33)
FERTILIZATION	PLANOCOCCUS (1) CITRI	COCOA	NG	(−)				CAMPBELL 1984 (33)
FERTILIZATION	GASCARDIA SP. NR. (1) ZONATA	COCOA	NG	(=)				CAMPBELL 1984 (33)
FERTILIZATION	PHENACOCCUS (1) HARGREAVESI	COCOA	NG	(=)				CAMPBELL 1984 (33)
WOODY WILD PLANTS:								
FERTILIZATION	PISSODES STROBUS (2)	PINUS STROBUS	NG	(+)				XYDIUS & LEAF 1964 (268)
<u>EXPERIMENTAL WATER STRESS:</u>								
HERBACEOUS CULTIVATED PLANTS:								
WATER STRESS	APHIS FABAE (1)	BROADBEAN	WILTING	(−)				KENNEDY ET AL. 1958 (112)
WATER STRESS	APHIS FABAE (1)	BEAN	WILTING	(=)				WEARING & VAN EMDEN 1967 (225)
WATER STRESS	APHIS FABAE (1)	SUGAR BEET	WILTING	(−)	(−)			KENNEDY & BOOTH 1959 (111)
WATER STRESS	APHIS FABAE (1)	MARIGOLD	WILTING	(=)				WEARING & VAN EMDEN 1967 (255)
WATER STRESS	BREVICORYNE BRASSICAE (1)	RAPE	(+) AA'S (−) H_2O, CARBOS		(=)	(+)	(=)	MILES ET AL. 1982b (159)
WATER STRESS	BREVICORYNE BRASSICAE (1)	BRUSSELS SPROUTS	WILTING	(−)				WEARING & VAN EMDEN 1967 (255)
WATER STRESS	BREVICORYNE BRASSICAE (1)	BRUSSELS SPROUTS	WILTING	YOUNG, MATURE LEAVES OLD LEAVES			(+) (−)	WEARING 1967 (252) WEARING 1972a,b (253,254)
WATER STRESS	MYZUS PERSICAE (1)	BRUSSELS SPROUTS	WILTING	YOUNG, MATURE LEAVES OLD LEAVES			(+) (−)	WEARING 1967 (252) WEARING 1972a,b(253,254)

TREATMENT	INSECT (GUILD)	PLANT	PLANT RESPONSE	HERBIVORE RESPONSE				REFERENCE
				#	SURV	GROW	FECUND	
WATER STRESS	MYZUS PERSICAE (1)	BRUSSELS SPROUTS	WILTING	(*)				WEARING & VAN EMDEN 1967 (255)
WATER STRESS	MYZUS PERSICAE (1)	MARIGOLD	WILTING	(*)				WEARING & VAN EMDEN 1967 (255)
WATER STRESS	RHOPALOSIPHUM MAIDIS (1)	WINTER WHEAT	(–) H_2O POTENTIAL		(–)		(–)	SUMNER ET AL. 1986b (231)
WATER STRESS	SCHIZAPHIS GRAMINUM (1)	WHEAT (RESISTANT)	(–)OSMOREGULATION (+) AA'S (–) GROWTH	(+)				DORSCHNER ET AL. 1986 (59)
WATER STRESS	SCHIZAPHIS GRAMINUM (1)	WHEAT (SUSCEPTIBLE)	(–) OSMOREGULATION (+) AA'S (–) GROWTH	(+)				DORSCHNER ET AL. 1986 (59)
WATER STRESS	SCHIZAPHIS GRAMINUM (1)	WHEAT (RESISTANT)	(–)H_2O POTENTIAL	(–)	(–)	(–)	(–)	SUMNER ET AL. 1986a (230), SUMNER ET AL. 1983 (232)
WATER STRESS	SCHIZAPHIS GRAMINUM (1)	WHEAT (SUSCEPTIBLE)	(–)H_2O POTENTIAL	(–)	(–)	(–)	(–)	SUMNER ET AL. 1986a (230), SUMNER ET AL. 1983 (232)
WATER STRESS	SCHIZAPHIS GRAMINUM (1)	BARLEY, WHEAT, RYE	(–) GROWTH	(–)				BEHLE & MICHAELS 1988 (12)
WATER STRESS	SCHIZAPHIS GRAMINUM (1)	SORGHUM	(–) H_2O POTENTIAL				(–)	MICHELS & UNDERSANDER 1986 (157)
WATER STRESS	SITOBION AVENAE (1)	WHEAT	(–) LEAF H_2O POTENTIAL		(=)		(=)	FERERES ET AL. 1988 (68)
WATER STRESS	THERIOAPHIS SP. (1)	ALFALFA (RESISTANT)	(–) H_20 POTENTIAL (–) AA'S (PROTEIN) (=) AA'S (FREE)			(=)	(=)	LORENZ-DEFRIES & MANGLITZ 1982 (129)
WATER STRESS	THERIOAPHIS SP. (1)	ALFALFA (SUSCEPTIBLE)	(–) H_2O POTENTIAL (–) AA'S (PROTEIN) (=) AA'S (FREE)			(=)	(=)	LORENZ-DEVRIES & MANGLITZ 1982 (129)
WATER STRESS	THERIOAPHIS MACULATA (1)	ALFALFA (RESISTANT)	WILTING				(=)	MCMURTRY 1962 (148)
WATER STRESS	THERIOAPHIS MACULATA (1)	ALFALFA (SUSCEPTIBLE)	WILTING				(=)	MCMURTRY 1962 (148)
WATER STRESS	SACCHAROSYDNE (1) SACCHARIVORA	SUGARCANE	(+) N				(=)	METCALFE 1970 (156)
WATER STRESS	EMPOASCA (1) KERRI	GROUNDNUT	(–) GROWTH, REPROD, (+) LEAF TEMP	(–)				WHEATLEY ET AL. 1989 (258)

TREATMENT	INSECT (GUILD)	PLANT	PLANT RESPONSE	HERBIVORE RESPONSE #	SURV	GROW	FECUND	REFERENCE
WATER STRESS	EMPOASCA FABAE (1)	ALFALFA	(–) H_20 POTENTIAL		(–)			HOFFMAN ET AL. 1990 (95)
WATER STRESS	EMPOASCA SPP. (1)	COTTON	(–) LEAF AREA	(–)				LEIGH ET AL. 1974 (123)
WATER STRESS	LYGUS HESPERUS (1)	COTTON	(–) LEAF H_20, (=) TANNIN	(–)				GUINN & EIDENBOCK 1982 (78)
WATER STRESS	LYGUS HESPERUS (1)	COTTON	(–) YIELD	(–)				LEIGH ET AL. 1970 (124), LEIGH ET AL. 1974 (123)
WATER STRESS	FRANKLINIELLA (1) SCHULTZEI	GROUNDNUT	(–) GROWTH, REPROD, (+) LEAF TEMP	(*)				WHEATLEY ET AL. 1989 (258)
WATER STRESS	EPILACHNA VARUVESTUS (2)	SOYBEAN	(+) A.A.'S	(*)				MCQUATE & CONNOR 1991a (150)
WATER STRESS	EPILACHNA VARUVESTUS (2)	SOYBEAN	(+) A.A.'S		(–)	(–)		MCQUATE & CONNOR 1991b (151)
WATER STRESS	HELIOTHIS SPP. (2)	COTTON	(–) LEAF H_20 (=) TANNIN	(–)				GUINN & EIDENBOCK 1982 (78)
WATER STRESS	APROAEREMA (2) MODICELLA	GROUNDNUT	(–) GROWTH, REPROD, (+) LEAF TEMP	(+)				WHEATLEY ET AL. 1989 (258)
WATER STRESS	PIERIS RAPAE (2)	RAPE	(+) AMINO ACIDS,N (–) H_2O, CARBO'S,			(=)	(=)	MILES ET AL. 1982a (158)
WATER STRESS	PHTHORIMAEA (2) OPERCULELLA	TOMATO, POTATO	WILTING	(+)	(+)			YATHOM 1986, MEISNER 1969 (270,152)
WATER STRESS	ELDANA SACCHARINA (2)	SUGARCANE	(+) N	(+)	(+)	(+)		ATKINSON & NUSS 1989 (8)
WATER STRESS	OLIGONYCHUS PRATENSIS (3)	CORN	(–) H_20, SENESCENCE				(*)	PERRING ET AL. 1986 (180)
WATER STRESS	OLIGONYCHUS PRATENSIS (3)	CORN	NG	(+)				FEESE & WILDE 1977 (67)
WATER STRESS	OLIGONYCHUS PRATENSIS (3)	CORN	NG	(+)				CHANDLER ET AL. 1979 (42)
WATER STRESS	TETRANYCUS (3) CINNABARINUS	CORN	NG	(+)				CHANDLER ET AL. 1979 (42)
WATER STRESS	TETRANYCHUS PACIFICUS (3)	COTTON	(*) YIELD	(*)				LEIGH ET AL. 1970 (124)
WATER STRESS	TETRANYCHUS URTICAE (3)	CHRYSANTHEMUM	NG	(+)				PRICE ET AL. 1982 (191)

TREATMENT	INSECT (GUILD)	PLANT	PLANT RESPONSE	HERBIVORE RESPONSE				REFERENCE
				#	SURV	GROW	FECUND	
WATER STRESS	TETRANYCHUS URTICAE (3)	ALFALFA	NG	(–)				BUTLER 1955 (30)
WATER STRESS	TETRANYCHUS URTICAE (3)	CUCUMBER (RESISTANT)	WILTING		(–)			GOULD 1978 (76)
WATER STRESS	TETRANYCHUS URTICAE (3)	CUCUMBER (SUSCEPTIBLE)	WILTING		(=)			GOULD 1978 (76)
WATER STRESS	TETRANYCHUS URTICAE (3)	BEAN	(–) LEAF EXPANSION	(*)	(=)	(+)	(*)	ENGLISH-LOEB 1989, 1990 (64,65)
WATER STRESS	TETRANYCHUS URTICAE (3)	SOYBEAN	(–) GROWTH	(–)				MELLORS ET AL. 1984 (153)
WATER STRESS	TETRANYCHUS URTICAE (3)	SOYBEAN	(–) LEAF H_20 POT., BIOMASS	(–)				OLOUMI-SADEGHI ET AL. 1988 (175)
WATER STRESS	TETRANYCHUS URTICAE (3)	SOYBEAN	(–)REPRODUCTION	(+)				KLUBERTANZ ET AL. 1990 (117)
WATER STRESS	TETRANYCHUS URTICAE (3)	RADISH	(–) GROWTH	(+)				MELLORS & PROPTS 1983 (154)
WATER STRESS	TETRANYCHUS URTICAE (3)	PEPPERMINT	(–) GROWTH	(+)				HOLLINGSWORTH & BERRY 1982 (96)
WATER STRESS	APHIDIDAE SPP. (1)	HERB	(–) BIOMASS			(=)	(=)	GANGE & BROWN 1989 (72)
HERBACEAOUS WILD PLANTS:								
WATER STRESS	PEMPHIGUS BETAE (1)	RUMEX CRISPUS	NG	(–)	(–)	(–)	(–)	MORAN & WHITHAM 1988 (166)
WATER STRESS	PEMPHIGUS BETAE (1)	CHENOPODIUM ALBUM	NG	(–)	(–)	(–)	(–)	MORAN & WHITHAM 1988 (166)
WATER STRESS	LEPTINOTARSA (2) DECEMLINEATA	SOLANUM BERTHAULTII	(+) TRICHOMES		(=)			PELLETIER 1990 (178)
WATER STRESS	3 SPP. LEPIDOPTERA (2)	VARIOUS FORBS	(–) LEAF H_20			(–)		SCRIBER 1979 (208)
WATER STRESS	MELANOPLUS (2) DIFFERENTIALIS	HELIANTHUS ANNUUS	(–) H_2O, (+) PHENOLS	(+)	(+)	(+)	(+)	LEWIS 1984 (126)
WATER STRESS	GIRAUDIELLA INCLUSA (4)	PHRAGMITES SP.	NARROWER SHOOTS, (–) SILICA, PROTEIN	(+)				TSCHARNTKE 1989 (241)

TREATMENT	INSECT (GUILD)	PLANT	PLANT RESPONSE	HERBIVORE RESPONSE #	SURV	GROW	FECUND	REFERENCE
WOODY CULTIVATED PLANTS:								
WATER STRESS	APHIS FABAE (1)	SPINDLE	WILTING	(–)	(–)		(–)	KENNEDY ET AL. 1958 (112), KENNEDY & BOOTH 1959 (111)
WATER STRESS	ZONOCERUS VARIEGATUS (2)	CASSAVA	WILTING, (–) HCN		(+)			BERNAYS ET AL. 1977 (14)
WATER STRESS	CERATITIS CAPITATA (2)	COFFEE	(–) FRUIT SIZE (=) FRUIT N AND H_20	(+)				HARRIS & LEE 1989 (84)
WATER STRESS	MONONYCHELLUS SP. (3) TANAJOA	CASSAVA	(–) LEAF PRODUCTION	(–)				YANNICK ET AL. 1989 (269)
WATER STRESS	PANONYCHUS CITRI (3)	CITRUS	(–) YIELD (=) NUTRIENTS	(=)				HARE ET AL. 1989 (82)
WATER STRESS	TETRANYCHUS (3) PACIFICUS	ALMOND	(+) LEAF TEMP. (–) N, H_20		(=)		(*)	YOUNGMAN ET AL. 1988 (271)
WATER STRESS	TETRANYCHUS PACIFICUS (3)	ALMOND	(+) LEAF TEMP, (–) LEAF H_20 POT.			(+)		OI ET AL. 1989 (173)
WOODY WILD PLANTS:								
WATER STRESS	APHIDIDAE SP. (1)	CHRYSOTHAMNUS NAUSEOSUS	(–) H_20 POTENTIAL	(–)				FERNANDES & PRICE, UNPUBL.
WATER STRESS	APHIS FABAE (1)	EUONYMUS EUROPEUS	WILTING	(–)	(–)			KENNEDY & BOOTH 1959 (111)
WATER STRESS	UNASPIS EUONYMI (1)	EUONYMUS FORTUNEI	(–) LEAF H_20, LEAF GROWTH	(–)				COCKFIELD & POTTER 1986 (47)
WATER STRESS	CORYTHUCA ARCUATA (1)	QUERCUS ALBA	(–) TRANSPIRATION RATE			(–)		CONNOR 1989 (49)
WATER STRESS	MIRIDAE SPP. (1)	CHRYSOTHAMNUS NAUSEOSUS	(–) H_20 POTENTIAL	(–)				FERNANDES & PRICE, UNPUBL.
WATER STRESS	ACRIDIDAE SP. 1 (2)	CHRYSOTHAMNUS NAUSEOSUS	(–) H_20 POTENTIAL	(–)				FERNANDES & PRICE, UNPUBL.

TREATMENT	INSECT (GUILD)	PLANT	PLANT RESPONSE	HERBIVORE RESPONSE				REFERENCE
				#	SURV	GROW	FECUND	
WATER STRESS	ACRIDIDAE SP. 2 (2)	CHRYSOTHAMNUS NAUSEOSUS	(–) H_20 POTENTIAL	(–)				FERNANDES & PRICE, UNPUBL.
WATER STRESS	ACRIDIDAE SP. 3 (2)	CHRYSOTHAMNUS NAUSEOSUS	(–) H_20 POTENTIAL	(–)				FERNANDES & PRICE, UNPUBL.
WATER STRESS	ACRIDIDAE SP. 4 (2)	CHRYSOTHAMNUS NAUSEOSUS	(–) H_20 POTENTIAL	(–)				FERNANDES & PRICE, UNPUBL.
WATER STRESS	AGRILUS BILINEATUS (2)	QUERCUS ALBA	(–)XYLEM POTENTIAL	(+)				DUNN ET AL. 1986 (61)
WATER STRESS	AGRILUS BILINEATUS (2)	QUERCUS SP.	(–) ROOT STARCH	(+)				HAACK & BENJAMIN 1982 (79)
WATER STRESS	CHRYSOMELA KNABI (2)	SALIX SP.	(–) LEAF H_20		(–)	(–)	(–)	HORTON 1989 (98)
WATER STRESS	PAROPSIS ATOMARIA (2)	EUCALYPTUS CAMALDULENSIS	(+) AA'S, N (–) PHENOLS, H_2O CONTENT, H_2O POTENTIAL		(=)	(=)	(=)	MILES ET AL. 1982a (158)
WATER STRESS	EPILACHNA DODECASTIGMA (2)	MOMORDICA CHARANTIA	(+) AA'S (–) H_20, CARBOS, PROTEIN			(–)	(–)	MANDAL ET AL. 1984 (133)
WATER STRESS	HYALOPHORA CECROPIA (2)	PRUNUS SEROTINA	(–) LEAF H_20			(–)		SCRIBER 1977 (207)
WATER STRESS	LYMANTRIA DISPAR (2)	BETULA PAPYRIFERA	(+) LEAF SUCROSE			(+)		TALHOUK ET AL. 1990 (233)
WATER STRESS	13 SPP. LEPIDOPTERA (2) TREE FEEDERS	VARIOUS	(–) LEAF H_20			(–)		SCRIBER 1979 (208)
WATER STRESS	CAMERARIA NOV. SP. A (2)	QUERCUS EMORYI	(–) FOLIAGE H_2O		(–)	(=)		BULTMAN & FAETH 1987 (29)
WATER STRESS	CAMERARIA NOV. SP. B (2)	QUERCUS EMORYI	(–) FOLIAGE H_2O		(–)			BULTMAN & FAETH 1987 (29)
WATER STRESS	CHORISTONEURA FUMIFERANA (2)	ABIES BALSAMEA	(–) H_20 POTENTIAL	(=)				MATTSON ET AL. 1983 (141)
WATER STRESS	PANOLIS FLAMMEA (2)	PINUS CONTORTA	NG		(–)	(–)		WATT 1986 (251)
WATER STRESS	GEOMETRIDAE SP. 1 (2)	CHRYSOTHAMNUS NAUSEOSUS	(–) H_20 POTENTIAL	(–)				FERNANDES & PRICE, UNPUBL.

TREATMENT	INSECT (GUILD)	PLANT	PLANT RESPONSE	HERBIVORE RESPONSE #	SURV	GROW	FECUND	REFERENCE
WATER STRESS	GEOMETRIDAE SP. 2 (2)	CHRYSOTHAMNUS NAUSEOSUS	(–) H_20 POTENTIAL	(–)				FERNANDES & PRICE, UNPUBL.
WATER STRESS	GEOMETRIDAE SP. 3 (2)	CHRYSOTHAMNUS NAUSEOSUS	(–) H_20 POTENTIAL	(–)				FERNANDES & PRICE, UNPUBL.
WATER STRESS	NEODIPRION (2) AUTUMNALIS	PINUS PONDEROSA	(–) H_2O POTENTIAL		(–)	(–)	(–)	MCCULLOUGH & WAGNER 1987 (144)
WATER STRESS	NEODIPRION (2) FULVICEPS	PINUS PONDEROSA	(–) H_2O POTENTIAL		(=)			CRAIG ET AL. 1990, TISDALE 1988 (51,239)
WATER STRESS	LEPIDOPTERA SP. (4)	CHRYSOTHAMNUS NAUSEOSUS	(–) H_20 POTENTIAL	(–)				FERNANDES & PRICE, UNPUBL.
WATER STRESS	CECIDOMYIIDAE SP. 1 (4)	CHRYSOTHAMNUS NAUSEOSUS	(–) H_20 POTENTIAL	(+)				FERNANDES & PRICE, UNPUBL.
WATER STRESS	CECIDOMYIIDAE SP. 2 (4)	CHRYSOTHAMNUS NAUSEOSUS	(–) H_20 POTENTIAL	(–)				FERNANDES & PRICE, UNPUBL.
WATER STRESS	CECIDOMYIIDAE SP. 3 (4)	CHRYSOTHAMNUS NAUSEOSUS	(–) H_20 POTENTIAL	(–)				FERNANDES & PRICE, UNPUBL.
WATER STRESS	PROCECIDOCHARIS SP. (4)	CHRYSOTHAMNUS NAUSEOSUS	(–) H_20 POTENTIAL	(–)				FERNANDES & PRICE, UNPUBL.
WATER STRESS	PAKTULOPSPHAIRA (4) VITIFOLIAE	VITIS ARIZONICA	(–) GROWTH, XYLEM H_20 POTENTIAL	(–)				KIMBERLING ET AL. 1990 (113), KIMBERLING, UNPUBL.
WATER STRESS	PEMPHIGUS BETAE (4)	POPULUS ANGUSTIFOLIA	(–) GROWTH		(+)			LARSON 1989 (119)
WATER STRESS	EUURA LASIOLEPIS (4)	SALIX LASIOLEPIS	(–) GROWTH, (+) PROTEIN, (=) PHENOLS	(–)	(–)			WARING & PRICE 1988 (248)
DROUGHT & CHRONIC WATER STRESS								
HERBACEOUS CULTIVATED PLANTS:								
DROUGHT	ACYRTHOSIPHUM PISUM (1)	PEA	NG	(–)				MAITEKI ET AL. 1986 (132)
DROUGHT	MIRIDAE SP. (1)	LUCERNE	NG	(–)				ERDELYI ET AL. 1983 (66)

TREATMENT	INSECT (GUILD)	PLANT	PLANT RESPONSE	HERBIVORE RESPONSE				REFERENCE
				#	SURV	GROW	FECUND	
DROUGHT	CHEILOSIA FASCIATA (2)	ALLIUM URISINUM	NG	(–)	(–)			HOVEMEYER 1987 (100)
DROUGHT	SITONA DISCOIDEUS (2)	LUCERNE	NG	(–)				GOLDSON ET AL. 1985 (75)
DROUGHT	TYCHIUS FLAVUS (2)	LUCERNE	NG	(+)				ERDELYI ET AL. 1983 (66)
DROUGHT	BRUCHOPHAGUS RODDI (2)	LUCERNE	NG	(=)				ERDELYI ET AL. 1983 (66)
DROUGHT	SCIRPOPHAGA (2) INCERTULAS	RICE	NG	(–)				CATLING ET AL. 1984 (41)
HERBACEOUS WILD PLANTS:								
DROUGHT	MELANOPLUS SPP. (2)	FIELD PLANTS	NG	(+)				EDWARDS 1960 (62)
DROUGHT	TELEOGRYLLUS (2) COMMODUS	PASTURE GRASS	NG	(+)				BLANK ET AL. 1986 (17)
DROUGHT	HETERONYCHUS ARATOR (2)	PASTURE GRASS	NG	(+)				BLANK ET AL. 1986 (17)
DROUGHT	ORTHOPTERA SPP. (2)	GRASSLAND	(–) PRECIPITATION	(–)				FIELDING & BRUSVEN 1990 (70)
DROUGHT	EUPHYDRYAS EDITHA (2)	PLANTAGO ERECTA & ORTHOCARPUS DENSIFLORA	WILTING	(–)				EHRLICH ET AL. 1980 (63)
DROUGHT	EUPHYDRYAS EDITHA (2)	PEDICUALRIS DENSIFLORA	NG	(–)				EHRLICH ET AL. 1980 (63)
DROUGHT	EUPHYDRYAS EDITHA (2)	COLLINSIA TINCTORIA	NG	(–)				EHRLICH ET AL. 1980 (63)
DROUGHT	EUPHYDRYAS (2) CHALCEDONA	CASTILLEJA SP. & PEDICULARIS SP.	NG	(–)				EHRLICH ET AL. 1980 (63)
DROUGHT	EUPHYDRYAS (2) CHALCEDONA	SCROPHULARIACEAE	NG	(–)				EHRLICH ET AL. 1980 (63)
DROUGHT	EUPHYDRYAS (2) CHALCEDONA	SCROPHULARIACEAE & PLANTAGINACEAE	NG	(=)				EHRLICH ET AL. 1980 (63)
WOODY CULTIVATED PLANTS:								
DROUGHT	PANONYCHUS ULMI (3)	APPLE	NG	(–)				SPECHT 1965 (225)

TREATMENT	INSECT (GUILD)	PLANT	PLANT RESPONSE	HERBIVORE RESPONSE #	SURV	GROW	FECUND	REFERENCE
WOODY WILD PLANTS:								
WATER STRESS (CHRONIC)	APHIS POMI (1)	CRATAEGUS SP.	(–) H_20	(+)				BRAUN & FLUCKIGER 1984 (22)
WATER STRESS (CHRONIC)	MATSUCOCCUS (1) ACALYPTUS	PINUS EDULIS	(–) GROWTH, RESINS	(+)				COBB ET AL. UNPUBL.
DROUGHT	CARDIASPINA (1) DENSITEXTA	EUCALYPTUS FASCICULOSA	NG	(+)				WHITE 1969 (259)
DROUGHT	CARDIASPINA SPP. (1)	EUCALYPTUS SPP.	NG	(+)				WHITE 1969 (259)
DROUGHT	CARDIASPINA (1) ALITEXTURA	EUCALYPTUS CAMALDULENSIS	NG	(+)				WHITE 1969 (259)
DROUGHT	CARDIASPINA (1) ALBITEXTURA	EUCALYPTUS BLAKELYI	NG	(+)				WHITE 1969 (259)
DROUGHT	INGLISIA FAGI (1)	NOTHOFAGUS FUSCA	(–) GROWTH	(+)				HOSKING & KERSHAW 1985 (99)
DROUGHT	HEMIPTERA SPP. (1)	ECUALYPTUS SP.	(–) GROWTH	(–)				BELL 1985 (13)
DROUGHT	AGRILUS BILINEATUS (2)	QUERCUS SPP.	(+) DIEBACK	(+)				LEWIS 1981 (127)
DROUGHT	BARK BEETLE (2)	PINUS PONDEROSA	NG	(+)				SHERMAN & WARREN 1988 (216)
DROUGHT	DENDROCTONUS (2) FRONTINALIS	PINUS SP.	NG	(+)				CRAIGHEAD 1925 (52)
DROUGHT	DENDROCTONUS (2) PONDEROSAE	LODGEPOLE PINE	(–) GROWTH	(+)				THOMSON & SHRIMPTON 1984 (237)
DROUGHT	IPS CALLIGRAPHIS (2)	PINUS SP.	NG	(+)				ST. GEORGE 1930 (203)
DROUGHT	IPS GRANDICOLLIS (2)	PINUS SP.	NG	(+)				ST. GEORGE 1930 (203)
DROUGHT	SCOLYTUS (2) QUADRISPINOSA	CARYA SP.	(–) GROWTH	(+)				ST. GEORGE 1929 (202)
DROUGHT	SCOLYTUS VENTRALIS (2)	ABIES SP.	NG	(+)				BERRYMAN 1973 (16)

TREATMENT	INSECT (GUILD)	PLANT	PLANT RESPONSE	HERBIVORE RESPONSE				REFERENCE
				#	SURV	GROW	FECUND	
DROUGHT	SCOLYTUS VENTRALIS (2)	ABIES CONCOLOR	(–) RADIAL GROWTH	(+)				FERRELL & HALL 1975 (69)
DROUGHT	CORTHYLUS (2) COLUMBIANUS	ACER RUBRUM	NG	(+)				MCMANUS & GIESE 1968 (147)
DROUGHT	TETROPIUM ABIETIS (2)	ABIES CONCOLOR	(–) RADIAL GROWTH	(+)				FERRELL & HALL 1975 (69)
DROUGHT	LAMBDINA (2) FISCELLARIA	ABIES BALSAMEA	(–) RADIAL GROWTH	(+)				CARROLL 1956 (37)
DROUGHT	SELIDOSEMA SUAVIS (2)	PINUS RADIATA	NG	(+)				WHITE 1974 (260)
WATER STRESS (CHRONIC)	CHORISTONEURA (2) OCCIDENTALIS	PSEUDOSTUGA MENZIESII	(–) H_20 POTENTIAL, TERPENE DIFFERENCES		(+)	(+)		CATES ET AL. 1983 (40)
WATER STRESS (CHRONIC)	CHORISTONEURA (2) FUMIFERANA	FOREST SPECIES	(–) MINERAL CONC.	(+)				KEMP & MOODY 1984 (110)
WATER STRESS (CHRONIC)	CHORISTONEURA (2) FUMIFERANA	PICEA MARIANA	NG	(+)				HIX ET AL. 1987 (94)
DROUGHT	CHORISTONEURA (2) FUMIFERANA	FOREST SPECIES	(–) GROWTH	(+)				THOMSON ET AL. 1984 (236)
DROUGHT	CHORISTONEURA (2) FUMIFERANA	FOREST SPECIES	NG	(+)				SHEPARD 1959 (214)
DROUGHT	CHORISTONEURA PINUS (2)	PINUS BANKSIANA	NG	(+)				CLANCY ET AL. 1980 (46)
WATER STRESS (CHRONIC)	DIORYCTRIA (2) ALBOVITELLA	PINUS EDULIS	(–) GROWTH, RESINS	(+)				COBB ET AL. UNPUBL.
DROUGHT	LYMANTRIA DISPAR (2)	GENERAL	NG	(+)				MILLER ET AL. 1989 (160)
DROUGHT	LYMANTRIA DISPAR (2)	GENERAL	NG	(+)				CAMPBELL & SLOAN 1977 (34)
DROUGHT	LYMANTRIA (2) DISPAR	FOREST SPECIES	NG	(+)				SKALLER 1985 (218), MILLER ET AL. 1989 (160)
DROUGHT	LYMANTRIA MONACHA (2)	CONIFERS	NG	(+)				SHEPARD ET AL 1988 (215)

TREATMENT	INSECT (GUILD)	PLANT	PLANT RESPONSE	HERBIVORE RESPONSE				REFERENCE
				#	SURV	GROW	FECUND	
DROUGHT	ORGYIA PSEUDOSUGATA (2)	PSEUDOTSUGA MENSIEZII	NG	(+)				SHEPARD ET AL. 1988 (215)
DROUGHT	PHLOEOSINUS ARMATUS (2)	CYPRESSUS SEMPERVIRENS	(–) RESIN FLOW	(+)				MENDEL 1984 (155)
DROUGHT	PHLOEOSINUS ACIBEI (2)	CYPRESSUS SEMPERVIRENS	(–) RESIN FLOW	(+)				MENDEL 1984 (155)
DROUGHT	HETEROCAMPA (2) GUTTIVITTA	BEECH, MAPLE, BIRCH	NG	(=)				MARTINAT & ALLEN 1987 (137)
DROUGHT	CHRYSOPHTHARTA (2) BIMACULATA	EUCALYPTUS OBLIQUA & EUCALYPTUS REGNANS	(–) GROWTH	(+)				WEST 1979 (257)
DROUGHT	PAROPSIS ATOMARIA (2)	EUCALYPTUS BLAKELYI	(–) LEAF PRODUCTION	(–)				LARSSON & OHMART 1988 (121)
DROUGHT	LARVAE (2)	EUCALYPTUS SP., ACACIA SP.	(–) GROWTH	(–)				BELL 1985 (13)
DROUGHT	CREISS PERICULOSA (2)	EUCALYPTUS RUDIS	NG	(+)				WHITE 1976 (261)
DROUGHT	NEODIPRION SERTIFER (2)	PINUS SYLVESTRIS	NG	(+)				LARSSON & TENOW 1984 (122)
WATER STRESS (CHRONIC)	NEODIPRION SP. (2)	PINUS EDULIS	(–) GROWTH, RESINS	(+)				COBB ET AL. UNPUBL.
DROUGHT	EPINOTIA TEDELLA (2)	PICEA ABIES	(–) GROWTH	(+)				MUNSTER-SWENDSEN 1984 (169)
DROUGHT	GLYCASPIS SPP. (2)	NG	NG	(+)				WHITE 1976 (261)
WATER STRESS (CHRONIC)	NEODIPRION EDULICOLIS(2)	PINUS EDULIS	(–) GROWTH, RESINS	(+)				COBB ET AL. UNPUBL.
WATER STRESS (CHRONIC)	ORGYIA (2) PSEUDOSUGATA	PSEUDOTSUGA MENSIEZII	NG	(+)				STOSZEK ET AL. 1981 (228)
WATER STRESS (CHRONIC)	GNORIMOSCHEMA (4) TETRADYMIELLA	TETRADYMIA SP.	DIFFUSE GROWTH	(+)				HARTMAN 1984 (86)
DROUGHT	ASPHONDYLIA SPP. (4)	ATRIPLEX SP.	NG	(–)				HAWKINS ET AL. 1986 (87)

TREATMENT	INSECT (GUILD)	PLANT	PLANT RESPONSE	HERBIVORE RESPONSE #	SURV	GROW	FECUND	REFERENCE
WATER STRESS (CHRONIC)	ASPHONDYLIA CLAVATA (4)	LARREA TRIDENTATA	DIFFUSE GROWTH, (+) PROTEIN, N, RESINS; (–) GROWTH, H_20 POT., (–) LEAF PHENOLS	(+)				WARING & PRICE 1990 (249)
WATER STRESS (CHRONIC)	ASPHONDYLIA PILOSA (4)	LARREA TRIDENTATA	" "	(=)				WARING & PRICE 1990 (249)
WATER STRESS (CHRONIC)	ASPHONDYLIA VILLOSA (4)	LARREA TRIDENTATA	" "	(+)				WARING & PRICE 1990 (249)
WATER STRESS (CHRONIC)	ASPHONDYLIA SILICULA (4)	LARREA TRIDENTATA	" "	(+)				WARING & PRICE 1990 (249)
WATER STRESS (CHRONIC)	ASPHONDYLIA FABALIS (4)	LARREA TRIDENTATA	" "	(+)				WARING & PRICE 1990 (249)
WATER STRESS (CHRONIC)	ASPHONDYLIA DIGITATA (4)	LARREA TRIDENTATA	" "	(–)				WARING & PRICE 1990 (249)
WATER STRESS (CHRONIC)	ASPHONDYLIA FLOREA (4)	LARREA TRIDENTATA	" "	(–)				WARING & PRICE 1990 (249)
WATER STRESS (CHRONIC)	ASPHONDYLIA ROSETTA (4)	LARREA TRIDENTATA	" "	(–)				WARING & PRICE 1990 (249)
DROUGHT	EUURA LASIOLEPIS (4)	SALIX LASIOLEPIS	(–) GROWTH	(–)	(–)			PRICE & CLANCY 1986 (192)

REFERENCES

1. **Abrahamson, W. G., Anderson, S. S., and McCrea K. D.,** Effects of manipulation of plant carbon/nutrient balance on tall goldenrod resistance to a gallmaking herbivore, *Oecologia*, 77, 302, 1988.
2. **Adkisson, P. L.,** The influence of fertilizer applications on populations of *Heliothis zea* (Boddie) and certain insect predators, *J. Econ. Entomol.*, 51, 757, 1958.
3. **Allen, M. D. and Selman, I. W.,** Egg-production in the mustard beetle, *Phaedon cochleariae* (F.) in relation to diets of mineral-deficient leaves, *Bull. Entomol. Res.*, 46, 393, 1955.
4. **Allen, M. D. and Selman, I. W.,** The response of larvae of the large white butterfly (*Pieris brasssicae* (L.)) to diets of mineral-deficient leaves, *Bull. Entomol. Res.*, 48, 229, 1957.
5. **Al-Zubaidi, F. S. and Capinera, J. L.,** Utilization of food and nitrogen by the beet armyworm, *Spodoptera exigua* (Hubner) (Lepidoptera: Noctuidae), in relation to food type and dietary nitrogen levels, *Environ. Entomol.*, 13, 1604, 1984.
6. **Arant, F. S. and Jones, C. M.,** Influence of lime and nitrogenous fertilizers on populations of greenbug infesting oats, *J. Econ. Entomol.*, 44, 121, 1951.
7. **Archer, T. L., Onken, A. B., Matheson, R. L., and Bynum, E. D.,** Nitrogen fertilizer influence on greenbug (Homoptera: Aphididae) dynamics and damage to sorghum, *J. Econ. Entomol.*, 75, 695, 1982.
8. **Atkinson, P. R. and Nuss, K. J.,** Associations between host-plant nitrogen and infestations of the sugarcane borer, *Eldana saccharina* Walker (Lepidoptera: Pyralidae), *Bull. Entomol. Res.*, 79, 489, 1989.
9. **Auerbach, M. J. and Strong, D. R.,** Nutritional ecology of *Heliconia* herbivores: experiments with plant fertilization and alternative hosts, *Ecol. Mon.*, 51, 63, 1981.
10. **Baker, G. L.,** The ecology of mermithid nematode parasites of grasshoppers and locusts in south-east Australia, in *Fundamental and Applied Aspects of Invertebrate Pathology,* Samson, R. A., Vlak, J. M., and Peters, D., Eds., Foundation 4th Int. Colloq. of Invertebrate Pathology, Wageningen, The Netherlands, 1986, 277.
11. **Barker, J. S. and Tauber, O. E.,** Fecundity of and plant injury by the pea aphid as influenced by nutritional changes in the garden pea, *J. Econ. Entomol.*, 44, 1010, 1951.
12. **Behle, R. W. and Michaels, G. J.,** Resposnes of greenbug to drought stressed small grain hosts, *Southwest. Entomol.*, 13, 55, 1988.
13. **Bell, H. L.,** Seasonal variation and the effects of drought on the abundance of arthropods in savanna woodland on the Northern Tablelands of New South Wales, *Australian J. of Ecol.*, 10, 207, 1985.
14. **Bernays, E. A., Chapman, R. F., Leather, E. M., and McCaffery, A. R.,** The relationship of *Zonocerus variegatus* (L.) (Acridoidea: Pyrgomorphidae) with cassava (*Manihot esculenta*), *Bull. Entomol. Res.*, 67, 391, 1977.
15. **Bernays, E. A. and Lewis, A. C.,** The effect of wilting on palatability of plants to *Schistocera gregaria*, the desert locust, *Oecologia,* 70, 132, 1986.
16. **Berryman, A. A.,** Population dynamics of the fir engraver, *Scolytus ventralis* (Coleoptera: Scolytidae). I. Analysis of population behavior and survival from 1964 to 1971, *Can. Entomol.*, 105, 1465, 1973.
17. **Blank, R. H., Bell, D. S., and Olson, M. H.,** Differentiating between black field cricket and black beetle damage in northland pastures under drought conditions, *New Zealand J. Experim. Agric.*, 14, 361, 1986.
18. **Bogenschutz, H. and Konig, E.,** Relationships between fertilization and tree resistance to forest insect pests, in *Fertilizer Use and Plant Health,* Proc. 12th Colloq. Int. Potash Inst., CH-3048 Worblaufen-Bern, Switzerland, 1976, 281.
19. **Bowling, C. C.,** Effect of nitrogen levels on rice water weevil populations, *J. Econ. Entomol.*, 56, 826, 1963.
20. **Boyer, J. S.,** Plant productivity and environment, *Science*, 218, 443, 1982.
21. **Branson, T. F. and Simpson, R. G.,** Effects of a nitrogen-deficient host and crowding on the corn leaf aphid, *J. Econ. Entomol.,* 59, 290, 1966.
22. **Braun, S. and Fluckiger, W.,** Increased population of the aphid *Aphis pomi* at a motorway. Part 2.The effect of drought and deicing salt, *Environ. Pollution,* (Series A), 36, 261, 1984.
23. **Breukel, L. M. and Post, A.,** The influence of the manurial treatment of orchards on the population density of *Metatetranychus ulmi* (Koch) (Acari: Tetranychidae), *Entomol. Exp. Appl.*, 2, 38, 1959.

24. **Brewer, J. W., Capinera, J. L., Deshon, R. E., and Walmsley, M. L.,** Infuence of foliar nitrogen levels on survival, development, and reproduction of western spruce budworm, *Choristoneura occidentalis* (Lepidoptera:Tortricidae), *Can. Entomol.*, 117, 23, 1985.
25. **Brix, H. and Mitchell, A. K.,** Thinning and nitrogen fertilization effects on soil and tree water stress in a Douglas-fir stand, *Can. J. For. Res.*, 16, 1334, 1986.
26. **Broadbeck, B. and Strong, D. R.,** Amino acid nutrition of herbivorous insects and stress to host plants, in *Insect Outbreaks: Ecological and Evolutionary Perspectives*, Barbosa, P. and Schultz, J. Eds., Academic Press, New York, 1987, chap. 14.
27. **Broadbent, L., Gregory, P. H., and Tinsley, T. W.,** The influence of planting date and manuring on the incidence of virus diseases in potato crops, *Ann. Appl. Biol.*, 39, 509, 1952.
28. **Bryant, J. P., Clausen, T. P., Reichardt, P. B., McCarthy, M. C., and Warner, R. A.,** Effect of nitrogen fertilization upon the secondary chemistry and nutritional value of quaking aspen (*Populus tremuloides* Michx.) leaves for the large aspen tortrix (*Choristoneura conflictana* (Walker)), *Oecologia*, 73, 513, 1987.
29. **Bultman, T. L. and Faeth, S. H.,** Impact of irrigation and experimental drought stress on leaf-mining insects of Emory oak, *Oikos*, 48, 5, 1987.
30. **Butler, G. D.,** The effect of alfalfa irrigation treatments on the two-spotted mite in alfalfa, *J. Econ. Entomol.*, 48, 221, 1955.
31. **Cagle, L. R.,** Life history of the European red mite, *Virginia Agric. Stat. Tech. Bull.*, 98, 1946.
32. **Cagle, L. R.,** Life history of the two-spotted spider mite, *Virginia Agric. Stat. Tech. Bull.*, 113, 1949.
33. **Campbell, C. A. M.,** The influence of overhead shade and fertilizers on the Homoptera of mature upper Amazon cocoa trees in Ghana, *Bull. Entomol. Res.*, 74, 163, 1984.
34. **Campbell, R. W. and Sloan, R. J.,** Release of gypsy moth populations from innocuos levels, *Environ. Entomol.*, 6, 323, 1977.
35. **Cannon, W. N. and Connell, W. A.,** Populations of *Tetranychus atlanticus* (Acarina:Tetranychidae) on soybean supplied with various levels of nitrogen, phosphorus, and potassium, *Entomol. Exp. Appl.*, 8, 153, 1965.
36. **Cannon, W. N. and Ortega, A.,** Studies of *Ostrinia nubilalis* larvae (Lepidoptera: Pyraustidae) on corn plants supplied with various amounts of nitrogen and phosphorus. I. Survival, *Ann. Entomol. Soc. Am.*, 59(4), 631, 1966.
37. **Carroll, W. J.,** History of the hemlock looper, *Lambdina fiscellaria fiscellaria* (Guen.) (Lepidoptera: Geometridae), in Newfoundland, and notes on its biology, *Can. Entomol.*, 88, 587, 1956.
38. **Carrow, J. R. and Betts, R. E.,** Effects of different foliar applied nitrogen fertilizers on balsam wooly aphid, *Can. J. For. Res.*, 3, 122, 1973.
39. **Carrow, J. R. and Graham, K.,** Nitrogen fertilization of the host tree and population growth of the balsam woolly aphid *Adelges piceae* (Homoptera: Adelgidae), *Can. Entomol.*, 100, 478, 1968.
40. **Cates, R. G., Redak, R., and Henderson, C. B.,** Natural product defensive chemistry of Douglas-fir, Western Spruce Budworm success and forest management practices, *Z. Ang. Ent.* 96, 173, 1983.
41. **Catling, H D., Islam, Z., and Pattrasudh, R.,** Seasonal occurrence of the yellow stem borer *Scirpophaga incertulas* (Walker) on deepwater rice in Bangledesh and Thailand, *Agric., Ecosyst. Environ.*, 12, 47, 1984.
42. **Chandler, L. D., Archer, T. L., Ward, C. R., and Lyle, W. M.,** Influences of irrigation practices on spider mite densities on field corn, *Environ. Entomol.*, 8, 196, 1979.
43. **Clancy, K. M.,** Douglas-fir nutrients and terpenes as potential factors influencing western spruce budworm defoliation, in *Proceedings, Forest Insect Guilds: Patterns of Interaction with Host Trees, USDA Forest Service Gen. Tech. Rep.*, 1990.
44. **Clancy, K. M. and Price, P. W.,** Rapid herbivore growth enhances enemy attack: sublethal plant defenses remain a paradox, *Ecology*, 68(3), 733, 1987.
45. **Clancy, K. M. and Price, P. W.,** Temporal variation in three-trophic-level interactions among willows, sawflies, and parasites, *Ecology*, 67, 1601, 1986.
46. **Clancy, K. M., Giese, R. L., and Benjamin, D. M.,** Predicting jack-pine budworm infestations in northwestern wisconsin, *Environ. Entomol.*, 9, 743, 1980.
47. **Cockfield, S. D. and Potter, D. A.,** Interaction of *Euonymus* scale (Homoptera: Diaspididae) feeding damage and severe water stress on leaf abscission and growth of *Euonymus fortunei*, *Oecologia* (*Berlin*), 71, 41, 1986.

48. **Cohen, A. C.,** Effects of water stress on hemolymph volume, osmotic potential and chemical composition in *Megetra cancellata*, *Comp. Biochem. Physiol.*, 79A, 547, 1984.
49. **Connor, E. F.,** Plant water deficits and insect responses: the preference of *Corythuca arcuata* (Heteroptera:Tingidae) for the foliage of white oak, *Quercus alba*, *Ecol. Ent.*, 13, 375, 1989.
50. **Coon, B. F.,** Aphid populations on oats grown in various nutrient solutions, *J. Econ. Entomol.*, 52, 624, 1959.
51. **Craig, T. P., Wagner, M. R., McCullough, D. G., and Frantz, D.,** Effects of experimentally altered plant moisture stress on the performance of *Neodiprion* sawflies, *Forest Ecology and Management*, Special Issue: XVIII International Congress of Entomology Towards Integrated Pest Management of Forest Defoliators, 1990.
52. **Craighead, F. C.,** Bark-beetle epidemics and rainfall deficiency, *J. Econ. Entomol.*, 18, 577, 1925.
53. **Cram, W. T.,** Fecundity of the root weevils *Brachyrhinus sulcatus* and *Sciopithes obscurus* on strawberry at different conditions of host plant nutrition, *Can. J. Plant Sci.*, 45, 219, 1965.
54. **Cunliffe, N.,** Studies on *Oscinella frit* L.: observations on infestations, yield, susceptibility, recovery power, the influence of variety on the rate of growth of the primary shoot of the oat and the reaction to manurial treatment, *Ann. Appl. Biol.*, 15, 473, 1928.
55. **Dahms, R. G.,** Oviposition and longevity of chinch bugs on seedlings growing in nutrient solutions, *J. Econ. Entomol.*, 40, 841, 1947.
56. **Dale, D.,** Plant-mediated effects of soil mineral stresses on insects, in *Plant Stress — Insect Interactions*, Heinrichs, E. A.,Ed., John Wiley & Sons, New York, 1988, chap. 2.
57. **Daniels, N. E.,** Greenbug populations and their damage to winter wheat as affected by fertilizer applications, *J. Econ. Entomol.*, 50, 793, 1957.
58. **Denno, R. F., Douglass, L. W., and Jacobs, D.,** Crowding and host plant nutrition: environmental determinants of wingform in *Prokelisia marginata*, *Ecology*, 66, 1588, 1985.
59. **Dorschner, K. W., Johnson, R. C., Eikenbary, R. D., and Ryan, J. D.,** Insect-plant interactions: greenbugs (Homoptera: Aphididae) disrupt acclimation of winter wheat to drought stress, *Environ. Entomol.*, 15, 118, 1986.
60. **Douglas, W. A. and Eckhardt, R. C.,** The effect of nitrogen in fertilizers on earthworm damage to corn, *J. Econ. Entomol.*, 46, 853, 1953.
61. **Dunn, J. P., Kimmerer, T. W., and Nordin, G. L.,** The role of host tree condition in attack of white oaks by the twolined chestnut borer, *Agrilus bilineatus* (Weber) (Coleoptera: Buprestidae), *Oecologia*, 70, 596, 1986.
62. **Edwards, R. L.,** Relationship between grasshopper abundance and weather conditions in Saskatchewan, *Can. Entomol.*, 92, 619, 1960.
63. **Ehrlich, P. R., Murphy, D. D., Singer, M. C., Sherwood, C. B., White, R. R., and Brown, I. L.,** Extinction, reduction, stability and increase: the responses of checkerspot butterfly (*Euphydryas*) populations to the california drought, *Oecologia*, 46, 101, 1980.
64. **English-Loeb, G. M.,** Nonlinear responses of spider mites to drought stressed host plants, *Ecol. Entomol.*, 14, 45, 1989.
65. **English-Loeb, G. M.,** Plant drought stress and outbreaks of spider mites: a field test, *Ecology*, 7, 1401, 1990.
66. **Erdelyi, C., Manninger, S., Manninger, K., and Dobrovolszky, A.,** Approximate forecasting of the main damage areas of some lucerne seed pests, and of probable years of high damage caused by them, by means of climatic and weather factors, *P. Int. Conf. Integr. Plant Prot.*, 1, 144, 1983.
67. **Feese, H. and Wilde, G.,** Factors affecting survival and reproduction of the Banks grass mite, *Oligonychus pratensis*, *Environ. Entomol.*, 6(1), 53, 1977.
68. **Fereres, A., Gutierrez, C., Del Estal, P., and Castanera, P.,** Impact of the English grain aphid, *Sitobion avenae* (F.) (Homoptera: Aphididae), on the yield of wheat plants subjected to water deficits, *Environ. Entomol.*, 17, 596, 1988.
69. **Ferrell, G. T. and Hall, R. C.,** Weather and tree growth associated with white fir mortality caused by fir engraver and roundheaded fir borer, *U.S. For. Serv. Res. Pap.*, PSW-109, 1975.
70. **Fielding, D. J. and Brusven, M. A.,** Historical analysis of grasshopper (Orthoptera:Acrididae) population responses to climate in southern Idaho, 1950–1980, *Environ. Entomol.*, 19, 1876, 1990.
71. **Fitter, A. H. and Hay, R. K. M.,** *Environmental Physiology of Plants*, 2nd ed., Academic Press, New York, 1987.

72. **Gange, A. C. and Brown, V. K.,** Effects of root herbivory by an insect on a foliar-feeding species, mediated through changes in the host plant, *Oecologia*, 81, 38, 1989.
73. **Garman, P. and Kennedy, B. H.,** Effect of soil fertilization on the rate of reproduction of the two-spotted spider mite, *J._Econ. Entomol.*, 42, 157, 1949.
74. **Gershenzon, J.,** Changes in the levels of plant secondary metabolites under water and nutrient stress, *Recent Adv. Phytochem.*, 18, 273, 1984.
75. **Goldson, S. L., Dyson, C. B., Proffitt, J. R., Frampton, E. R., and Logan, J. A.,** The effect of *Sitona discoideus* Gyllenhal (Coleoptera: Curculionidae) on lucerne yields in New Zealand, *Bull. Entomol. Res.*, 75, 429, 1985.
76. **Gould, F.,** Resistance of cucumber varieties to *Tetranychus urticae*: genetic and environmental determinants, *J. Econ. Entomol.*, 71, 680, 1978.
77. **Goyer, R. A. and Benjamin, D. M.,** Influence of soil fertility on infestation of jack pine plantations by the pine root weevil, *For. Sci.*, 18, 139, 1972.
78. **Guinn, G. and Eidenbock, M. P.,** Catechin and condensed tannin contents of leaves and bolls of cotton in relation to irrigation and boll load, *Crop Science*, 22, 614, 1982.
79. **Haack, R. A. and Benjamin, D. M.,** The biology and ecology of the two-lined chestnut borer, *Agrilus biliniatus* (Coleoptera: Buprestidae), on oaks, *Quercus* spp., in Wisconsin, *Can. Entomol.*, 114, 385, 1982.
80. **Hale, M. G. and Orcutt, D. M.,** *The Physiology of Plants Under Stress*, Wiley Interscience, New York, 1987, 206.
81. **Hamstead, E. O. and Gould, E.,** Relation of mite populations to seasonal leaf nitrogen levels in apple orchards, *J. Econ. Entomol.*, 50, 109, 1957.
82. **Hare, J. D., Morse, J. G., Menge, J. L., Pehrson, J. E., Coggins, C. W., Embeton, T. W., Jarrell, W. M., and Meyer, J. L.,** Population responses of the citrus red mite and citrus thrips to 'navel' orange practices, *Environ. Ent.*, 18(3), 481, 1989.
83. **Harrewijn, P.,** Reproduction of the aphid *Myzus persicae* related to the mineral nutrition of potato plants, *Entomol. Exp. Appl.*, 13, 307, 1970.
84. **Harris, E. J. and Lee, C. Y.,** Development of *Ceratitis captitata* (Diptera:Tephritidae) in coffee in wet and dry habitats, *Environ. Entomol.*, 18, 1042, 1989.
85. **Harris, M. K. and Ring, D. R.,** Adult pecan weevil emergence related to soil moisture, *J. Econ. Entomol.*, 73, 339, 1980.
86. **Hartman, H. A.,** Ecology of gall-forming Lepidoptera on *Tetradymia* II. Plant stress effects of infestation intensity, *Hilgardia*, 52, 17, 1984.
87. **Hawkins, B. A., Goedon, R. D., and Gagne, R. J.,** Ecology and taxonomy of the *Asphondylia* spp. (Diptera: Cecido-myiidae) forming galls on *Atriplex* spp. (Chenopodiaceae) in Southern California, *Entomography*, 4, 55, 1986.
88. **Heathcote, G. D.,** The effect of plant spacing, nitrogen fertilizer and irrigation on the appearance of symptoms and spread of virus yellows in sugarbeet crops, *J. Agric. Sci.*, 82, 53, 1974.
89. **Heinrichs, E. A.,** Global food production and plant stress, in *Plant Stress — Insect Interactions*, Heinrichs, E. A., Ed., John Wiley & Sons, New York, 1988, chap. 1.
90. **Henderson, I. F. and Clements, R. O.,** Stem-boring Diptera in grassland in relation to management practice, *Ann. Appl. Biol.*, 87, 524, 1977.
91. **Henderson, I. F. and Perry, J. N.,** Some factors affecting the build-up of cereal aphid infestations in winter wheat, *Ann. Appl. Biol.*, 89, 177, 1978.
92. **Henneberry, T. J.,** The effect of host-plant nitrogen supply and age of leaf tissue on the fecundity of the two-spotted spider mite, *J. Econ. Entomol.*, 55, 799, 1962.
93. **Henneberry, T. J. and Shriver, D.,** Two-spotted spider mite feeding on bean leaf tissue of plants supplied with various levels of nitrogen, *J. Econ. Entomol.*, 57, 377, 1964.
94. **Hix, D. M., Barnes, B. V., Lynch, A. M., and Witter, J. A.,** Relationships between Spruce budworm damage and site factors in spruce-fir dominated ecosystems of Western upper Michigan, *Forest Ecol. Manage.*, 21, 129, 1987.
95. **Hoffman, G. D., Hogg, D. B., and Boush, G. M.,** The effect of plant-water stress on potato leafhopper, *Empoasca fabae*, egg developmental period and mortality, *Entomol. Exp. Appl.*, 57, 165, 1990.
96. **Hollingsworth, C. S. and Berry, R. E.,** Two-spotted sider mite (Acari: Tetranychidae) in peppermint: population dynamics and influence of cultural practices, *Environ. Entomol.*, 11, 1280, 1982.
97. **Holtzer, T. O., Archer, T. L., and Norman, J. M.,** Host plant suitability in relation to water stress, in *Plant Stress — Insect Interactions*, Heinrichs, E. A., Ed., John Wiley & Sons, New York, 1988.

98. **Horton, D. R.,** Performance of a willow-feeding beetle, *Chrysomela Knabi* Brown, as affected by host species and dietary moisture, *Can. Entomol.*, 121, 777, 1989.
99. **Hosking, G. P. and Kershaw, D. J.,** Red beech death in the Maruia valley South Island, New Zealand, *New Zealand J. Bot.*, 23, 201, 1985.
100. **Hovemeyer, K.,** The population dynamics of *Cheilosia fasciata* (Diptera, Syrphidae): significance of environmental factors and behavioral adaptations in a phytophagous insect, *Oecologia,* 73, 537, 1987.
101. **Ingrisch, S.,** The plurennial life cycles of the European tettigoniidae (Insecta: Orthoptera), *Oecologia,* 70, 624, 1986.
102. **Israel, P. and Prakasa Rao, P. S.,** Influence of potash on gall-fly incidence on rice, *Oryza,* 4, 85, 1967.
103. **Ivanovici, A. M. and Wiebe, W. J.,** Towards a working definition of stress: a review and critique, in *Stress Effects on Natural Ecosystems*, Barrett, G. W. and Rosenberg, R., Eds., John Wiley & Sons, New York, 1981.
104. **Jackson, P. R. and Hunter, P. E.,** Effects of nitrogen-fertilizer level on development and populations of pecan leaf scorch mite (Acarina: Tetranychidae), *J. Econ. Entomol.*, 76, 432, 1983.
105. **Jansson, R. K. and Smilowitz, Z.,** Influence of nitrogen on population parameters of potato insects: abundance, development, and damage of the Colorado potato beetle, *Leptinotarsa decemlineata* (Coleoptera: Chrysomelidae), *Environ. Entomol.*, 14, 500, 1985.
106. **Jansson, R. K. and Smilowitz, Z.,** Influence of nitrogen on population parameters of potato insects: abundance, population growth, and within-plant distribution of the green peach aphid, *Myzus persicae* (Homoptera: Aphididae), *Environ. Entomol.*, 15, 49, 1986.
107. **Johnstone, I. M.,** Consumption of leaves by herbivores in mixed mangrove stands, *Biotropica,* 13, 252, 1981.
108. **Jones, C. G. and Coleman, J. S.,** Plant stress and insect herbivory: toward an integrated perspective, in *Integrated Responses of Plants to Stress*, Mooney, H. A., Winner, W. E., and Pell, E. J., Eds., Academic Press, New York, 1990.
109. **Joyce, R. J. V.,** Effect on the cotton plant in the Sudan Gezira of certain leaf feeding pests, *Nature,* 182, 1463, 1958.
110. **Kemp, W. P. and Moody, U. L.,** Relationships between regional soils and foliage characteristics and western spruce budworm (Lepidoptera: Tortricidae) outbreak frequency, *Environ. Entomol.*, 13, 1291, 1984.
111. **Kennedy, J. S. and Booth, C. O.,** Response of *Aphis fabae* Scop. to water shortage in host plants in the field, *Entomol. Exp. Appl.*, 2, 1, 1959.
112. **Kennedy, J. S., Lamb, K. P., and Booth, C. O.,** Responses of *Aphis fabae* Scop. to water shortage in host plants in pots, *Entomol. Exp. Appl.*, 1, 274, 1958.
113. **Kimberling, D.N., Scott, E., and Price, P. W.,** Testing a new hypothesis: plant vigor and Phylloxera distribution on wild grape leaves in Arizona, *Oecologia,* 84, 1, 1990.
114. **Kindler, S. D. and Staples, R.,** Nutrients and the reaction of two alfalfa clones to the Spotted alfalfa aphid, *J. Econ. Entomol.*, 63(3), 938, 1970.
115. **Kirchner, T. B.,** The effects of resource enrichment on the diversity of plants and arthropods in a shortgrass prairie, *Ecology,* 58, 1334, 1977.
116. **Klostermeyer, E. C.,** Effect of soil fertility on corn earworm damage, *J. Econ. Entomol.*, 43, 427, 1950.
117. **Klubertanz, T. H., Pedigo, L. P., and Carlson, R. E.,** Effects of plant moisture stress and rainfall on population dynamics of the two-spotted spider mite (Acari: Tetranychidae), *Env. Entomol.*, 19, 1773, 1990.
118. **Landsberg, J., Morse, J., and Khanna, P.,** Tree dieback and insect dynamics in remnants of native woodlands on farms, *Proc. Ecol. Soc. Aust.*, 16, 149, 1990.
119. **Larson, K. C.,** Sink-source Interactions between a Galling Aphid and its Narrowleaf Cottonwood Host: within and between Plant Variation, dissertation, Northern Arizona University, Flagstaff, Arizona, 1989.
120. **Larsson, S.,** Stressful times for the plant stress — insect performance hypothesis, *Oikos,* 56, 277, 1989.
121. **Larsson, S. and Ohmart, C. P.,** Leaf age and larval performance of the leaf beetle, *Paropsis atomaria, Ecol. Entomol.*, 13, 19, 1988.
122. **Larsson, S. and Tenow, O.,** Areal distribution of a *Neodiprion sertifer* (Hym., Diprionidae) outbreak on Scots pine as related to stand condition, *Holarctic Ecology,* 7, 81, 1984.
123. **Leigh, T. F., Grimes, D. W., Dickens, W. L., and Jackson, C. E.,** Planting pattern, plant population, irrigation, and insect interactions in cotton, *Environ. Entomol.,* 3, 492, 1974.

124. **Leigh, T. F., Grimes, D. W., Yamada, H., Bassett, D., and Stockton, J. R.,** Insects in cotton as affected by irrigation and fertilization practices, *California Agric.*, 24, 12, 1970.
125. **Leuck, D. B.,** Induced fall armyworm resistance in pearl millet, *J. Econ. Entomol.*, 65, 1608, 1972.
126. **Lewis, A. C.,** Plant quality and grasshopper feeding: effects of sunflower condition on preference and performance in *Melanoplus differentialis*, *Ecology*, 65, 836, 1984.
127. **Lewis, R.,** *Hypoxylon* spp., *Ganoderma lucidum* and *Agrilus bilineatus* in association with drought related oak mortality in the South, *Phytopathology*, 71, 890, 1981.
128. **Lightfoot, D. C. and Whitford, W. G.,** Variation in insect densities on desert creosotebush: is nitrogen a factor?, *Ecology*, 68, 547, 1987.
129. **Lorenz-DeVries, N. E. and Manglitz, G. R.,** Spotted alfalfa aphid (*Therioaphis maculata* (Buckton)) (Homoptera: Aphididae): water stress, amino acid content and plant resistance, *J. Kansas Entomol. Soc.*, 55(1), 57, 1982.
130. **Lynch, R. E.,** Effects of "coastal" bermudagrass fertilization levels and age of regrowth on fall armyworm (Lepidoptera: Noctuidae): larval biology and adult fecundity, *J. Econ. Entomol.*, 77, 948, 1984.
131. **Maddox, D. M. and Rhyne, M.,** Effects of induced host-plant mineral deficiencies on attraction, feeding and fecundity of the alligatorweed flea beetle, *Environ. Entomol.*, 4, 682, 1975.
132. **Maiteki, G. A., Lamb, R. J., and Ali-Khan, S. T.,** Seasonal abundance of the Pea aphid, *Acyrthosiphon pisum* (Homoptera: Aphididae), in Manitoba field peas, *Can. Entomol.*, 118, 601, 1986.
133. **Mandal, S, Kukherjee, S. P., Choudhuri, M. A., and Choudhuri, D. K.,** Water stress-induced plant metabolism and its effect on the insect pest. 1. *Mormordica charantia* cultivar korola vs. *Epilachna dodecastigma*, *Proc. Indian Natl. Sci. Acad.*, Part B Biol. Sci., 50, 163, 1984.
134. **Manglitz, G. R., Gorz, H. J., Haskins, F. A., Akeson, W. R., and Beland, G. L.,** Interactions between insects and chemical components of sweetclover, *J. Environ. Qual.*, 5, 347, 1976.
135. **Marino, P. C.,** Activity patterns and microhabitat selection in a desert tenebrionid beetle (Coleoptera: Tenebrionidae), *Ann. Entomol. Soc. Am.*, 79, 468, 1986.
136. **Markkula, M. and Tiittanen, K.,** Effect of fertilizers on the reproduction of *Tetranychus telarius* (L.), *Myzus persicae* (Sulz.) and *Acyrthosiphon pisum* Harris, *Ann. Agric. Fenn.*, 8, 9, 1969.
137. **Martinat, P. J. and Allen, D. C.,** Relationship between outbreaks of Saddled Prominent, *Heterocampa guttivitta* (Lepidoptera: Notodontidae), and drought, *Environ. Entomol.*, 16, 246, 1987.
138. **Mattson, W. J.,** Herbivory in relation to plant nitrogen content, *Annu. Rev. Ecol.*, 11, 119, 1980.
139. **Mattson, W. J. and Haack, R. A.,** The role of drought in outbreaks of plant-eating insects, *BioScience*, 37, 110, 1987a.
140. **Mattson, W. J. and Haack, R. A.,** The role of drought stress in provoking outbreaks of phytophagous insects, in *Insect Outbreaks*, Barbosa, P. and Schultz, J., Eds., Academic Press, New York, 1987b, chap. 15.
141. **Mattson, W. J., Slocum, S. S., and Koller, C. N.,** Spruce budworm performance in relation to foliar chemistry of its host plants, *USDA For. Serv. Gen. Tech. Rep.*, NE-85, 55, 1983.
142. **McClure, M. S.,** Foliar nitrogen: a basis for host suitability for elongate hemlock scale, *Fiornia externa* (Homoptera: Diaspididae), *Ecology*, 61, 72, 1980.
143. **McClure, M. S.,** Dispersal of the scale *Fiorinia externa* (Homoptera: Diaspididae) and effects of edaphic factors on its establishment on hemlock, *Environ. Entomol.*, 6, 539, 1977.
144. **McCullough, D. G. and Wagner, M. R.,** Influence of watering and trenching Ponderosa pine on a pine sawfly, *Oecologia*, 71, 382, 1987.
145. **McDonald, G. and Smith, A. M.,** The incidence and distribution of the armyworms *Mythimna convecta* (Walker) and *persectania* spp. (Lepidoptera: Noctuidae) and their parasitoids in major agricultural districts of Victoria, south-eastern Australia, *Bull. Ent. Res.*, 76, 199, 1986.
146. **McGarr, R. L.,** Relation of fertilizers to the development of the cotton aphid, *J. Econ. Entomol.*, 35, 482, 1942.
147. **McManus, M. L. and Giese, R. L.,** The columbia timber beetle, *Corthylus columbianus*. VII. The effect of climatic integrants on historic density fluctuations, *For. Sci.*, 14, 242, 1968.
148. **McMurtry, J. A.,** Resistance of alfalfa to spotted alfalfa aphid in relation to environmental factors, *Hilgardia*, 32, 501, 1962.

149. **McNeill, S. and Southwood, T. R. E.,** The role of nitrogen in the development of insect/plant relationships, in *Biochemical Aspects of Plant and Animal Coevolution,* Harborne, J. B., Ed., Academic Press, New York, 1978, chap. 4.
150. **McQuate, G. T. and Connor, E. F.,** Insect responses to plant water deficits. I. Effect of water deficits in soybean plants on the feeding preference of Mexican bean beetle larvae, *Ecol. Entomol.*, 15, 419, 1990a.
151. **McQuate, G. T. and Connor, E. F.,** Insect responses to plant water deficits. II. Effect of water deficits in soybean plants on the growth and survival of Mexican bean beetle larvae, *Ecol. Entomol.*, 15, 433, 1990b.
152. **Meisner, J.,** Attraction and repellence of the potato tuber moth, *Gnorimoschema operculella* Zell.: phagostimulants and antifeedants for the larvae: some factors of the attraction to oviposition, Dissertation, Hebrew University of Jerusalem, Israel, 1969.
153. **Mellors, W. K., Allegro, A., and Hsu, A. N.,** Effects of carbofuran and water stress on growth of soybean plants and two-spotted spider mite (Acari: Tetranychidae) populations under greenhouse conditions, *Environ. Entomol.,* 13, 561, 1984.
154. **Mellors, W. K. and Propts, S. E.,** Effects of fertilizer level, fertility balance, and soil moisture on the interaction of two-spotted spider mites (Acari: Tetranychidae) with radish plants. *Environ. Entomol.,* 12, 1239, 1983.
155. **Mendel, Z.,** Life history of *Phloeosinus armatus* reiter and *P. aubei perris* (Coleoptera: Scolytidae) in Israel, *Phytoparasitica*, 12, 89, 1984.
156. **Metcalfe, J. R.,** Studies on the effect of nutrient status of sugar-cane on the fecundity of *Saccarosydne saccharivora* (Westw.) (Hom., Delphacidae), *Bull. Entomol. Res.*, 60, 309, 1970.
157. **Michels, G. J. Jr. and Undersander, D. J.,** Temporal and spatial distribution of the greenbug (Homoptera: Aphididae) on sorghum in relation to water stress, *J. Econ. Entomol.*, 79, 1221, 1986.
158. **Miles, P. W., Aspinall, D., and Correll, A. T.,** The performance of two chewing insects on water-stressed food plants in relation to changes in their chemical composition, *Aust. J. Zool.,* 30, 347, 1982a.
159. **Miles, P. W., Aspinall, D., and Rosenburg, L.,** Performance of the cabbage aphid, *Brevicoryne brassicae* (L.) on water-stressed rape plants, in relation to changes in their chemical composition, *Aust. J. Zool.*, 30, 337, 1982b.
160. **Miller, D. R., Mo, T. K., and Wallner, W. E.,** Influence of climate on gypsy moth defoliation in Southern New England, *Environ. Entomol.*, 18(4), 646, 1989.
161. **Mistric, J. R.,** Effects of nitrogen fertilization on cotton under boll weevil attack in North Carolina, *J. Econ. Entomol.*, 61, 282, 1968.
162. **Mitchell, R. G. and Paul, H. G.,** Field fertilization of Douglas Fir and its effect on *Adelges cooleyi* populations, *Environ. Entomol.*, 3, 501, 1974.
163. **Moore, D. and Clements, R. O.,** Stem-boring Diptera in perennial ryegrass in relation to fertilizer. I. Nitrogen level and form, *Ann. Appl. Biol.*, 105, 1, 1984.
164. **Mopper, S. and Whitham, T. G.,** The stress paradox: effects on pinyon sawfly sex ratios and fecundity, *Ecology,* in press, 1992.
165. **Moran, N. and Hamilton, W. D.,** Low nutritive quality as defense against herbivores, *J. Theor. Biol.,* 86, 247, 1980.
166. **Moran, N. A. and Whitham, T. G.,** Population fluctuations in complex life cycles: an example from *Pemphigus* aphids, *Ecology*, 69, 1214, 1988.
167. **Morgan, J. M.,** Osmoregulation and water stress in higher plants, *Annu. Rev. Plant Physiol.*, 35, 299, 1984.
168. **Morris, O. N.,** The development of the clover mite *Bryobia praetiosa* (Acarina: Tetranychidae) in relation to nitrogen, phosphorus, and potassium nutrition of its plant host, *Ann. Entomol. Soc. Am.*, 54, 551, 1961.
169. **Munster-Swendsen, M.,** The effect of precipitation on radial increment in Norway spruce (*Picea abies* Karst.) and on the dynamics of a Lepidopteran pest insect, *J. Appl. Ecol.*, 24, 563, 1984.
170. **Munster-Swendsen, M.,** Index of vigour in Norway spruce (*Picea abies* Karst.), *J. Appl. Ecol.*, 24, 551, 1987.
171. **Myers, J. H.,** Effect of physiological condition of the host plant on the ovipositional choice of the cabbage white butterfly, *Pieris rapae, J. Anim. Ecol.*, 54, 193, 1985.
172. **Ohmart, C. P., Stewart L. G., and Thomas J. R.,** Effects of food quality, particularly nitrogen concentrations, of *Eucalyptus blakelyi* foliage on the growth of *Paropsis atomaria* larvae (Coleoptera: Chrysomelidae), *Oecologia,* 65, 543, 1985.
173. **Oi, D. H., Sanderson, J. P., Youngman, R. R., and Barnes, M. M.,** Developmental times of the pacific spider mite (Acari: Tetranychidae) on water-stressed almond trees, *Environ. Entomol.,* 18, 208, 1989.

174. **Okigbo, B. N. and Gyrisco, G. G.,** Effects of fertilizers on Hessian fly infestation, *J. Econ. Entomol.*, 55, 753, 1962.
175. **Oloumi-Sadeghi, H, Helm, C. G., Kogan, M., and Schoenweiss, D. F.,** Effect of water stress on abundance of two-spotted spider mite on soybeans under greenhouse conditions, *Entomol. Exp. Appl.*, 48, 85, 1988.
176. **Onuf, C. P., Teal, J. M., and Valiela, I.,** Interactions of nutrients, plant growth and herbivory in a mangrove ecosystem, *Ecology,* 58, 514, 1977.
177. **Patch, L. H.,** Manual infestation of dent corn to study resistance to European corn borer, *J. Econ. Entomol.*, 40, 667, 1947.
178. **Pelletier, Y.,** The effect of water stress and leaflet size on the density of trichomes and the resistance to Colorado potato beetle larvae (*Leptinotarsa decemlineata* (Say)) in *Solanum berthaultii* Hawkes, Can. Entomol., 122, 1411, 1990.
179. **Perrenoud, S.,** Contribution to the discussion: the effect of K on insect and mite development, in *Fertilizer Use and Plant Health*, Int. Potash Inst., CH-3048 Worblaufen-Bern, Switzerland, 317, 1976.
180. **Perring, T. M., Holtzer, T. O., Toole, J. L., and Norman, J. M.,** Relationships between corn-canopy microenvironments and Banks grass mite (Acari: Tetranychidae) abundance, *Environ._Entomol.*, 15, 79, 1986.
181. **Pfeiffer, D. G. and Burts, E. C.,** Effect of tree fertilization on numbers and development of pear psylla (Homoptera: Psyllidae) and on fruit damage, *Environ. Entomol.*, 12, 895, 1983.
182. **Pfeiffer, D. G. and Burts, E. C.,** Effect of tree fertilization on protein and free amino acid content and feeding rate of pear psylla (Homoptera: Psyllidae), *Environ. Entomol.*, 13, 1487, 1984.
183. **Posey, C. E. and Benjamin, D. M.,** Pine sawfly infestation related to cultural treatments, *Tree Plant Notes*, 20, 28, 1969.
184. **Potter, D. A.,** Effect of soil moisture on oviposition, water absorption, and survival of southern masked chafer (Coleoptera: Scarabaeidae) eggs, *Environ. Entomol.*, 12, 1223, 1983.
185. **Potter, D. A. and Gordon, F. C.,** Susceptibility of *Cyclocephala immaculata* (Coleoptera: Scarabaeidae) eggs and immatures to heat and drought in turf grass, *Environ. Entomol.* 13, 794, 1984.
186. **Power, A. G.,** Plant community diversity, herbivore movement and an insect-transmitted disease of maize, *Ecology,* 68, 1658, 1987.
187. **Power, A. G.,** Influence of plant spacing and nitrogen fertilization in maize on *Dalbulus maidis* (Homoptera: Cicadellidae), vector of cornstunt, *Environ. Entomol.*, 18, 494, 1989.
188. **Prestidge, R. A.,** Instar duration, adult consumption, oviposition and nitrogen utilization efficiencies of leafhoppers feeding on different quality food (Auchenorryncha: Homoptera), *Ecol. Entomol.*, 7, 91, 1982a.
189. **Prestidge, R. A.,** The influence of nitrogenous fertilizer on the grassland Auchenorrhyncha (Homoptera), *J. Appl. Ecol.*, 19, 735, 1982b.
190. **Prew, R. D., Church, B. M., Dewar, A. M., Lacey, J., Penny, A., Plumb, R. T., Thorne, G. N., Todd, A. D., and Williams, T. D.,** Effects of eight factors on the growth and nutrient uptake of winter wheat and on the incidence of pests and diseases, *J. Agri. Sci.*, 100, 363, 1983.
191. **Price, J. F., Harbough, B. K., and Stanley, C. D.,** Response of mites and leaf miners to trickle irrigation rates in spray chrysanthemum production, *Hortscience,* 17, 895, 1982.
192. **Price, P. W. and Clancy, K. M.,** Multiple effects of precipitation on *Salix lasiolepis* and populations of the stem-galling sawfly, *Euura lasiolepis*, *Ecol. Res.*, 1, 1, 1986.
193. **Rawat, R. R. and Sahu, H. R.,** Effect of date of sowing and nitrogen application on the incidence of the wheat stem fly, *Atherigona bituberculata* Malloch, *Indian J. Entomol.,* 31, 152, 1969.
194. **Reagan, T. E., Rabb, R. L., and Collins, W. K.,** Selected cultural practices as affecting production of tobacco hornworms on tobacco, *J. Econ. Entomol.*, 71, 79, 1978.
195. **Rhoades, D. F.,** Evolution of plant chemical defense against herbivores, in *Herbivores, Their Interaction with Secondary Plant Matabolites*, G. A. Rosenthal and D. H. Janzen, Eds., Academic Press, New York, 1979, chap. 1.
196. **Rhoades, D. F.,** Offensive-defensive interactions between herbivores and plants: their relevance in herbivore population dynamics and ecological theory, *Am. Nat.*, 125, 205, 1985.
197. **Rhoades, D. F.,** Herbivore population dynamics and plant chemistry, in *Variable Plants and Herbivores in Natural and Managed Systems,* R. F. Denno and M. S. McClure, Eds., Academic Press, New York, 1983, chap. 6.

198. **Rodriguez, J. G.,** Mineral nutrition of the two-spotted spider mite, *Tetranychus bimaculatus* Harvey, *Ann. Entomol. Soc.*, 44, 511, 1951.
199. **Rodriguez, J. G.,** The comparative NPK nutrition of *Panonychus ulmi* (Koch.) and *Tetranychus telarius* (L.) on apple trees, *J. Econ. Entomol.*, 51, 369, 1958.
200. **Rodriguez, J. G. and Rodriguez, L. D.,** The relation between minerals, B-complex vitamins and mite populations on tomato foliage, *Ann. Ent. Soc. Am.*, 45, 331, 1952.
201. **Room, P. M., Julien, M. H., and Forno, I. W.,** Vigorous plants suffer most from herbivores: latitude, nitrogen and biological control of the weed *Salvinia molesta*, *Oikos*, 54, 92, 1989.
202. **St. George, R. A.,** Weather, a factor in outbreaks of the hickory bark beetle, *J. Econ. Entomol.*, 22, 573, 1929.
203. **St. George, R. A.,** Drought affected and injured trees attractive to bark beetles, *J. Econ. Entomol.*, 23, 825, 1930.
204. **Salama, H. S., Amin, A. H., and Hawash, M.,** Effect of nutrients supplied to citrus seedlings on their susceptibility to infestation with the scale insects *Aonidiella aurantii* (Maskell) and *Lepidosaphes beckii* (Newman) (Coccoidea), *Z. Angew. Entomol.*, 71, 395, 1972.
205. **Salama, H. S., El-Sherif, A. F., and Megahed, M.,** Soil nutrients affecting the population density of *Parlatoria zizyphus* (Lucas) and *Icerya purchasi* Mask. (Homopt., Coccoidea) on citrus seedlings, *Z. Angew. Entomol.*, 99, 471, 1985.
206. **Schoene, W. J.,** Plant food and mealybug injury, *J. Econ. Entomol.*, 31, 241, 1941.
207. **Scriber, J. M.,** Limiting effects of low leaf-water content on the nitrogen utilization, energy budget, and larval growth of *Hyalophora cercropia* (Lepidoptera: Saturniidae), *Oecologia* (Berlin), 28, 269, 1977.
208. **Scriber, J. M.,** Effects of leaf-water supplementation upon post-ingestive nutritional indices of forb-, shrub-, vine-, and tree-feeding Lepidoptera, *Entomol. Exp. Appl.*, 25, 240, 1979.
209. **Scriber, J. M.,** Host-plant suitability, in *Chemical Ecology of Insects*, Bell, W. J. and Carde, R. T., Eds., Chapman and Hall, London, 1984, chap. 7.
210. **Scriber, J. M. and Slansky, F., Jr.,** The nutritional ecology of immature insects, *Annu. Rev. Entomol.*, 26, 183, 1981.
211. **Semtner, P. S., Rasnake, M., and Terrill, T. R.,** Effect of host plant nutrition on the occurrence of tobacco hornworms and tobacco flea beetles on different types of tobacco, *J. Econ. Entomol.*, 73, 221, 1980.
212. **Shaw, G. G. and Little, C. H. A.,** Effects of high urea fertilization of balsam fir trees on spruce budworm development, in *Insect and Mite Nutrition: Significance and Implications in Ecology and Pest Management*, Rodriguez, J. G., Ed., North Holland, Amsterdam, 1972.
213. **Shaw, G. G., Little, C. H. A., and Durzan, D. J.,** Effect of fertilization of balsam fir trees on spruce budworm nutrition and development, *Can. J. For. Res.*, 8, 364, 1978.
214. **Shepard, R. F.,** Phytosociological and environmental characteristics of outbreak and non-outbreak areas of the two-year cycle spruce budworm, *Choristoneura fumiferana*, *Ecology*, 40, 608, 1959.
215. **Shepard, R. F., Van Sickle, G. A., and Clarke, D. H. L.,** Spatial relationships of Douglas-fir tussock moth defoliation within habitat and climatic zones, in *Proceedings Lymantriidae: A Comparison of Features of New and Old World Tussock Moths*, Northeast Forest Experiment Station, Hamden, 1988.
216. **Sherman, R. J. and Warren, R. K.,** Factors in *Pinus ponderosa* and *Calocedrus decurrens* mortality in Yosemite valley, USA, *Vegetation*, 77, 79, 1988.
217. **Singh, R. and Agarwal, R. A.,** Fertilizers and pest incidence in India, *Potash Rev.*, 23, 1, 1983.
218. **Skaller, P. M.,** Patterns in the distribution of gypsy moth (*Lymantria dispar*) (Lepidoptera: Lymantriidae) egg masses over an 11 year population cycle, *Environ. Entomol.*, 14, 106, 1985.
219. **Slansky, F. and Feeny, P.,** Stabalization of the rate of nitrogen accumulation by larvae of the cabbage butterfly on wild and cultivated food plants, *Ecol. Mon.*, 47, 209, 1977.
220. **Slosser, J. E.,** Irrigation timing for bollworm management in cotton, *J. Econ. Entomol.*, 73, 346, 1980.
221. **Smirnoff, W. A. and Bernier, B.,** Increased mortality of swaine jack-pine sawfly, and foliar nitrogen concentrations after fertilization, *Can. J. For. Res.*, 3, 112, 1973.
222. **Smith, D. S. and Northcott, F. E.,** The effects on the grasshopper, *Melanoplus mexicanus mexicanus* (Sauss) (Orthoptera: Acrididae) of varying the nitrogen content in its food plant, *Can. J. Zool.*, 29, 297, 1951.

223. **Smith, E. H. and Harris, R. W.,** Influence of tree vigor and winter injury on the lesser peach tree borer, *J. Econ. Entomol.*, 45, 607, 1952.
224. **Southwood, T. R. E., and Reader, P. M.,** Population census data and key factor analysis for the viburnum whitefly, *Aleurotrachelus jelinekii* (Frauenf.), on three bushes, *J. Anim. Ecol.*, 45, 313, 1976.
225. **Specht, H. B.,** Effect of water-stress on the reproduction of European red mite, *Panonychus ulmi* (Koch), on young apple trees, *Can. Entomol.*, 97, 82, 1965.
226. **Stark, R. W.,** Recent trends in forest entomology, *Ann. Rev. Entomol.*, 10, 303, 1965.
227. **Storms, J. J. H.,** Observations on the relationship between mineral nutrition of apple root stocks in gravel culture and the reproduction rate of *Tetranychus urticae* (Acarina: Tetranychidae), *Entomol. Exp. Appl.*, 12, 297, 1969.
228. **Stoszek, K. J., Mika, P. G., Moore, J. A., and Osborne, H. L.,** Relationships of Douglas-fir tussock moth defoliation to site and stand characteristics in northern Idaho, *For. Sci.*, 27, 431, 1981.
229. **Strauss, S. Y.,** Direct and indirect effects of host-plant fertilization on an insect community, *Ecology*, 68, 1670, 1987.
230. **Sumner, L. C., Dorschner, K. W., Ryan, J. D., Eikenbary, R. D., Johnson, R. C., and McNew, R. W.,** Reproduction of *Schizaphis graminum* (Homoptera: Aphididae) on resistant and susceptible wheat genotypes during simulated drought stress induced with polyethylene glycol, *Environ. Entomol.*, 15, 756, 1986a.
231. **Sumner, L. C., Eikenbary, R. D., and Johnson, R. C.,** Survival and reproduction of *Rhopalosiphum maidis* (Fitch) (Homoptera: Aphididae) on winter wheat during simulated drought stress, *J. Kansas Entomol. Soc.*, 59, 561, 1986b.
232. **Sumner, L. C., Need, J. T., McNew, R. W., Dorschner, R. D., and Eikenberry, R. D.,** Response of *Schizaphis graminum* (Homoptera:Aphididae) to drought-stressed wheat, using polyethylene glycol as a matricum, *Environ. Entomol.*, 12, 919, 1983.
233. **Talhouk, S. N., Nieslen, D. G., and Montgomery, M. E.,** Water deficit, defoliation and birch clones: short-term effect on gypsy moth (Lepidoptera:Lymantriidae) performance, *Environ. Entomol.*, 19, 937, 1990.
234. **Taylor, C. E.,** The population dynamics of aphids infesting the potato plant with particular reference to the susceptibility of certain varieties to infestation, *Eur. Pot. J.*, 3, 204, 1962.
235. **Taylor, L. F., Apple, J. W., and Berger, K. C.,** Response of certain insects to plants grown on varying fertility levels, *J. Econ. Entomol.*, 45, 843, 1952.
236. **Thomson, A. J., Sheperd, R. F., Harris, W. E., Silversides, R. H.,** Relating weather to outbreaks of western spruce budworm, *Choristoneura occidentalis* (Lepidoptera: Tortricidae), in British Columbia, *Can. Entomol.*, 116, 375, 1984.
237. **Thomson, A. J. and Shrimpton, D. M.,** Weather associated with the start of mountain pine beetle outbreaks, *Can. J. For. Res.*, 14, 255, 1984.
238. **Tingey, W. M. and Singh, S. R.,** Environmental factors influencing the magnitude and expression of resistance, in *Breeding Plants Resistant to Insects*, Maxwell F. G. and Jennings, P. R., Eds., John Wiley & Sons, New York, 1980, 89.
239. **Tisdale, R. A.,** Influence of Water Stress, Photoperiod, Temperature, and Humidity on Oviposition Behavior and Egg Development of *Neodiprion fulviceps* (Cresson) (Hymenoptera: Diprionidae), M.S. thesis, Northern Arizona University, Flagstaff, 1988.
240. **Todd, J. W., Parker, M. B., and Gaines, T. P.,** Populations of Mexican bean beetles in relation to leaf protein of nodulating and non-nodulating soybeans, *J. Econ. Entomol.*, 65, 729, 1972.
241. **Tscharntke, T.,** Attack by a stem-boring moth increases susceptibility of *Phragmites australis* to gall-making by a midge: mechanisms and effects on midge population dynamics, *Oikos*, 54, 93, 1989.
242. **Turchin, P., Lorio Jr., P. L., Taylor, A., and Billings, R. F.,** Why do populations of southern pine beetles (Coleoptera: Scolytidae) fluctuate?, *Environ. Entomol.*, 20, 401, 1991.
243. **Van Emden, H. F.,** Studies on the relations of insect and host plant. III. A comparison of the reproduction of *Brevicoryne brassicae* and *Myzus persicae* (Hemiptera: Aphididae) on brussels sprouts plants supplied with different rates of nitrogen and potassium, *Entomol. Exp. Appl.*, 9, 444, 1966.
244. **Van Emden, H. F. and Bashford, M. A.,** A comparison of the reproduction of *Brevicoryne brassicae* and *Myzus persicae* in relation to soluble nitrogen concentration and leaf age (leaf position) in brussels sprouts plants, *Entomol. Exp. Appl.*, 12, 351, 1969.

245. **Vince, S. W. and Valiela, I.,** An experimental study of the structure of herbivorous insect communities in a salt marsh, *Ecology*, 62, 1662, 1981.
246. **Wagner, M. R. and Frantz, D. P.,** Influence of induced water stress in ponderosa pine on pine sawflies, *Oecologia*, 83, 452, 1990.
247. **Waller, G. D., Carpenter, E. W., and Ziehl, O. A.,** Potassium in onion nectar and its probable effect on attractiveness of onion flowers to honey bees, *Am. Soc. Hortic. Sci.*, 97, 535, 1972.
248. **Waring, G. L. and Price, P.,** Consequences of host plant chemical and physical variability to an associated herbivore, *Ecol. Res.*, 3, 205, 1988.
249. **Waring, G. L. and Price, P.,** Plant water stress and gall formation (Cecidomyiidae: *Asphondylia* spp.) on creosote bush, *Ecol. Entomol.*, 15, 87, 1990.
250. **Waring, R. H. and Pitman, G. B.,** Modifying lodgepole pine stands to change susceptibility to mountain pine beetle attack, *Ecology*, 66, 889, 1985.
251. **Watt, A. D.,** The performance of the pine beauty moth on water-stressed lodgepole pine plants: a laboratory experiment, *Oecologia*, 70, 578, 1986.
252. **Wearing, C. H.,** Studies on the relations of insect and host plant: II. Effects of water stress in host plants on the fecundity of *Myzus persicae* (Sulz.) and *Brevicoryne brassicae* (L.), *Nature*, 213, 1052, 1967.
253. **Wearing, C. H.,** Responses of *Myzus persicae* and *Brevicoryne brassicae* to leaf age and water stress in brussels sprouts grown in pots, *Entomol. Exp. Appl.*, 15, 61, 1972a.
254. **Wearing, C. H.,** Selection of brussels sprouts of different water status by apterous and alate *Myzus persicae* and *Brevicoryne brassicae* in relation to the age of leaves, *Entomol. Exp. Appl.*, 15, 139, 1972b.
255. **Wearing, C. H. and van Emden, H. F.,** Studies on the relations of insect and host plant. I. Effects of water stress in host plants on infestation by *Aphis fabae* Scop., *Myzus persicae* (Sulz.) and *Brevicoryne brassicae* (L.), *Nature*, March, 1051, 1967.
256. **Wermelinger, B., Oertli, J.J., and Delucchi, V.,** Effect of host plant nitrogen fertilization on the biology of the two-spotted spider mite, *Tetranychus urticae*, *Entomol. Exp. Appl.*, 38, 23, 1985.
257. **West, P. W.,** Date of onset of regrowth dieback and its relation to summer drought in eucalypt forest of southern Tasmania, *Ann. Appl. Biol.*, 93, 337, 1979.
258. **Wheatley, A., Wightman, J. A., Williams, J. H., and Wheatley, S. J.,** The influence of drought stress on the distribution of insects on four groundnut genotypes grown near Hyderabad, India, *Bull. Entomol. Res.*, 79, 567, 1989.
259. **White, T. C. R.,** An index to measure weather-induced stress of trees associated with outbreaks of psyllids in Australia, *Ecology*, 50, 905, 1969.
260. **White, T. C. R.,** A hypothesis to explain outbreaks of looper caterpillars with special reference to populations of *Selidosema suavis* in a plantation of *Pinus radiata* in New Zealand, *Oecologia*, 16, 279, 1974.
261. **White, T. C. R.,** Weather, food, and plagues of locusts, *Oecologia*, 22, 119, 1976.
262. **White, T. C. R.,** The abundance of invertebrate herbivores in relation to the availability of nitrogen in stressed food plants, *Oecologia*, 63, 90, 1984.
263. **Whitham, T. G.,** Habitat selection by *Pemphigus* aphids in response to resource limitation and competition, *Ecology*, 59, 1164, 1978.
264. **Wiseman, B. R., Leuck, D. B., and McMillian, W. W.,** Effects of fertilizers on resistance to antigua corn to fall armyworm and corn earworm, *Florida Entomol.*, 56, 1, 1973.
265. **Wolfson, J. L.,** Ovipositional response of *Pieris rapae* to environmentally induced variation in *Brassica nigra*, *Entomol. Exp. Appl.*, 27, 223, 1980.
266. **Wolfson, J. L.,** Developmental responses of *Pieris rapae* and *Spodoptera eridania* to environmentally induced variation in *Brassica nigra*, *Environ. Entomol.*, 11, 207, 1982.
267. **Woolridge, A. W. and Harrison, F. P.,** Effects of soil fertility on abundance of green peach aphids on Maryland tobacco, *J. Econ. Entomol.*, 61, 387, 1968.
268. **Xydius, G. K., and Leaf, A. L.,** Weevil infestation in relation to fertilization of white pine, *For. Sci.*, 10, 428, 1964.
269. **Yannick, J. S., Herron, H. R., and Gutierrez, A. P.,** Dynamics of *Mononychellus tanajoa* (Acari: Tetranychidae) in Africa: seasonal factors affecting phenology and abundance, *Environ. Entomol.*, 18, 625, 1989.
270. **Yathom, S.,** Phenology of the potato tuber moth (*Phthorimaea operculella*), a pest of potatoes and of processing tomatoes in Israel, *Phytoparasitica*, 14, 313, 1986.

271. **Youngman, R. R., Sanderson, J. P., and Barnes, M. M.,** Life history parameters of *Tetranychus pacificus* McGregor (Acari: Tetranychidae) on almonds under differential water stress, *Environ. Entomol.*, 17, 488, 1988.

272. **Zar, J. H.,** *Biostatistical Analysis*, Prentice-Hall, Englewood Cliffs, New Jersey, 1984.

INDEX

A

D

E

G

H

M

N

O

P

Q

R

S

T

U

V

W

X

Y

Z

Milton Keynes UK
Ingram Content Group UK Ltd.
UKHW020821141024
449569UK00008B/513